Network Security and Wireless Technology

Network Security and Wireless Technology

Edited by **Bernhard Ekman**

Willford Press

New York

Published by Willford Press,
118-35 Queens Blvd., Suite 400,
Forest Hills, NY 11375, USA
www.willfordpress.com

Network Security and Wireless Technology
Edited by Bernhard Ekman

International Standard Book Number: 978-1-68285-166-1 (Hardback)

Printed in China.

Contents

Preface

In my initial years as a student, I used to run to the library at every possible instance to grab a book and learn something new. Books were my primary source of knowledge and I would not have come such a long way without all that I learnt from them. Thus, when I was approached to edit this book; I became understandably nostalgic. It was an absolute honor to be considered worthy of guiding the current generation as well as those to come. I put all my knowledge and hard work into making this book most beneficial for its readers.

There has been a considerable increase in the use of internet across the world. This digitalization has made data access very simple and available to all. Thus, there is a need to secure relevant data. Network security restricts unauthorized access, modification and misuse of data. This book traces the development of this field by elucidating significant topics like wireless network security, cryptography, routing, algorithms, clustering, etc. This book will prove beneficial to students and professionals alike.

I wish to thank my publisher for supporting me at every step. I would also like to thank all the authors who have contributed their researches in this book. I hope this book will be a valuable contribution to the progress of the field.

Editor

1

A Weakest Chain Approach to Assessing the Overall Effectiveness of the 802.11 Wireless Network Security

Berker Tasoluk[1] and Zuhal Tanrikulu[2]

[1]Department of Informatics, Istanbul University, Istanbul, Turkey
berker@berker.cc

[2]Department of Management Information Systems, Bogazici University, Istanbul, Turkey
tanrikul@boun.edu.tr

Abstract

This study aims to assess wireless network security holistically and attempts to determine the weakest link among the parts that comprise the 'secure' aspect of the wireless networks: security protocols, wireless technologies and user habits. The assessment of security protocols is done by determining the time taken to break a specific protocol's encryption key, or to pass an access control by using brute force attack techniques. Passphrase strengths as well as encryption key strengths ranging from 40 to 256 bits are evaluated. Different scenarios are planned and created for passphrase generation, using different character sets and different number of characters. Then each scenario is evaluated based on the time taken to break that passphrase. At the end of the study, it is determined that the choice of the passphrase is the weakest part of the entire 802.11 wireless security system.

Keywords

Brute Force Attack, Password Security, Wireless Network Protocols, Wireless Network Security

1. Introduction

The 802.11 wireless Internet and network usage is increasing daily for the benefits and ease of use offered to its users. Almost everywhere that people go, including airports, cafes, their homes and workplaces, and hotels, they use wireless networks. This increasing usage also brings some security concerns. Users require that their sensitive information not be seen or altered by anyone else. In addition, users require that any security measure taken will not interfere with their working habits; they want the security to be transparent to their normal network usage.

Many wireless network protocols exist today. Some of them were used in networks a few years ago and are no longer used, and some of them are still used in today's networks. The standards are historically outlined below.

802.11: This is the first standard published by IEEE in 1997. It operates on a 2.4 Ghz frequency band and can increase up to 2 Mbps. It is not used at the moment, since the 2 Mbps speed does not meet users' needs.

802.11a: This is the second standard published in 1999. It operates on a 5 Ghz band and can operate at speeds up to 54 Mbps. The main advantage of 802.11a is that it operates at a 5 Ghz frequency band, which makes it less prone to interference. It also has more non-overlapping channels, which also reduces the possibility of interference. The main drawback of this protocol is that it is incompatible with other wireless protocols such as 802.11b and 802.11g. It is seldom used at the moment [1].

802.11b: Before the 802.11g protocol, this was the protocol that was most commonly used. It operates on a crowded 2.4 Ghz frequency band and can reach speeds up to 11 Mbps.

802.11g: This protocol is similar to the 802.11b, as it operates on a 2.4 Ghz band, but with speeds of up to 54 Mbps. It is compatible with 802.11b, and is the most widely used protocol for new installations.

802.11n: It is designed to produce speeds up to 600 Mbps. This standard operates at 2.4 Ghz and/or 5 Ghz. At the moment, some 802.11n compatible laptops and wireless access points can be seen in the market [2].

There are numerous wireless security protocols that can be deployed in wireless networks. Wireless network security can be divided into three basic layers: [3] Wireless LAN Layer, Access Control Layer and Authentication Layer.

Wireless LAN Layer Security Protocols is responsible for the encryption and decryption of data. These protocols are as follows:

WEP (Wired Equivalent Privacy): Despite its name, this is the weakest form of encryption security for wireless networks. WEP can use a 64-bit or 128-bit stream RC4 cipher with a 24-bit initialization vector (IV), which leads to an effective key size of 40 or 104 bits. RC4 is also a weak cipher that can be easily cracked within minutes or hours.

WPA (Wi-Fi Protected Access): WPA still uses RC4 algorithm with doubled initialization vector (IV) and supports key lengths up to 152 bits. Recent attacks have targeted the WPA algorithm and achieved success on certain specific WPA implementations [4].

WPA2: WPA2 uses AES algorithm with varying lengths of 128, 192 and 256 bits. The AES algorithm is considered more secure than RC4 and uses block cipher, which involves 128 bits of data [5].

Access Control Layer Security Protocols is responsible for allowing only legitimate users access to network resources. For this to happen, this layer of protocols talks to authentication layer protocols. The protocols are outlined below.

802.1X: Originally developed for wired network access control, it is now also widely deployed for wireless networks. Many institutions integrate dot1X into their existing local area networks and wireless networks. LAN switches and wireless access points must support dot1X in order to use this protocol. 802.1X simply obtains user credentials and asks an authentication server whether to allow this user or not. It typically asks a RADIUS server, although it is not mandatory to use RADIUS.

RADIUS (Remote Access Dial-In User Service): As its name suggests, RADIUS was originally designed for dial-in user access, but it can be used for any access control purpose, including dot1X integration. It is a central database of usernames and passwords, as well as some user attributes such as VLAN ID [6].

EAP (Extensible Authentication Protocol): This is an intermediary protocol between the access control layer and authentication layer security protocols. It can work with a variety of authentication protocols; future authentication protocol support is embedded into the protocol (the word 'extensible' is used for this behavior).

Authentication Layer Security Protocols responds to access control layer security protocols. It decides who can join the network, and the decision is implemented by access control layer security protocols. These protocols are as follows:

TLS: This is the IEEE standardized version of Netscape's well-known, widely used SSL protocol. It relies on using certificates and symmetric/asymmetric ciphers. The Wi-Fi alliance required vendors to support at least the TLS protocol. Therefore, it is widely deployed with WPA, and the interoperability between different vendors is high [7]. For wireless security, TLS is generally used more often than EAP.

Kerberos: Invented at MIT during the 1980s, this protocol has undergone some changes, been standardized by IETF and reached its fifth version. It relies on the ticket granting concept and symmetric ciphers, and is still used in some deployments.

LEAP (Cisco Light EAP): This is a Cisco devised protocol that is well suited to the 802.1X scheme. It is also based on MS-CHAPv1, which is considered an unsecure protocol.

PEAP (Protected EAP): The PEAP's main goal is to bring confidentiality to an EAP session and safeguard the credentials. If a secure protocol like TLS is not used within EAP, it is strongly recommended to use PEAP instead of EAP.

Other than these protocols, there is another component of the entire system: The users. Often, weak passwords are used by them; either short number of characters or dictionary words are used. In a work performed by Klein, even using a small 62,727 word dictionary; 25% of 13,797 passwords were cracked [8].

2. Problem Statement

The afore mentioned wireless technologies and protocols provide different levels of security to users. In a typical wireless network environment, a blend of these protocols is used. Although some protocols and security mechanisms provide adequate security, others may not. Using a strong security mechanism with a weak one may decrease the overall security posture of the wireless network. It may also provide its users with a false sense of safety. Most of the time, only a few aspects of the wireless network security are considered, while other aspects are omitted. Is it possible to analyze all pieces and components of the entire system? Do people realize that they are only as secure as the weakest component of the entire system? This study aims to find answers to these problems.

3. Objectives

This study aims to explore the 802.11 wireless network security protocols; identifying the underlying ciphering algorithms' strengths based on used key's bit length, and identifying the used passphrases' strength based on the number and type of passphrase characters used. Then, the weakest chain among those two components will be determined and a best practice approach to increase the overall security of a wireless network system will be suggested. This is the main objective of the study.

4. Method

The assessment of security protocols is done by determining the time taken to break a specific protocol's encryption key, or to pass an access control. This time is directly proportional to all possible keys that an encryption algorithm may have, or all possible values that an authentication protocol may require from the user. It is assumed that application specific integrated circuits (ASIC) are used to break any given encryption key.

In this study, the brute force resistance of security protocols is evaluated by assuming a 10,000 ASIC chip setup is established, and 1×10^{12} keys per second are tried. By using 10,000 ASIC circuits, 100M x 10,000 = 1×10^{12} keys per second can be tried. A 100 MHz ASIC can try 100M keys per second [9]. Schäfer [10] stated that it is feasible to achieve 10^{6} encryptions / μ second in today's technology, which is equivalent to 1×10^{12} keys per second.

Other cryptanalysis methods are not considered, as they are beyond the scope of this study.

5. Results

The assessment of a wireless security protocol can be found below by investigating the time required to break that specific protocol.

Table 1: WEP / WPA / WPA2 Key Strength: Cipher Strength

Key Length	Typically Found in	Duration of Cracking
40-bits	WEP	$2^{40} / 10^{12}$ = **1.09951163 seconds**
104-bits	WEP	$2^{104} / 10^{12}$ = **6.42724504 × 10^{11} years**
256-bits	WPA / WPA2	$2^{256} / 10^{12}$ = **3.67174306 × 10^{57} years**

The above table reveals that, without considering the cryptanalysis methods, breaking 40-bit keys takes just a second, but breaking 104-bit and 256-bit keys requires a considerable amount of time. Here, the weaknesses of the WEP algorithm and other attack methods may further reduce the key-breaking times.

Theoretically, ten totally random hexadecimal characters as a passphrase for 40-bit WEP, 26 totally random hexadecimal characters as a passphrase for 104-bit WEP, and 64 totally random hexadecimal characters as a passphrase for WPA/WPA2 Phrase Shared Key Mode can be used. However, in practice, it is known that this not the case. In their study, Morris and Thompson [11] found that among 3,289 user passwords investigated, eighty-six percent of all passwords were equal to or less than six characters long or were easily found in a dictionary or name list.

WPA2 passphrases are another kind of 'user password' and have the same vulnerabilities as traditional user passwords. A brute force attack analysis of the mentioned passphrases is outlined in Table 2.

In the table, the duration of crack times is calculated using two variables: the first variable being the encryption algorithm's passphrase strength (second column of the table), and the second variable being the character sets chosen to generate the passphrase (first line of the table).

With a 256-bit encryption algorithm, there are 2^{256} different key possibilities, and this leads to 3.67174306 × 10^{57} years of cracking time (Table 1). When using 63-character passphrases, which are created using 26-character sets or higher, even higher calculated values are possible. These values have no practical significance and are reduced to 3.67174306 × 10^{57} years on the last line of Table 2.

Table 2: WEP / WPA / WPA2 Key Strength: User Given Passphrase Strength

Encryption Algorithms' Key & Passphrase Lengths		**Passphrase Character Sets**				
Encryption Key Length (Bits)	**Passphrase Length (Characters)**	**Using 10 character set**	**Using 26 character set**	**Using 36 character set**	**Using 52 character set**	**Using 62 character set**
40-bits WEP	5 Characters	100 nano-seconds	11.881376 micro-seconds	60.466176 micro-seconds	380.204032 micro-seconds	916.132832 micro-seconds
256-bits WPA/WPA2 PSK	8 Characters	100 micro-seconds	0.208827065 seconds	2.82110991 seconds	53.4597285 seconds	3.63900176 minutes
104-bits WEP	13 Characters	10 seconds	28.7170471 days	5.40552424 years	644.093292 years	6 338.6573 years
256-bits WPA/WPA2 PSK	16 Characters	2.77777778 hours	1 381.90719 years	252 200.139 years	90 564 669.6 years	1.51067952 × 10^{9} years
256-bits WPA/WPA2 PSK	32 Characters	3.16887646 × 10^{12} years	6.02632354 × 10^{25} years	2.00717544 × 10^{30} years	2.58828625 × 10^{35} years	7.20177208 × 10^{37} years
256-bits WPA/WPA2 PSK	63 Characters	3.16887646 × 10^{43} years	4.40784416 × 10^{69} years (calculated)	3.53152965 × 10^{78} years (calculated)	4.06551865 × 10^{88} years (calculated)	2.63986518 × 10^{93} years (calculated)
256-bits WPA/WPA2 PSK	63 Characters	3.16887646 × 10^{43} years	3.67174306 × 10^{57} years (practical)	3.67174306 × 10^{57} years (practical)	3.67174306 × 10^{57} years (practical)	3.67174306 × 10^{57} years (practical)

Considering the duration of crack times, it is assumed that using a combination of key strength and a chosen character set is safe if it takes longer than 89.78 years to decipher the key; this is the greatest "average life expectancy of a human at birth" among all the countries [12].

Here, it is assumed that the following character sets are used when producing passphrases. Usage of different character sets yields different levels of security.

- Only Arabic numbers:

  0 1 2 3 4 5 6 7 8 9 (10 characters)

- Only Latin lowercase letters:

  a b c d e f g h i j k l m n o p q r s t u v w x y z (26 characters)

- Only Latin uppercase letters:

  A B C D E F G H I J K L M N O P Q R S T U V W X Y Z (26 characters)

- Either lowercase or uppercase letters + numbers (36 characters)
- Both lowercase and uppercase letters (52 characters)
- Both lowercase and uppercase letters + numbers (alphanumeric set) (62 characters)

From the table, it can be concluded that using five characters (typical for 40-bit WEP) or eight characters (minimum character number required for WPA/WPA2 PSK mode) offers no security. Even when producing the keys completely randomly from the alphanumeric set, it can be cracked in a very short time frame.

Generating 13 character passphrases by using only numbers or only lowercase or uppercase letters offers poor security. Only using numbers and letters combined seems to provide adequate security at 13 character passphrases, given that the passphrase is randomly produced.

Using a 16-character scenario seems to be secure unless the passphrase uses only numbers.

Using 32 or 63 character seems safe even if the passphrase only contains numbers.

Table 3: Minimum Number of Characters in the Set to Ensure a 'Secure' Passphrase

Encryption Key Length (Bits)	Passphrase Length (Characters)	Minimum Number of Characters in the Set
40-bits WEP	5 Characters	**19 516.289**
256-bits WPA/WPA2 PSK	8 Characters	**480.284174**
104-bits WEP	13 Characters	**44.6840764**
256-bits WPA/WPA2 PSK	16 Characters	**21.9153867**
256-bits WPA/WPA2 PSK	32 Characters	**4.68138726**
256-bits WPA/WPA2 PSK	63 Characters	**2.19032075**

For each of the passphrase lengths, Table 3 gives the minimum number of character sets that must be used in order to ensure the key can be recovered in 89.78 years. These values are calculated by using the following formula:

$$x = \sqrt[y]{\left(89.78\,years * 365\,\frac{days}{years} * 86400\,\frac{seconds}{days} * 10^{12}\,\frac{characters}{seconds}\right)}$$

where y denotes the passphrase length and x denotes the minimum number of characters in the set.

6. Conclusions

From these findings, it can be concluded that even if WPA2 with a 256-bit AES algorithm is used and this algorithm's code breaking requires 3.67 x 10^{57} years, using weak passphrases may reduce this security level and code breaking times to days or even minutes. The choice of the passphrase seems to be the weakest part of the entire 802.11 wireless security system. It is advised to ensure that adequate procedures are in place that guarantee that chosen passphrases are generated completely randomly, and that passphrases include the minimum required characters and are generated from character sets containing the minimum number of elements.

The assessment is completed by assuming that 10,000 x 100 MHz ASIC circuits are used for deciphering the key. Using the CPUs of personal computers or distributed computing may be other alternatives.

This work only focuses on brute force attacks. Future studies may focus on dictionary attacks, or using word lists. Furthermore, the entropy of language may be considered. For example, Shannon estimates that the English language has entropy of approximately 2.3 bits per letter [13], which means that even using a non-dictionary word or phrase may yield a less secure key than normally expected.

This work does not take into account wireless security protocol weaknesses. The key length of any given protocol is assumed to be the ultimate criterion. Future studies may focus on protocol weaknesses and attacks targeted on them.

References

[1] Peikari, C., & Fogie, S. (2002). *Maximum Wireless Security*, Sams Publishing.

[2] Zimmerman, T., & King, M. J. (2009), *IEEE Ratifies 802.11n to Begin New Era of High-Speed WLANs*, Research ID Number G00171326, Gartner Publishing.

[3] Edney, J., & Arbaugh, W.A. (2003). *Real 802.11 Security: Wi-Fi Protected Access and 802.11i*, Addison Wesley.

[4] Morii, M., & Ohigashi, T. (2009). *A Practical Message Falsification Attack on WPA.*

[5] Chandra, P. (2005). *Bulletproof Wireless Security: GSM, UMTS, 802.11 and Ad Hoc Security*, Elsevier Inc.

[6] Khan, J., & Khwaja, A. (2003). *Building Secure Wireless Networks with 802.11*, Indianapolis, Indiana: Wiley Publishing, Inc.

[7] Vladimirov, A.A., Gavrilenko, K.V., & Mikhailovsky, A.A. (2004). *Wi-Foo*, Addison Wesley.

[8] Klein, D. V. (1990). "Foiling the cracker: A survey of, and improvements to, password security", Proceedings of the 2nd USENIX Security Workshop.

[9] Gutmann, P. (2001). *Encryption and Security Tutorial*, e-book.

[10] Schäfer, G. (2003). *Security in Fixed and Wireless Networks: An Introduction to Securing Data Communications,* translated by Hedwig Jourdan von Schmoeger, Wiley&Sons Inc.

[11] Morris, R., & Thompson, K. (1979). "Password Security: A Case History", *Communications of the ACM*, v. 22, November 1979, pp. 594-597.

[12] Central Intelligence Agency (2009). *The World Factbook 2009*, Washington DC.

[13] Shannon, C. E. (1950). "Prediction and Entropy of Printed English", *The Bell System Technical Journal*, v.30, January 1951, pp.50-64.

2

Telematics Based Security System

1. A.V.Prabu 2.Baburao Kodavati 3.Tholada Appa Rao 4. E.rambabu
5. Shyam sundar tripathy

1.Lecturer in AE & IE Dept, 2.Asst.Prof in ECE Depr 3.Asst. Prof in EE Dept

4. Asst Prof in AE & IE Dept 5. Senior Training officer in Electronics Dept

1,2,3,4- Gandhi Institute Of Engg & Technology , Gunupur,Rayagada,Orissa-765022,India

5-NTTF at J.N.TATA Technical Education Center , Gopalpur,Orissa,India..

1-prabu.deva@gmail.com , 2-baburaokodavati@gmail..com ,
3-nithiamar@yahoo.in ,4- edubilliramu98@gmail.com &
5- syamtripathy1@gmail.com

Abstract--*This paper describes a new way of providing security for objects; the object can either be a file or an automotive like car, etc. The method used for providing security to objects is by creating a virtual fence around the object in such a way that whenever the object is moved out of the fence it is considered as an event and the event is notified to the user. Encryption is one of the techniques for providing security to objects, and the key used for encryption plays major role in providing security. This paper explains a new way of key generation which makes the file to be decrypted at the same location and by the same person (who knows the password) where it is encrypted, and the decrypted file is deleted whenever the fence is exited. This paper also explains a method for providing security to automobile by creating a fence around the vehicle. The engine automatically locks whenever the fence is exited and when the vehicle is used by an unauthorized person.*

Keywords: *Encryption, , Data Encryption System, HASH, RFID & GPS.*

1. INTRODUCTION

A. *Security*

In the computer industry, security refers to technique for ensuring that data stored in a computer cannot be read or compromised by any individual without authorization. Most security mechanisms [1] involve data encryption and passwords. Data encryption is the technique of translation of data into a form that is unintelligible without a deciphering mechanism. A password is a secret word or phrase that gives a user access to a particular program or system.

B. *Need of Security*

Computer security is necessary as the information is a strategic resource and a siginificant portion of organization budget is spent on maintaining information. Rising crime rates highten the need of people with regard to their home and car security. More people are now looking for autocop (-electronic device doing some sort of policing), for protecting their property against theft [2] and burglary.

The main idea of this paper is to make the object functioning only in a specified region and the functionality of the object is destroyed whenever the object is exited from that region. This can be achieved by using the latitude and longitude of the region as the key for the object to do its function. A Geo-fence is a virtual boundary on a geographic area. When that boundary is entered or exited it can be recognized as an event and the user can be notified of that event. This event information can be sent to a mobile telephone as well as to an email account.

C. *Geofencing Applications*

- **Vehicle Tracking System:** It is an electronic device installed in a vehicle to enable the owner or a third party to track the vehicle's location. Most modern vehicle tracking system uses GPS modules for accurate location of the vehicle. Many systems also combine communication components such as satellite transmitters to communicate the vehicle's location to a remote user. Vehicle information can be viewed on electronic maps via internet or specialized software.
- **Stolen Vehicle Recovery:** A method of tracing stolen property uses a radio receiver for receiving encoded signals from a central station. When the received signal corresponds to a unique code stored at the receiver, a GPS receiver and a radio transmitter located with the stolen property are connected for a predetermined amount of time to a source of power that a GPS signal received by the receiver is processed and the position data is computed and it is transmitted to the central station.

The organization of the paper is as follows: In Section 2 we discuss the related work about the latitude and longitude information and how to create a geofence. Section 3 discusses how geofencing based security is applied for the files and Section 4 discusses how geofencing based security can be applied to automotives. In Section 5 we derive our conclusions.

II. RELATED WORK

Every region has a unique value (latitude and longitude). This unique value can be used to identify the region. The region's latitude and longitude value is unique and doesn't vary with time (the rotation of the earth).The latitude and longitude value is represented in terms of degrees, minutes and seconds. These values vary from one region to another region, but do not vary for the same location.The following graph of latitude namely Fig 2 as a function of distance shows that the latitude varies from one region to another region. In the following graph, a point 'A' is considered and from that point at a distance of nearly 100ft to right, at point 'C', we can see a change in the variation of latitude's minute field. Similarly as we move from the point 'A' to 'B' there will be a change in minute field of latitude. So the latitude (and longitude) values can be used as key to control an object.

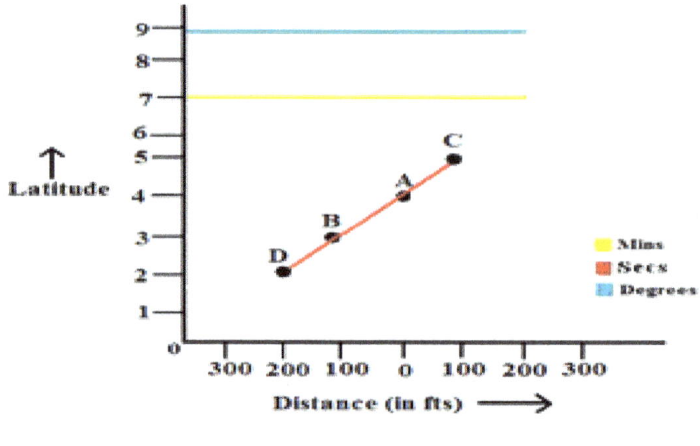

Fig 2-Variation of Latitude value with time

III. GEO FENCING BASED SECURITY TO FILES

Encryption is one of the techniques for providing the security to files. In encryption, the key used for encrypting the data plays an important role. The strength of the key increases the strength of the encryption. Hence the selection of key is vital.This paper proposes a new way for generating an input key which is given to a key generator algorithm. The key generator algorithm generates the key which is used for encrypting and decrypting the file.

A. GPS Receiver

The GPS Receiver is a device which continuously receives signal from the satellites about the latitude and longitude of a particular region, where the receiver is placed. It gets the information from the satellites in NMEA format. For latitude and longitude values to be accurate there must be at least three satellites from which the receiver should receive the information [3].

B. User Identity

User Identity is a password which uniquely identifies the person who encrypts the file. Generally it is about eight (8) alphanumeric characters to be kept secret for every individual.

C. Creating the virtual fence

The latitude and longitude values vary from place to place. Following table Table 1 gives detailed information on the variation of latitudes and longitudes from place to place.

Table 1 – Distance covered by Degree, Minute, Second values of Latitude and Longitude variation.

Latitude Longitude	Area covered (in kms)
1° of latitude	110.5743km (68.70768 miles)
1' of latitude	1.8429 km (1.1451 miles)
1"of latitude	0.0307151km (100.771 feet)
1°of longitude	111.3195 km (69.17073miles)
1' of longitude	1.8553 km (1.1528 miles)
1"of longitude	0.0309221km (101.45 feet)

From the above table it is clear that for a distance of about hundred feet, the latitude/ longitude values remain same. If we need better accuracy range, we need to have accurate GPS receiver(s). Inorder to create a fence either latitude or longitude or both should differ.

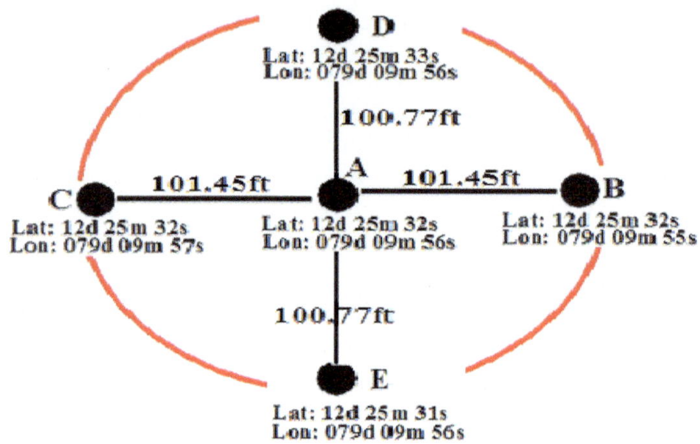

Figure 3 a- Creating a Virtual Geo fence

The above figure Fig 3a, explains the way to create a virtual fence around an object. Let us consider that an object is located at the point A (i.e., at Lat: 12d 25m 32s, Lon: 079d 09m 56s) on the earth. In order to create a virtual fence around this object, let us consider four points around this object in such a way that these points differ by only one second (1 sec) in their latitude and/or longitude values.

If we consider the point A as the object's position on the earth and consider four other points around A. Let these points be B, C, D and E. The distance between two points A and B (also A and C) is about 101 feet as we know that the distance between two longitudes when there is a variation of one second is 101 feet (approx) (from the above Table 1). Similarly, distance between the two points A and E (also A and D) is 100 feet (approx) as we see from the Table 1. So, if we join these four points we get an elliptical fence of approximately 90sq.mts area (~Π*r*r (area of circle, r is radius=100ft)) and, with in this area the value of the latitude and longitude will be same. Thus a virtual geographic fence can be created around the object, and if we use the latitude and longitude values as the key for locking of the object, the object can be localised with in an area of 90sq.mts.

D. Generation of Input Key

Once the latitude and longitude of the location is received, it must be verified whether the location lies inside or outside the fence. If the location lies inside the fence, the input key is generated by combining the latitude and longitude values received from the GPS receiver and the User Identity (password).The generated input key is given to a key generation algorithm [4], which in turn generates a key and it can be used for encrypting a file. A 64-bit input key is generated out of two keys namely LocationKey (generated from the latitude and longitude values) and UserIdentity key (generated from the User password called UIKey). For example,

Let the value of latitude be - 12 deg, 08 hrs, 15 sec

Let the value of longitude be - 79deg, 05hrs, 55sec

Let the user password is (maximum of 8 characters) - Anjali

We represent the digits of latitude and longitude using four bits. Thus, we get the following result:

12—00010010	079--000001111001
08—00001000	05--00000101
15—00010101	55--01010101

Concatenating these, yields a key (Location Key) of 52 bits (0001001000001000000101010000 01111 0010000010101010101). If we pad with twelve (12) 0s, we get a location key of length 64 bits (00010010000010000001010100000111001000001 01010101010000000000000). Let us convert each character of the password into ASCII value and then convert the ASCII values into binary as explained earlier. Hence the size of UIKey is given as:

UIKey = (number of characters in password*8) bits

'A'- 01000001	'a'- 01100001
'n' - 01101110	'l' - 01101100
'j' - 01101010	'i' - 01101001

UIKey= 0100000101101110011010100110000 10110110 0011010010000000000000000

If the length of the UIKey is less than 64-bits, let us pad it also with 0s, and then perform an XORing operation between the two keys namely the location key and the UIKey to get an input key.

Input key = (locationKey) XOR (UIKey).

=1010110010011001100000001001100 1000000111100001110101111111111111

The resultant input key can be directly used for encryption or it can be given as input key to another key generation algorithm which can generate the necessary keys.The File encrypted with this key can be decrypted only by the person who encrypted the file and the location at which the person encrypted the file.

E. Encryption Algorithm

F. For out implementation, we have the used the standard DES [5] algorithm for encryption.

As with any encryption scheme, there are two inputs to the encryption function; the plain text to be encrypted and the key. In this case, the plain text must be 64 bits in length and the key is 56-bits in length.

G. Architecture Diagram

i. File Encryption

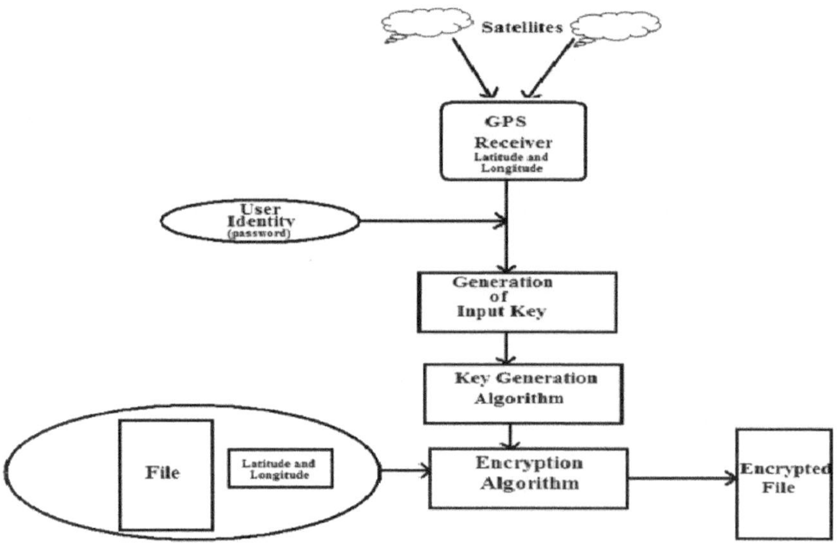

Fig 3b- File encryption using Geo fencing

During the encryption process (as shown in the Fig 3b) , the system first takes the position information (latitude, longitude) from the satellites using a GPS receiver and this position information is combined with the user password and an input key is generated. The generated input key is given to a Key Generator which generates key that can be given to the encryption algorithm. The encryption algorithm takes input a file that is to be encypted and the position information (latitude, longitude) and generates an encrypted file, which consists of the file data and the position information in encrypted form.

File Decrption:

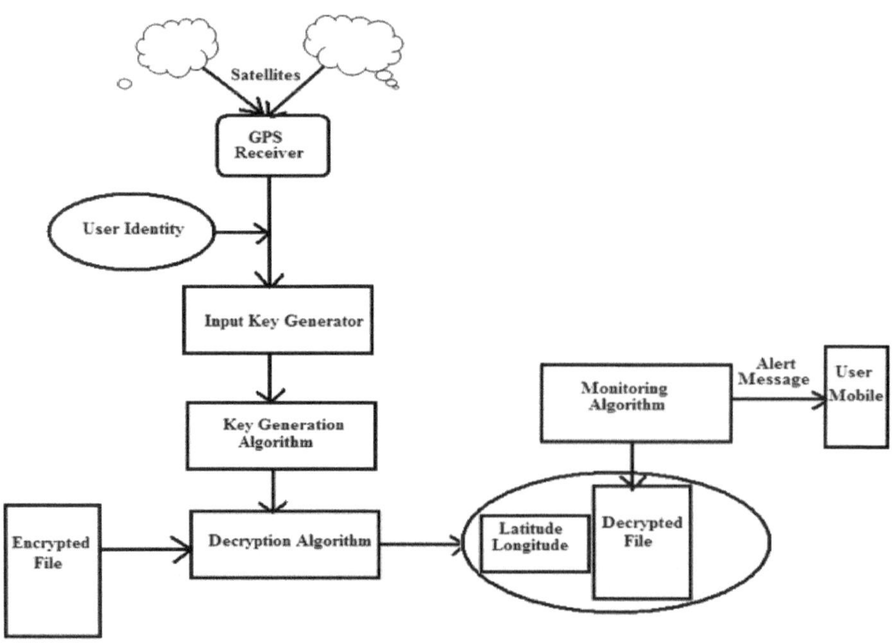

Fig 3c- File decryption using Geo fencing

H. Decryption Algorithm

The Decryption algorithm [6] transforms the cipher text into plain text and it is the reverse of the encryption algorithm.

During the decryption process, as shown in the Fig 3c, the system takes location information (latitude and longitude) from the satellites using a GPS receiver and this location information is combined with the user password and an input key is generated. The generated input key is given to a Key Generator which generates the key that can be given to the decryption algorithm. If the key generated is a valid key i.e, if the location where the file is encrypted and the person who is decrypting the file is same then the encrypted file is decrypted, and we get the original file and the location information. After the file is decrypted, the monitoring algorithm checks whether the file is within the fence by comparing the decrypted values of location information and the freshly received values from the GPS receiver. If the file exits the fence the decrypted file is deleted and an alert message is sent to the user mobile.

In the next section, we explain how the above mentioned Geofencing based technique can be extended to secure automobiles.

IV. GEO FENCING BASED SECURITY TO AUTOMOTIVES

In this section we eplain how the geofencing based security can be used in protecting parked automotives (particularly in open lawns, roadsides and streets).

Whenever a vehicle is parked at a particular location, this method takes the latitude and longitude values of the location of the parking place [7] and lock the engine. It also takes a password from the user and compares these values while unlocking the engine. The engine will be unlocked only when the both the values are matched. After unlocking the engine, this system continuously reads values from the GPS receiver and continuously compares it with the values where the vehicle was parked and locks the engine whenever there is a mismatch in the values. In order to avoid an unauthorized user guessing a password and trying to unlock an engine, we also use an RFID system. The user is expected to be tied with an RFID tag and the RFID reader is attached with the engine locking system. If the RFID tag details do not match then the engine is locked again.

A.RFID Receiver

RFID receiver is basically a radio frequency (RF) transmitter and receiver, controlled by microprocessor or Digital Signal Processor (DSP). The RFID reader, using an attached antenna, captures data from RFID tags, and then passes the data to a computer for processing.

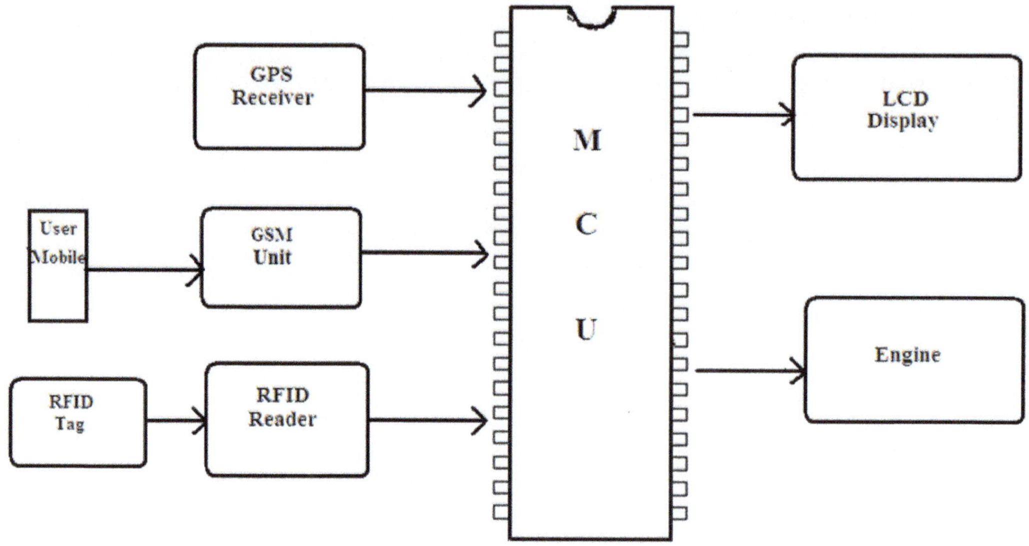

Fig 4a- Geofencing based security to Automotives

As shown in the above figure Fig 4a, while locking or unlocking, the system takes the latitude and longitude information of the place where the vehicle is parked, and also it takes password from the user. The user sends password from his mobile using an SMS (Short Messaging Service). The System allows the user to move out of the fence only if the user posses the correct RFID tag.

B. Locking Phase

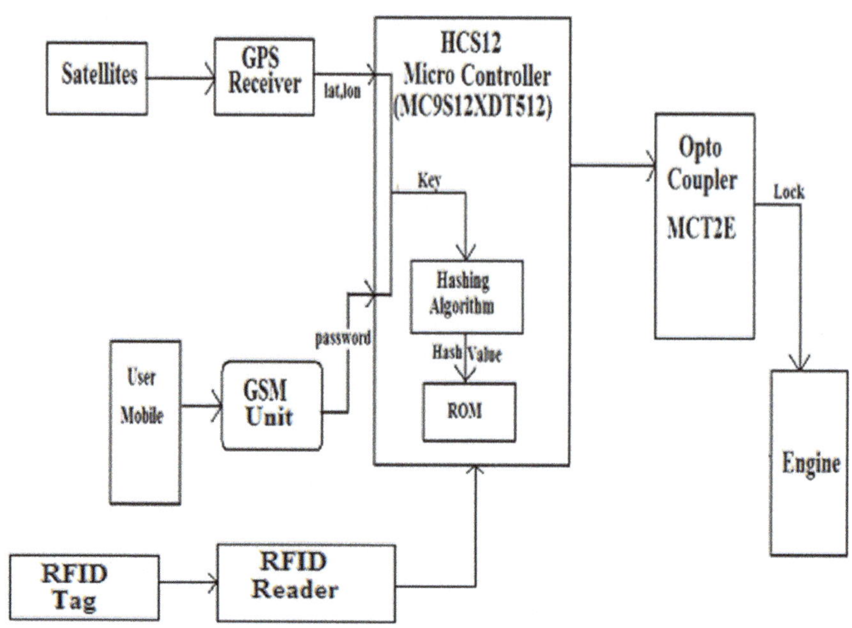

Fig 4b- Locking Engine using Geofencing

As shown in the Fig 4b while locking the car, the GPS receiver present in the car receives the latitude and longitude values of the car location [8] from the satellites and these values are sent to the microcontroller (MC9S12XDT512) (in our case). Simultaneously the user of the car sends a password through his mobile, using the SMS (Short Messaging Service) service provided, to the GSM present in the car, and then the password is retrieved from it by microcontroller. In the microcontroller the above two values are mixed and a key is generated. This generated key is then passed to a hashing algorithm which generates a hash value. The generated hash value is stored in the read only memory (ROM) of the microcontroller and using an opto-coupler (MCT2E), it converts the voltage signal to light, and switch off the battery, subsequently the engine is locked.

C. Hashing Algorithm

A Hashing algorithm generates a hash value 'h' by a function H (Hash function) of the form h = H (M); where M is the variable-length message and H (M) is the fixed length hash value.

D. Unlocking Phase

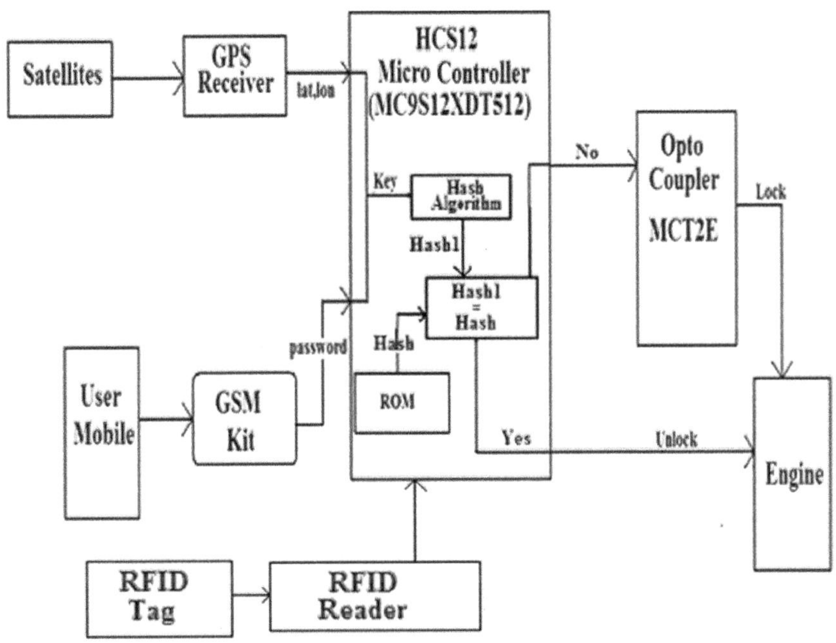

Fig 4c- Unlocking Engine using Geofencing

As shown in the above figure Fig 4c while unlocking the car, the GPS receiver present in the car receives the latitude and longitude values of the car location [7] from the satellites and these values are sent to the microcontroller (MC9S12XDT512). Simultaneously the user of the car sends a password from his mobile, using the SMS service provided, to the GSM unit present in the car. Then the password is retrieved from it by microcontroller.

In the microcontroller the above two values are mixed and a key is generated (key). This generated key is then passed to a hashing algorithm which generates a hash value (Hash1) and then compared with the hash value (Hash) stored in the read only memory (ROM) of the microcontroller and if there is no mismatch in the values then we understand that the car is within the fence then the engine is unlocked but when there is mismatch i.e., when the car is moved out of the fence (identified by latitude and longitude values) or change in the password then using a opto-coupler (MCT2E) we send low voltage which switch off the battery and subsequently the engine will be locked.

V. CONCLUSION

This paper proposes a new way of providing security to protect objects. Usually, the basic mechanism for securing the objects is to protect it by using a password but the security mechanism fails once the password is cracked. The proposed method in this paper generates a key by using both the Geo-positioning data and user password which is used for protecting the object. Applying the Geo-position data and user password as a key is a better approach than the existing mechanisms as it creates a virtual fence around the object and the object loses its functionality whenever it exits the fence. Even if the password is cracked the geo-positioning data adds security to the object. Providing the security to automotive like car by using the proposed method is better approach as it makes the car to be unlocked only at the place where it has been locked. The RFID module ensues that only the person who parked the car can unlock the engine. If any unauthorized person attempts to use the car then it locks the engine.

REFERENCES

[1] MarkoWolf, AndreWeimerskirch, and ThomasWollinger, "State of the Art: Embedding security in vehicles", Journal on Embedded Systems, Volume 2007, Article ID 74706.

[2] J.-P. Hubaux, S. C Apkun, and J. Luo, "The security and privacy of smart vehicles," *IEEE Security & Privacy Magazine*, vol. 2, no. 3, pp. 49–55, 2004.

[3] Alison Brown, JacobGriesbach and Bruce Bockius, "GPS traking location based service using wrist watch GeoZigBee Sensors", Proceddings of ION NTM, 2007, Som Diego, pp 1-10, December 2007.

[4] W.Stallings, "Cryptography and Network Security", Prentice-Hall, Englewood Cliffs, NJ, USA, 4th edition, 2005.

[5] Ingrid Verbourwhede, Frank Hoornaert, Joos Vandewalle, Hugo J. Deman "Security and performance optimization of a new DES", IEEE Journal on Solid State ciruits, vol. 23, no.3, pp 647-656, 1999.

[6] National Institute of Standards & Technology, "FIPS-46-3: Data Encryption Standard (DES)," October 1977, reaffirmed in October 1999.

[7] Sinpyo Hong, Man Hyung Lee, Sun Hong Kwon, and Ho Hwan Chun, "A Car test for the estimation of GPS/INS alignment errors", IEEE TRANSACTIONS ON INTELLIGENT TRANSPORTATION SYSTEMS, VOL. 5, NO. 3, pp 208-218, SEPTEMBER 2004.

[8] Youjing Cui and Shuzhi Sam Ge,"Autonomous vehicle positioning with GPS in urban canyon environments", IEEE TRANSACTIONS ON ROBOTICS AND AUTOMATION, vol. 19, NO. 1, pp 15-25, February 2003.

3

ENHANCED THREE TIER SECURITY ARCHITECTURE FOR WSN AGAINST MOBILE SINK REPLICATION ATTACKS USING MUTUAL AUTHENTICATION SCHEME

Linciya.T[1] and Anandkumar. K.M[2]

[1]P.G Scholar, Department of Computer Science and Engineering, Easwari Engineering College, Chennai, India
linciya.thomas@gmail.com

[2]Assistant Professor (Sl.Gr.), Department of Computer Science and Engineering, Easwari Engineering College, Chennai, India
kmanandmss@gmail.com

ABSTRACT

Recent developments on Wireless Sensor Networks have made their application in a wide range such as military sensing and tracking, health monitoring, traffic monitoring, video surveillance and so on. Wireless sensor nodes are restricted to computational resources, and are always deployed in a harsh, unattended or unfriendly environment. Therefore, network security becomes a tough task and it involves the authorization of admittance to data in a network. The problem of authentication and pair wise key establishment in sensor networks with mobile sink is still not solved in the mobile sink replication attacks. In q-composite key pre distribution scheme, a large number of keys are compromised by capturing a small fraction of sensor nodes by the attacker. The attacker can easily take a control of the entire network by deploying a replicated mobile sinks. Those mobile sinks which are preloaded with compromised keys are used authenticate and initiate data communication with sensor node. To determine the above problem the system adduces the three-tier security framework for authentication and pair wise key establishment between mobile sinks and sensor nodes. The previous system used the polynomial key pre distribution scheme for the sensor networks which handles sink mobility and continuous data delivery to the neighbouring nodes and sinks, but this scheme makes high computational cost and reduces the life time of sensors. In order to overcome this problem a random pair wise key pre distribution scheme is suggested and further it helps to improve the network resilience. In addition to this an Identity Based Encryption is used to encrypt the data and Mutual authentication scheme is proposed for the identification and isolation of replicated mobile sink from the network.

KEYWORDS:

Wireless Sensor Network, Pair wise Key Establishment, Random Key Pre distribution, Mobile Sink, ID Encryption, Mutual Authentication Scheme.

1. INTRODUCTION

Recent development in electronic and computer technologies made the way for emergent of Wireless Sensor Networks (WSN). The WSN consists of wide distributed sensors to monitor physical or environmental conditions, such as temperature, sound, pressure, etc. and to pass their data through the network to a destination. The advancement of wireless communication technologies and rooted computing, are being widely adapted into many applications through sensor networks and many active researches on related subject are being carried out. Several sensor nodes are connected to build the wireless sensor networks. The large number of small autonomous devices are embedded in the sensors which are interconnected to form a sensor network. The WSN make use of a number of sensor nodes within or neighbouring the area of event to not only collect and integrate but also process and relay the information. Sensor nodes are equipped with integrated sensors that has the capability to process the collected data and for short range radio communication. The randomly deployed sensor nodes in the area of scrutiny collect the sensor data in typical application scenarios.

In application such as military sensing and tracking, environment monitoring, patient monitoring and tracking, smart environment, etc are sensed by the deployed sensor. Those sensors are interconnected to form the sensor network and are deployed in hostile environment where they are prone to different malicious attack. There are various attacks in the WSN. For example, messages can me spoofed during the transit and then the copy of the message can be sent to the recipient in the altered form. One can also intrude with a node and alter its behaviour. The types of counter measures can be addressed by different types of attacks.

The most formidable attack in wireless sensor network is the replication attack. The attacker compromises a node and uses its secret cryptographic key materials to effectively colonize the network with the clones of it. Foe collects all the credentials like keys, identity etc once the node has been captured. Those captured credentials are reprogrammed by the attacker and the nodes are replicated using those credentials in order to eavesdrop the transmitted messages or compromise the functionality of the network. The replication attack is carried in two ways: mobile sink replication attack and stationary access node replication attack. There are some solutions available for stationary access point replication attack compare with mobile sink replication attack. Many important functions of the sensor network such as routing, resource allocation, misbehaviour detection, network resilience, throughput etc are remarkably injurious by replication attack. The above problem of replication attack can be solved through polynomial pool pr distribution scheme.

There are number of key pre distribution schemes for solving the problem of authentication and pair wise key distribution which do not exhibit desirable network resilience. The breach of security in one node makes the entire sensor network unsafe. The new security challenge for data collection is by avocation of pair wise key establishment and authentication in the mobile sinks. In the basic probabilistic and q-composite key pre distribution schemes, the attacker gain the control of entire network by deploying a replicated mobile sink preloaded with some compromised keys. Those compromised keys can be achieved by an attacker by obtaining large number of keys by capturing a small fraction of nodes.

To handle the afore-said problem, we use pair wise key pre distribution scheme that provide authentication and pair wise key establishment amid sensor nodes and Mobile Sink. To expedite the study of a new security technique, a general three-tier security agenda is established for authentication and pair wise key establishment, based on the polynomial pool-based key pre distribution scheme.

In polynomial pool-based key pre distribution scheme both mobile sink and stationary access point generates the separate subset of keys which results in the high computational cost. In order to overcome this problem Random pair wise pre distribution scheme is used to reduce the computational cost and provides security against the replication attacks in the proposed system. In Random pair wise pre distribution scheme only mobile sink generates a key with key identifiers and broadcasted to stationary access point and sensor node. Fig 1 represents the system architecture of the prospective system.

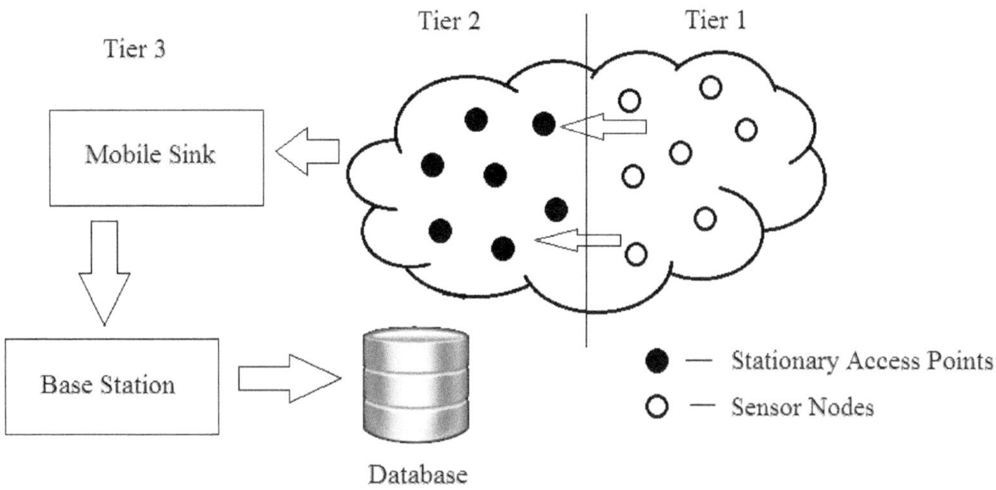

Fig 1. System Architecture

The rest of the paper is organized as follows. Section 2 discusses the previous work on related topics and Section 3 gives the models and assumptions. Section 4 describes proposed Random Pair wise Pre distribution in detail. Section 5 discusses the encryption algorithm. Section 6 discusses the analysis results and Finally, Section 7 concludes the paper.

2. Related Works

A comprehensive literature review was carried out on node replication attacks and other security issue in static and mobile wireless sensor networks. This literature survey revealed versions mathematical and operational methods for solving the problems. The key management problem is a progressive research area in wireless sensor networks. Eschenauer and Gilgor [4] proposed a probabilistic key pre distribution scheme which helps to bootstrap the initial trust between the sensor nodes. The discriminative distribution and countermanding of keys to sensor nodes and node re-keying without substantial computation and communication capabilities are briefly explained in this scheme. For the random subset of keys from a large key pool before deployment, the probabilistic key pre distribution scheme is used. As an outcome, two nodes have assured a prospect to share at least one key after deployment.

The characteristic of network connectivity and security can be poised by key management scheme. The core idea is the sensor node randomly picks a set of keys from a key pool before deployment, so that any two sensor nodes had a certain prospect to share at least one common key. The problem existing in this key pre distribution is the probability of establishing a common key between communicating nodes, and the ability to tolerate compromised nodes is highly dependent on the memory available on the sensor nodes. In other words, these schemes still require each node to be assigned a large number of keys for a high-performance WSN, and some solutions for this problem focus on pre deployment knowledge, post deployment knowledge, the state of the sensors, and overlapping key strings. However, these solutions are not quite viable since their key sharing mechanism is not enough to be efficient. One of the most important information is the signal range of the sensor node that might significantly improve the performance of the key sharing mechanism however, this is not exploited in these schemes.

Liu et al.[3] The sensor nodes are communicated securely with each other using cryptographic techniques by enabling bi variate key polynomial. This is one of the most primary security services. The resource constraints of the sensor nodes makes it not feasible for sensors to use key polynomial scheme. The scheme assures a direct key established between any two neighbour sensors in any deployment group. Better trade-off is achieved between communication overhead, network connectivity and security against node capture compared to the existing key pre-distribution schemes. Ultimately, it also supports dynamic node addition after the initial deployment of the nodes in the network.

Chan et al. [2] further unmitigated this proposal and developed two key predistribution schemes: the q-composite key predistribution scheme and the random pairwise keys scheme. In q-composite key predistribution scheme two sensor nodes are required to compute compute a pairwise key from the shared q predistributed keys. In random pairwise keys scheme pair of sensor nodes are randomly picked and assign each pair a unique random key. Comparing to the basic probabilistic key predistribution scheme the above schemes improved the security. They developed a general scaffold for pairwise key establishment using the polynomial-based key pre distribution protocol [5] and the probabilistic key distribution in [6] and [7]. Tolerating not more than t compromised node, where the value of t was limited by the memory available in the sensor nodes can be carried in this scheme.

Amar Rasheed et.al [1] developed for group key pre distribution. The pair wise key establishment in the context of sensor networks is the ultimate aspiration. This system is unreservedly secure and t-collusion resistant. The network resilience is significantly improved to the mobile sink replication attack. The small fraction of preselected sensor nodes are called as sensor nodes. In this new security agenda, stationary access nodes act as an authentication access point to a network to elicit the sensor node to transmit the collected to the mobile sink. The data request message from the mobile sink to sensor node is transmitted through stationary access node. These data request will elicit the sensor node to transmit the data from sensor node to mobile sink through stationary access point.

3. Models and Assumptions

3.1 Network Model

We consider a sensor networks, where a large number of sensor nodes are randomly distributed in the two dimensional area and remain stationary after deployment, and all of them have similar capabilities and equal significance. One sink which has the mobility named as mobile sink which collects the data from all other sensor nodes. Each sensor node sends its data packets to sink nodes through access point. For the secure data transmission we go three tier security scheme.

3.2 Three Tier Security Scheme

The two separate polynomial pools are used in three tier security scheme: the mobile polynomial pool and the static polynomial pool. For data gathering from the sensor through the mobile sink the polynomials from mobile polynomial pool are used. To gain access to the network for the sensor's data gathering, an attacker needs to compromise at least a single polynomial. For key setup between the sensor nodes and stationary access nodes the polynomials from static polynomial pool are used.

Afore mention to deployment, The subset of polynomials from the mobile polynomial pool is picked by each mobile sink. The randomly selected sensor nodes called stationary access nodes carry a polynomial from the mobile polynomial pool. These nodes acts as the authentication point for the network which triggers the sensor node to transmit the data. The data request is transmitted to sensor node from the mobile sink through stationary access node. This data request will initiate the sensor node to transmit the collected data. Every stationary access node may share a mobile polynomial with a mobile sink. The subset of polynomials from both static and mobile polynomial pool is randomly picked by stationary access point and sensor node. The main benefit to use separate pools is the mobile sink authentication is autonomous of the key distribution scheme used to unite the sensor network

3.2.1 Key Discovery between a Mobile Sink and a Sensor Node

To establish a direct pair wise key between sensor node u and mobile sink v, a sensor node u needs to find a stationary access node a in its neighbourhood, such that, node a can establish pair wise keys with both mobile sink v and sensor node u. Fig 2 shows a direct secure path establishment between nodes u and v, mobile sink v sends the pair wise key to node a in message encrypted and authenticated with the shared pair wise key ;a between v and a. If node a receives the above message and it shares a pair wise key with u, it sends the pair wise key to node u in a message encrypted and authenticated with pair wise key; u between a and u.

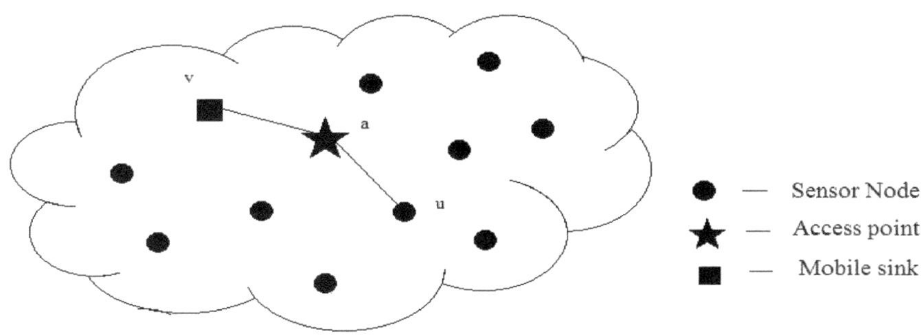

Fig 2. Direct Key Establishment

Fig 3 illustrates that the mobile sink and the sensor node will have to establish a pair wise key with the help of other sensor nodes using indirect key discovery. To establish a pair wise key with mobile sink u, a sensor node v has to find a stationary access node a in its neighbourhood such that node a can establish a pair wise key with both nodes u and v. If node (a) establishes a pair wise key with only node v and not with u. As the probability is high that the access node a can discover a common mobile polynomial with node v, sensor node u needs to find an intermediate sensor node i along the path u - i - a - v, such that intermediate node i can establish a direct pair wise key with node (a).

Fig 3. Indirect Key Discovery through Intermediate Stationary Access Node

Fig 4 shows that the mobile sink and the sensor node will have to establish a pair wise key with the help of other sensor nodes using indirect key discovery. To establish a pair wise key with mobile sink u, a sensor node v has to find a stationary access node a in its neighbourhood such that node a can establish a pair wise key with both nodes u and v. If node (a) establishes a pair wise key with only node v and not with u. As the probability is high that the sensor node a can discover a common mobile polynomial with node v, sensor node u needs to find an intermediate sensor node i along the path u - i - a - v, such that intermediate node i can establish a direct pair wise key with node (a).

Fig 4. Indirect Key Discovery through Intermediate Stationary Node

4. RANDOM PAIR WISE KEY PRE DISTRIBUTION

In this section, we first list phases for pair wise key setup in Random Key Pre distribution (RKP) schemes

A. Phases in Random Key Pre distribution Schemes

Four main phases for key setup in RKP schemes are presented as follows.

1. Key pre distribution phase:

   A mobile sink has a polynomial pool.

   - generate a large key pool of size ;
   - different keys for each sensor from the key pool to form a key ring are selected randomly;
   - In the memory of the sensor the key ring is loaded.
   - Each sensor is loaded with unique node identifier or key identifier.

2. Sensor deployment phase:

   Sensors are erratically picked and uniformly dispersed in a large area. Typically, the average number of neighbors of a sensor is much smaller than the total number of deployed sensors.

3. Key discovery phase:

   Two steps are involved in the key discovery phase. In the first step, each sensor attempts to discover shared key(s) with each of its neighbors. To accomplish this, the sensor can broadcast its key ring identifier to its neighbours. After the first step of the key discovery phase, the sensor knows all its neighbors. The set of all neighbors of sensor is represented by W_i and $|W_i|=n'$. The set of neighbours of sensor i who share at least one key with the sensor i is represented by R_i. Thus, we have $W_i = Q_i \cup R_i$ and $|Q_i|+|R_i|= n'$. In the second step, every sensor i broadcasts its set Q_i. Using the sets received from neighbors, a sensor can build a key graph based on the key-share relations among neighbors.

4. Pair wise key establishment phase:

   If sensor shares at least one key with a given neighbor, the shared key(s) can be used as their pair wise key(s).

In the proposed system the random pair wise key is used for the security and for reducing the computational cost. Figure 5 shows the functional architecture for the three tier security scheme. Sensor nodes collect the data from the remote area and send to the mobile sink. The data that are collected from sensor node is forwarded to access point which in turn sent to mobile sink. For secure transmission of data we generate keys from the polynomial pool. The key is generated in sensor node then it encrypt the collected data using the generated key and transfer it to the mobile sink.

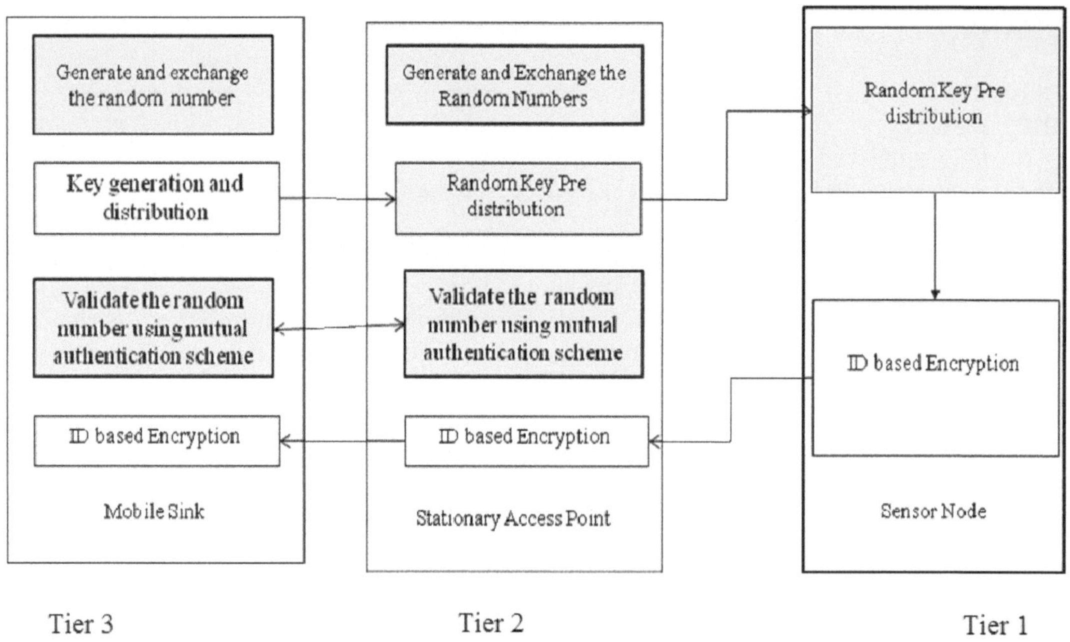

Fig 5. Functional Architecture of the Proposed System

For the encryption of the data using the key ID based encryption algorithm is used. For the ID based encryption algorithm its essential for the private key generator which is to be located in the base station from which the master key to be stored in the mobile sink.

5. ID BASED ENCRYPTION ALGORITHM

ID-based public key infrastructure involves a trusted Key Generation Center (KGC), and users. The basic operations consist of **Set Up** and **Private Key Extraction**. KGC runs BDH parameter generator to generate two groups G_1, G_2 and a bilinear pairing e: $G_1 \times G_1 \rightarrow G_2$. Itchooses an arbitrary generator $P \in G_1$ and defines two cryptographic hash functions:

$$H_1 : \{0,1\}^* \rightarrow G_1,$$
$$H_2 : \{0,1\}^* \rightarrow G_2.$$

- **Set Up:** KGC chooses a random number $s \in Z_q^*$ and set $P_{pub}=sP$. Then the KGC publishes system parameters params=$\{G_1,G_2,q,P,P_{pub},H_1,H_2\}$, and keep s as master-key.
- **Private Key Extraction:** An identity information ID to KGC is submitted by the user. The user's public key is computed by KCG as $Q_{ID}=H_1(ID)$ and reversion private key as $S_{ID}= sQ_{ID}$.

The holder of private key S_{ID} to decrypt a message sent to her under the public key $Q_{ID.}$
Let m denote the message to be encrypted.

- **Encryption:**
  Compute U=rP where $r \in_R Z_q^*$. Then compute
  $$V = m \oplus H_2(e(P_{pub},rQ_{ID})) \qquad (1)$$
  Output the cipher text (U,V)

- **Decryption:**
  Decryption is performed by computing

$$V \oplus H_2(e(U,S_{ID})) = V \oplus H_2(e(rP,sQ_{ID})) \quad (2)$$
$$= V \oplus H_2(e(sP,rQ_{ID})) \quad (3)$$
$$= V \oplus H_2(e(P_{pub},rQ_{ID})) \quad (4)$$
$$= m$$

The node ID is used for the creation of the private key and public key for the secure transmission of data from the sensor node to the mobile sink. Hence ID based encryption algorithm is adhibited for the encryption and decryption of the data.

6. Mutual Authentication Scheme

Challenge–response authentication is an authentication process by verifying an identity using the required authentication information. This authentication information is usually a value and that are to be computed in response for inevitable challenge. Challenge–response authentication helps to crack the problem of exchanging session keys for encryption. This is chiefly effective for man-in-the-middle attack, because without knowing the secret the attacker will not be able to derive the session key. Through this protocol we are able to state things other than the secret value. Every challenge–response sequence is unique to ensure a challenge by employing a cryptographic nonce in the authentication protocol.

Mutual authentication is percolated using a challenge–response handshake in both directions. Mutual authentication is the process to authenticate the both ends of transmission that is authentication between sender and receiver. This is the basic function for both private and public communication. It is also known as two-way authentication where two parties authenticating each other suitably. It refers to a one node authenticating themselves to another node and that node authenticating itself to the first node in such a way that both parties are assured of the others' identity.

Mutual authentication scheme is anticipated in this paper for identifying the replicated mobile sink. Figure 6 describes the pseudo code for mutual authentication scheme. Mobile Sink (MS) sends mobile sink challenge (MSC) to the Access point (AP). AP generates the challenge and Access point response (APR) calculates the hash function using APC, MSC and the random number. If MS receives is not equal to null then MS calculates the estimated value using APC and APR. If the estimated value equal to APR then mobile sink response (MSR) evaluates the hash function using APC, MSC and random number. If AP receives MSR and AP is not equal to null then AP calculates the estimated value using MSC and MSR. If the estimated value equals to MSR then MS is authenticated as a trusted node or else MS is the replicated node.

```
MS ←Mobile Sink, AP ←Access Point
MSC ←Mobile Sink Challenge, APC ←Access Point Challenge
MSR ←Mobile Sink Response, APR ←Access Point Response
        if (MS in AP range)
                AP ←recv (MSC, AP)
AP → gen(APC)
APR ←HashA(APC+MSC+random number)
        if (MS ←recv(APC,APR,MS)!=NULL)
                MS Calculate the estimated value using received APC and APR
                        if ( estimated value== APR)
                                MSR ←HashA(APC+MSC+random number)
        if (AP ←recv(MSR,AP)!=NULL)
                AP calculate the estimate value using MSC and MSR
                        if (estimated value==MSR)
                                MS is authenticated as a trusted node
                        else
                                MS is the replicated node
```

Fig 6. Pseudo code for Mutual Authentication Scheme

7. RESULT ANALYSIS

The proposed scheme has been implemented and analysed in the network simulator NS2. The possible outcomes when there is a change in the key size is to be determined. The various sets of key size is given and the change of key size broadcasted from mobile sink to stationary access node and stationary access node to sensor node is to be determined. Table 1 shows the calculation of broadcasting the key from the mobile sink to stationary access node and stationary access node to sensor node.

Table 1. Calculation of Broadcasting the Key of Various Size

Size of Key Generated in Mobile Sink	Size of Key Broadcasted to Stationary Access Node	Size of Key Generated in Stationary Access Node	Size of Key Broadcasted to Sensor Node
5	4	5	4
10	9	10	9
20	19	20	19
30	29	30	29
40	39	40	39

If the key size is larger it should be stored in a disk file which can be discovered by someone else. Thus large key size is not only convenient to use but also it is a security risk. It considerably takes longer time to encrypt and decrypt messages and broadcast the generated key. So it is convenient to use the smaller key size. The key size used in this project is 4 (32 bit). Thus the following analysis shows that as the size of the polynomial pool increases the probability of sharing the key size also increases

The key is generated in the static polynomial pool and it is broadcasted to the sensor node. Fig. 7 and Fig. 8 represent the sharing of the key from static polynomial pool to the stationary access node and sensor node.

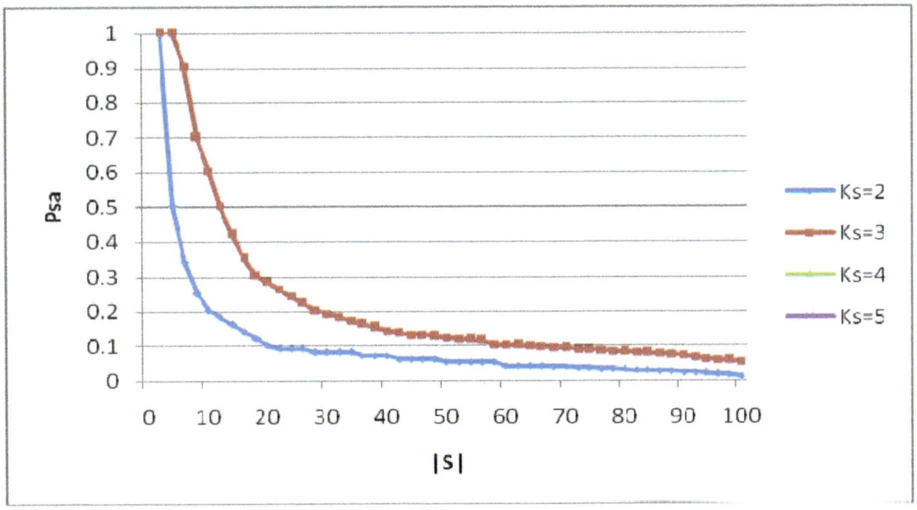

Fig 7. The probability P_{sa} that a sensor and stationary access node share a static polynomial versus the size $|S|$

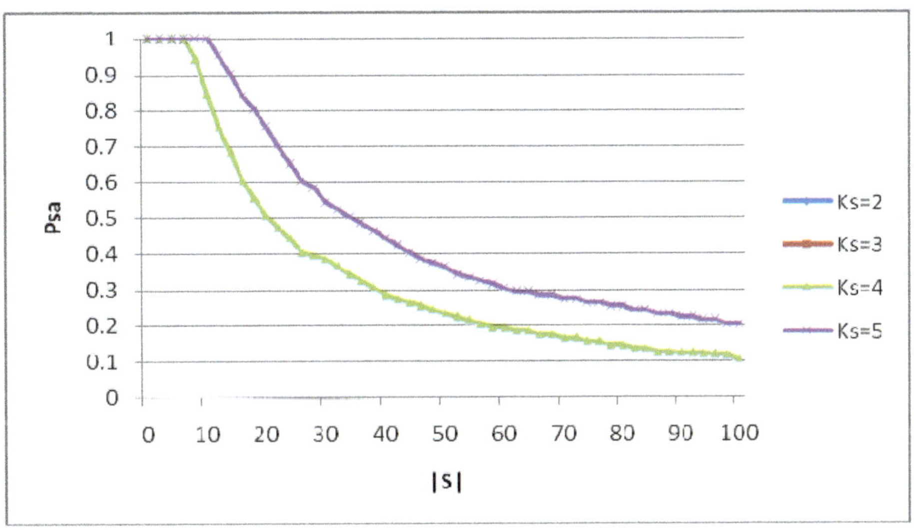

Fig 8. The probability P_{sa} that a sensor and stationary access node share a static polynomial versus the size $|S|$.

7. Conclusions

The security performance of the three tier scheme can be carried out by strengthening the authentication mechanism between stationary access nodes and sensor nodes. We used the one-way hash chains algorithm in conjunction with the static polynomial pool-based scheme.

The network resilience is significantly improved to the mobile sink replication attack using random pair wise key pre distribution compared to single polynomial pool-based key pre distribution approach. The use of key pools and stationary access nodes that carry polynomials from the mobile pool of the network will inhibit an attacker from collecting sensor data, by deploying a replicated mobile sink. The proposed key distribution scheme reduces the computational cost in the nodes.

REFERENCES

[1] Amr Rasheed, Rabi Mahapatra. N. (2012) 'The Three-Tier Security Scheme in Wireless Sensor Network with Mobile Sinks' IEEE Transactions on Parallel and Distributed system, IEEE Computer Society, VOL. 23, NO. 5, pp 958-965.

[2] H. Chan, A. Perrig, and D. Song(2003), "Random Key Pre-Distribution Schemes for Sensor Networks," Proc. IEEE Symp. Research in Security and Privacy.

[3] D. Liu, P. Ning, and R.Li(2003) " Establishing Pairwise Keys in Distributed Sensor Networks," Proc. 10th ACM Conf. Computers and Comm. Security, pp. 52-61.

[4] L. Eschenauer and V.D. Gligor(2002) "A Key-Management Scheme for Distributed Sensor Networks," Proc. ACM Conf. Computer Comm. Security, pp. 41-47.

[5] H. Chan, A. Perrig, and D. Song(2004), "Key Distribution Techniques for Sensor Networks," Wireless Sensor Networks, Kluwer Academic, pp. 277-303.

[6] D. Liu and P. Ning,(2003) "Location-Based Pairwise Key Establishments for Static Sensor Networks," Proc. First ACM Workshop Security Ad Hoc and Sensor Networks.

[7] A. Rasheed and R. Mahapatra(2009), "A Key Pre-Distribution Scheme for Heterogeneous Sensor Networks," Proc. Int'l Conf. Wireless Comm. and Mobile Computing Conf. (IWCMC '09), pp. 263-268.

[8] T. Gao, D. Greenspan, M. Welesh, R.R. Juang, and A. Alm,(2005) "Vital Signs Monitoring and Patient Tracking over a Wireless Network," Proc. IEEE 27th Ann. Int'l Conf. Eng. Medicine and Biology Soc. (EMBS).

[9] H. Deng, W. Li, and D.P. Agrawal,(2002) "Routing Security in Wireless Ad Hoc Networks," Proc. IEEE Comm. Magazine, pp. 70-75.

[10] Parno B, Perrig A, Gligor V.(2005) "Distributed Detection of Node Replication Attacks in Sensor Networks" In: Proceedings of the IEEE Symposium on Security and Privacy; pp.49-63.

[11] Y. Tirta, Z. Li, Y. Lu, and S. Bagchi, (2004)"Efficient Collection of Sensor Data in Remote Fields Using Mobile Collectors," Proc. 13th Int'l Conf. Computer Comm. and Networks (ICCCN '04).

[12] M.Conti, R. Di Pietro,L.V. Mancini and A. Mei (2007)"A randomized, efficient, and distributed protocol for the detection of node replication attacks in wireless sensor networks" In ACM MobiHoc, pages 80-89.

[13] Jun-Won Ho, Donggang Liu, Matthhew wright (2009) " Distributed Detection of replica node attacks with group deployment knowledge in wireless sensor networks", Ad Hoc Networks, pp:1476-1488.

[14] Chia-Mu, Chun-Shien, Lu., and Sy-Yen Kuo (2008) "Mobile Sensor Network Resilient Against Node Replication Attacks", SECON '08. 5th Annual IEEE communications Society conference, pp.597-599.

[15] Chia-Mu, Chun-Shien, Lu., and Sy-Yen Kuo (2009) "Efficient Distributed and Detection of Nodes Replication Attacks in Mobile Sensor Networks", IEEE 2009.

[16] A. Rasheed and R. Mahapatra,(2007) "An Energy-Efficient Hybrid Data Collection Scheme in Wireless Sensor Networks," Proc. Third Int'l Conf. Intelligent Sensors, Sensor Networks and Information Processing.

[17] W. Zhang, G. Cao, and T. La Porta(2003), "Data Dissemination with Ring-Based Index for Wireless Sensor Networks," Proc. IEEE Int'l Conf. Network Protocols (ICNP), pp. 305-314.

[18] L. Hu and D. Evans(2004), "Using Directional Antenna to Prevent Wormhole Attacks," Proc. Network and Distributed System Security Symp.

4

IMPLEMENTING A SECURE KEY ISSUING SCHEME FOR COMMUNICATION IN P2P NETWORKS

Mohammed Azharuddin , Annapurna P Patil

Department of Computer Science and Engineering, M.S.Ramiah Institute of Technology, Bangalore-560054, India
azharadhoni@yahoo.com, annapurnap2@yahoo.com

ABSTRACT

Key issuing scheme focuses on the confidentiality maintained in using the secret key for communication in p2p networks. Identity based cryptography (IBC) was introduced into peer-to-peer (P2P) networks recently for identity verification and authentication purposes. However, current IBC-based solutions were not addressing the problem of secure private key issuing. In this paper we propose a novel secure key issuing scheme for P2P networks using IBC. We present an IBC infrastructure setup phase, a peer registration solution using Shamir's (k, n) secret sharing scheme, and a secure key issuing scheme, which adopts key generate centre (KGC) and key privacy authorities (KPAs) to issue private keys to peers securely. This enables the IBC systems to be more acceptable and applicable in real-world P2P networks.

1. BACKGROUND

With its distributed, self-organization and self maintenance nature, P2P networks are extremely vulnerable to a large spectrum of attacks [1], mainly due to the lack of a certification service responsible for peers identity verification and for authentication purposes. Traditional certificate-based public key infrastructure (PKI) was used to solve some of the problems by verifying the authenticity of nodes' identities and issuing public key certificate to each node. However, as the node churn is highly frequent in the P2P network, many nodes that stored certificates may quickly become invalid, hence PKI based security protocol is difficult to be deployed. Besides, each node requires large amounts of space to store public key certificates, which can be difficult to implement in practice.

Furthermore, secured P2P overlay communication is efficient if the overlay nodes have a common, shared key for securing the communication. This is difficult to achieve in dynamic P2P overlay networks, as a new key must be generated every time an overlay node membership change occurs in order to preserve forward secrecy. Compared with the PKI technique, identity based cryptography (IBC) can simplify the key management process in P2P networks significantly. The identity of a peer (e.g., peer identifier or peer geometric coordinate) in P2P overlay networks is used to create its public key, thus avoiding the use of any certificates. These IBC-based systems are scalable, simple to administer, and each user can carry out anytime/anywhere encryption, establish secure communication channels, prove its identity to other nodes, verify protected messages and produce a form of signature with non-repudiation property.

2. INTRODUCTION

Modern cryptography follows a strongly scientific approach, and designs cryptographic algorithms around computational hardness assumptions, making such algorithms hard to break by an adversary. Such systems are not unbreakable in theory but it is infeasible to do so by any practical means. These schemes are therefore computationally secure. There exist information-theoretically secure schemes that provably cannot be broken but these schemes are more difficult to implement than the theoretically breakable but computationally secure mechanisms. Computer security can focus on ensuring the availability and correct operation of a computer system without concern for the information stored or processed by the computer. Governments, military, corporations, financial institutions, hospitals, and private businesses amass a great deal of confidential information about their employees, customers, products, research, and financial status. Most of this information is now collected, processed and stored on electronic computers and transmitted across networks to other computers.

Since the 1980s, public-key infrastructures(PKIs) have been widely anticipated as a primary means to make entities' keys available to others in a trusted fashion, thereby enabling a qualitative improvement in the protection and assurance of communications and transactions carried out over the Internet. Certificate-based authentication has become common practice in certain contexts, particularly in conjunction with SSLprotected web sites. In recent years, however, many commentators have lamented the fact that PKI has not achieved more pervasive adoption and deployment. Some have concluded that PKI is a failure or does not address users' primary security needs [19].Opinions differ on the reasons for these results, but most can be distilled into a few general categories.

The demand for the services offered by PKI, in terms of PKI-integrated applications and/or security-oriented use cases for those applications, has not yet emerged to a degree sufficient to motivate deployment of a trust infrastructure. A belief that characteristics of current PKI architectures and implementations make them unnecessarily difficult to deploy, and/or that those characteristics render them incapable of delivering value which alternate approaches could achieve[18].

2.1 Related Work

IBC uses the user's identity as the public key. The private keys of the users are issued by a key generate centre (KGC) after verifying the user's credentials. IBC was introduced in 1984 by Shamir [2]; however, the first practical encryption scheme (IBE) was not available until 2001 which was developed by Boneh and Franklin [3]. Though IBC overcomes the problems of the traditional PKI, it suffers from some inherent problems, one of which is the secure channel requirement: key issuing requires secure channel to avoid eavesdropping. In 2001, Boneh and Franklin [3] addressed secure key issuing problem using multiple key issuing authorities. After that, many key issuing protocols [4], [5], [6] without secure channels were proposed.

So far, several studies have been focused on introducing IBC into P2P security applications. Lu et al. in [7] combined distributed hash tables (DHTs [8]) and identity based encryption (IBE) to defend against man-in-the-middle attacks, however, the scheme assumed that each node has had a pre-assigned unique identifier, and has obtained the corresponding private key through a secure offline channel. This is expensive and difficult to achieve in a large scale P2P overlay network. In [9], Lua proposed a hybrid security protocol using IBE to resist the Sybil attacks, Ryu et al. in [10] proposed ID assignment protocols based on IBC to permit the acquisition of node IDs to be tightly regulated in order to mitigate the Sybil attacks, but these two schemes still suffered from the attack against key issuing phase. Likir [11] presented by Aiello et al. signs messages with IBS in Kademlia-based P2P networks, however the authors supposed every system user had already obtained a private key and did not consider the key issuing problem.

In real-world P2P networks, it is important to have a key issuing scheme in order to keep in secret whether the private key corresponding to a certain identity has been requested. In this paper, a secure key issuing scheme for P2P networks, which addresses the shortcomings of [7], [9], [10], [11], and makes IBC more applicable in the real world is presented.

2.2 Contribution

In this paper, a novel secure key issuing scheme for P2P networks is proposed along with the setup scheme of IBC infrastructure. A peer registration protocol which can register peers adopting Shamir's secret sharing scheme is introduced [12].Finally a secure key issuing protocol which can issue private keys securely without the requirement of secure channels is introduced. The protocol enables IBC more acceptable and applicable in real-world P2P networks.

3. DESIGN

We state the assumptions and requirements firstly, then we propose NOVEL KEY ISSUING SCHEME in the following four sections: system setup, peer registration, secure key issuing and system maintenance. System setup section describes how KGC and KPAs work at the beginning of the system. Peer registration section and secure key issuing section describe how a peer joins in the system. Adopting the threshold cryptography to register users, and using secure key issuing scheme to issue private keys. System maintenance section describes the maintenance mechanisms of KPAs .

ID_A :	Peer A's identity (ID)
K_A :	Peer A's private key
$Proof_A$:	Peer A's proof of the registration
$\cdot$	Concatenation
$SS(x, k)$	Secret share of secret x in Shamir's (k, n) threshold secret sharing scheme
$MAC(x, K)$	Keyed message authentication code of data x and key K
$\{X\}K_A$	A string X signed by peer A
$Thres_{KPA}$	Minimum number of KPAs system possesses
$PK_A(ID)$	Partial key of peer ID issued by A
$Pzl(x)$	A puzzle generated using Seed x
$Sln(x)$	Solution of Puzzle x

3.1 Terminology and assumptions

We present the entities involved and the security assumptions for the proposed scheme in this section.

KGC: There is a trusted core node which acts as KGC at the centre of the system, which provides peer registration and key issuing service. We assume that it has been highly fault tolerant and always available.

KPA: n nodes are selected as Key Privacy Authorities (KPAs) in order to provide the key privacy service in the key issuing phase, which are not required to be as reliable as KGC. In addition, malicious attackers can potentially compromise some of these nodes to perform insider attacks.

Peer: A peer is an ordinary node in P2P networks, which is vulnerable to all kinds of attacks.

3.2 Requirements

Secure peer registration: It must provide a method to mitigate attacks such as man-in-the-middle attacks, collusion attacks and DoS attacks during the peer registration phase.

Secure key issuing: It must provide a method to issue keys securely without secure channels during the key issuing phase, and defend against replay attacks, man-in-the-middle attacks and insider attacks.

3.3 System setup

There is one KGC node and n bootstrap KPA nodes at the setup phase. First, KGC selects a master key, publishes its identity (ID) and specifies the system parameters; Secondly, KGC assigns to each bootstrap KPA node an ID and a corresponding private key based on IBC scheme via a secure offline channel. Note that, the secure offline channel is only required in the system bootstrap phase, since with its ID and private key, a KPA can communicate with the KGC through a secure channel established based on IBC.

3.4 Peer registration

Before joining the network, a peer A should get registered to the KGC at first.
A simple protocol employed is as following. The peer A generates a request with a random nonce and sends it to KGC. After KGC receives the request, KGC issues (IDA,ProofA) for A. The proof of registration Proof A is a message that can prove whether the peer has been registered. Assigning ID by KGC can prevent a peer from choosing its own ID, and mitigate the Sybil attack in the system. However, the communication between KGC and A may be intercepted or modified by malicious peers in real-world P2P networks. We adopt Shamir's (k, n) threshold secret sharing scheme [12] to secure this process. To protect the registration data (IDA,ProofA), we divide it into many secret shares, so that if some of the secret shares have been intercepted, A can eventually recover the registration data if at least k secret shares are collected. On the other side, if the adversary also wants to recover the registration data, it has to get sufficient secret shares, which is difficult to achieve if the threshold k is appropriately configured.

The protocol is described as follows:

Step1: A -> KGC N
Step2: KGC ->KPA SS(IDA · ProofA, k),N
Step3: KPA ->A SS(IDA · ProofA, k),N

Request: When the peer A wishes to join the network, it must first get registered from KGC. Finding the KGC can be accomplished by consulting a bootstrap node or using an automatic service discovery mechanism. Then A sends a request to KGC. In order to avoid the replay attacks, A couples the request with a nonce N.

Distribution: After KGC receives the request, it generates IDA and ProofA for A. In particular, ProofA can be a keyed message authentication code of IDA, i.e., ProofA = MAC(IDA,KKGC). After that, KGC divides IDA and ProofA into n secret shares using Shamir's (k, n) threshold secret sharing scheme, and then KGC distributes those n secret shares to n KPAs respectively. KGC can divide IDA and ProofA into n pieces, IDA and ProofA are easily reconstructable from any k pieces, but even complete knowledge of k -1 pieces reveals absolutely no information about IDA and ProofA.

Thus, it is very difficult for the adversary to obtain sufficient secret shares in a P2P network if we divide the registration data and set an appropriate threshold k. Reconstruction: After receiving the secret share from KGC, KPAs send them to A. Upon receiving the messages from a KPA, A first checks the nonce N, if N is different from the one in its original request, it just ignores the message. Then A waits for collecting k secret shares from different KPAs within a predefined time window. After the peer A gets at least k different secret shares, IDA and ProofA can be reconstructed.

A detailed description of reconstruction process can be found in [12]. If the peer does not get sufficient secret shares, it may run the peer registration protocol again later. Since in real-world P2P networks, it is difficult to have all those paths node-disjoint, here KPAs can use approaches (such as [13]) to avoid paths between KPAs and A from node-joint, in order to mitigate eavesdropping and DoS attacks. In practice, many ISPs have a single connection to the Internet and if an eavesdropper is listening at that point, it will hear all of the n shares being transmitted. This kind of attack is usually a rare situation.

3.5 Secure key issuing

After registration phase, a peer obtains its ID. The next step is to describe how KGC issues a private key to a peer securely without the requirement of secure channels, and how a peer constructs its private key securely from the KGC and KPAs. Shamir's secret sharing scheme we used in Section 3.3 can also be utilized here, however, with KGC and KPAs in the system, we can make the key issuing phase more secure. We present a protocol which utilizes IBC secure key issuing schemes [4], [5], [6] below. Those schemes use one KGC and multiple KPAs for issuing the private keys to the users. KPAs participate in the key generation phrase, they assign the joining peer partial private keys. A registered peer can obtain its private key securely by collecting partial private key from KGC and KPAs. Those schemes avoid the need for secure channels, and the adversary who wants to obtain the private key must compromise not only KGC but also many KPAs. In our scheme, Saxena's scheme [6] is followed therefore it can easily be extended to other schemes [4], [5]. Our scheme is described as follows:

Step1: A ->KGC:	Request, IDA, ProofA,N
Step2: KGC ->A :	Partial key from KGC, N
Step3: A ->KPA :	Request, IDA, ProofA,N
Step4: KPA ->A :	Partial key from KPA, N

System setup: KGC selects its private key and specifies the system parameters. KPAs collaboratively run a key generation and distribution protocol [6], and share a secret s such that any k KPAs can construct it with their own secret shares.

Peer registration: As the system setup process is updated, in the peer registration process, IDA and ProofA are generated in a new way, but we can still utilize the protocol described in Section 3.3 to secure this process.

Request: A sends a request with its proof of registration as well as a nonce to KGC to obtain the partial private key;

KGC response: On receiving A's request, KGC checks the proof to verify whether A has been registered or not, if the result is positive, KGC responses with a partial private key;

Blind KPA request: After receiving the partial private key from KGC, A randomly selects some KPAs and requests them in parallel to provide key privacy service by sending a request;

KPA response: Each KPA authenticates A and issues a partial private key to it;

Key retrieval: On receiving at least k partial private keys from different KPAs, A combines them and then unbinds the resulting value to produce the private key; The scheme above is secure against replay attacks, man in-the-middle attacks and insider attacks, and more details can be found in [6]. It can easily be incorporated with other secure key issuing schemes such as [4], [5] that use KPAs to protect the private key.

4. IMPLEMENTATION

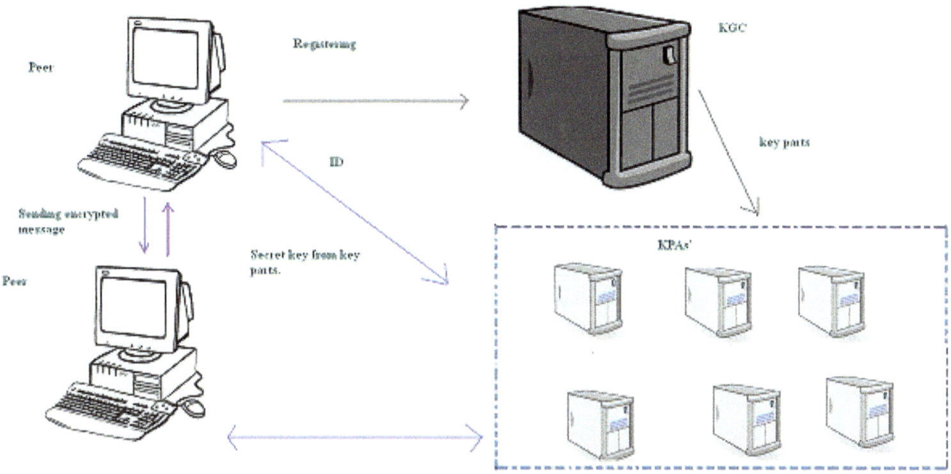

Figure 1: System architecture

The implementation of the key issuing scheme can be as shown by the system architecture in figure 1. The scheme is implemented with six KPAs. The detailed design of our work can be described by following UML design diagrams, which are documented below in Figure 2-5.

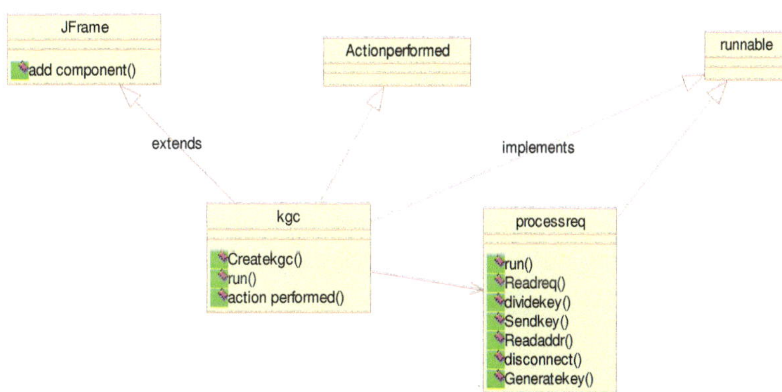

Figure 2: Class diagram for KGC

This is the Key Generation Center (KGC), central node operating in the network.This implements a form of location transparency.The nodes added to the network are unaware of actual location of this central node.This is possible with a level of abstraction provided by the underlying framework. KGC runs as thread on a peer and services the incoming requests by other nodes for registration .It also performs the key generation with the prerequistes of underlying algorithm.It also performs the part of look up work for synthesizing the whole key into parts and back parts into the whole with the mathematical background running over it.

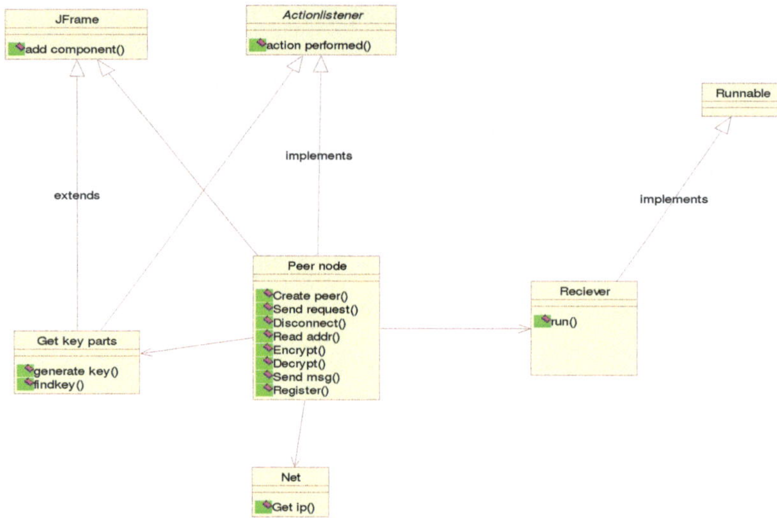

Figure 3: Class diagram for peer node

This is the peer node, a regular node seeking to enter the network and establish the communication with other peer in the network. The peer runs as a thread with different roles. The first form sends a request for registration of node for the communication and awaits the reply token for acquiring the parts of key. The second form is receiver ,which involves rendering the message sent it to it by another peer in the network after unwrapping up of the scrambled data.

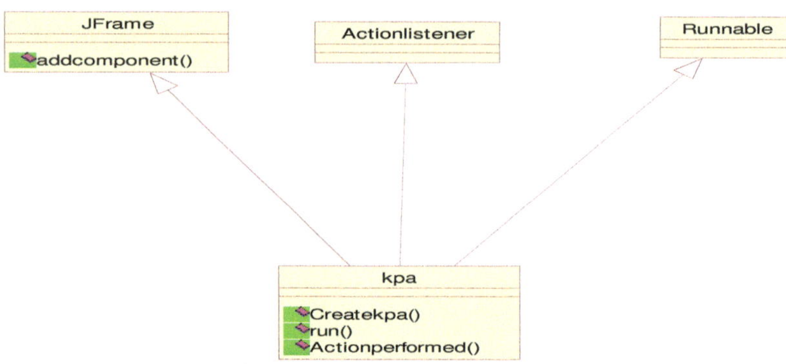

Figure 4: Class diagram for KPA

This is the Key Privacy Authority(KPA).This also runs as a thread. This involves a node in a network , which actually conceals a part of key after it is broken by KGC.The existence of KPAs is solely known to KGC with degree of location transparency and authentication involved.The other peers are unaware of the KPAs . KGC interacts with KPAs after breaking the key.And its upto the peers to render these parts (not all , but subset) with the usage of reference given by KGC.Even if peers come to know the keyparts directly,there is no way of rendering the key without the synthesis process involving the KGC references[2].

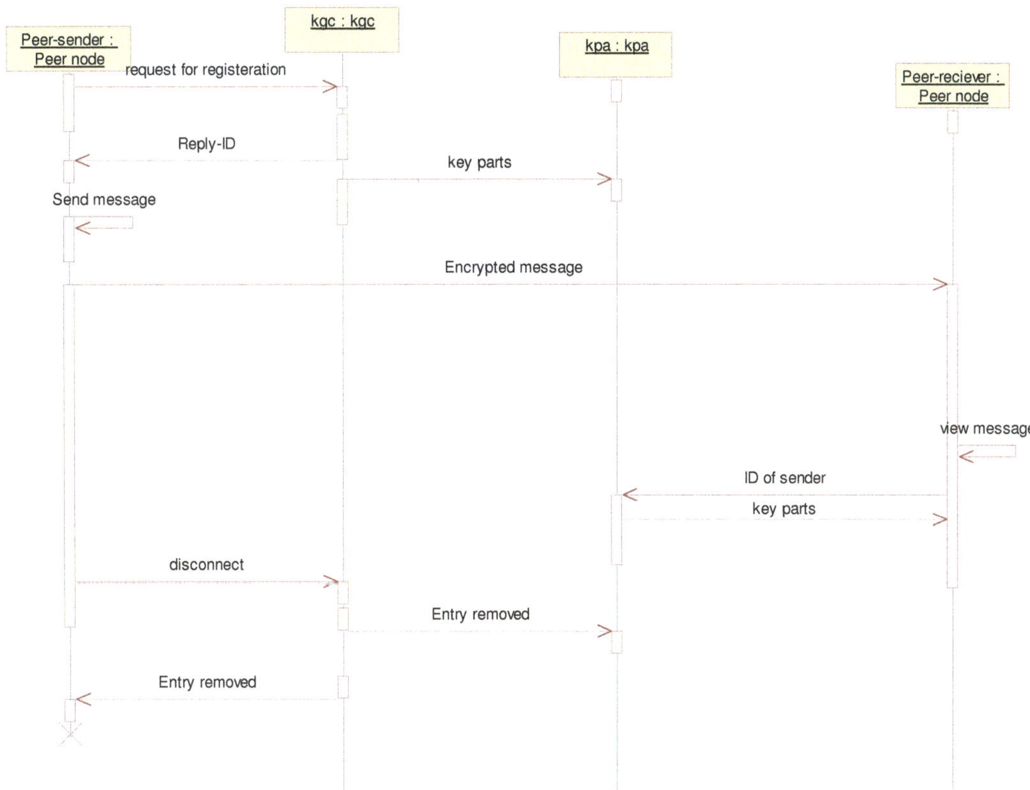

Figure 5: Sequence diagram

The sequence diagram shown above describes the interaction of various modules described with a timeline of events occuring across the network.The initiation of underlying cryptographic process is marked by addition of peer node (sender) in a network and completion by a successful transfer of message with underyling cryptographic process [18].The KGC,KPA are kind of server processes pre running as shown by their early entry in timeline.The process continues with arrival and departure of several peer nodes in a network.

5. CONCLUSION

As the networks are evolving, they also give rise to a requirement of strong security. This feature of networks related to security should be addressed properly. There have been a large number of ways to efficiently handle the security requirement in networks.

Peer to peer networks have a large role in today's fast evolving world and it becomes very essential to manage them efficiently.Security of these networks can be handled on various levels of cryptographic abstractions[1].All these levels have been explored with certain degree of threshold behaviour . In our work we focus on key exchanging phase of a cryptographic schema,where the keys are given to the authenticated nodes involved in communication[12].

In our work we have proposed a novel secure key issuing scheme which is used for basic cryptographic needs such as encryption and decryption. Here we have mainly concentrated on key issuing part of cryptography as previously proposed techniques do not address the issue effectively. Public Key Infrastructure(PKI) can be used for key issuing however there is problem of managing certificates in PKI as it becomes cumbersome. Hence we go for ID based cryptography, which is efficient compared to PKI.

In this contribution we have implemented a secure key issuing scheme for P2P networks using IBC and have provided a peer registration service using Shamir's secret sharing algorithm. We develop a secure key issuing protocol, which adopts KGC and KPAs to issue private keys to peers securely. The experiments are done on six peers and documented [18].

Our work can be extended to support large scale peer to peer networks of magnitude (10000 peers in simulation environment).It can address Sybil attacks with help of byzantine fault tolerance.

REFERENCES

[1] E. Sit and R. Morris, "Security considerations for peer-to-peer distributed hash tables," in IPTPS, 2002, pp. 261–269.

[2] A. Shamir, "Identity-based cryptosystems and signature schemes," in CRYPTO, 1984, pp. 47–53.

[3] D. Boneh and M. K. Franklin, "Identity-based encryption from the weil pairing," in CRYPTO, 2001, pp. 213–229.

[4] B. Lee, C. Boyd, E. Dawson, K. Kim, J. Yang, and S. Yoo, "Secure key issuing in id-based cryptography," in ACSW Frontiers, 2004, pp. 69–74.

[5] R. Gangishetti, M. C. Gorantla, M. L. Das, A. Saxena, and V. P. Gulati, "An efficient secure key issuing protocol in idbased cryptosystems," in ITCC (1), 2005, pp. 674–678.

[6] A. Saxena, "Threshold ski protocol for id-based cryptosystems," in IAS, 2007, pp. 65–70.

[7] Z.-L. Lu, G.-H.; Zhang, "Wheel of trust: A secure framework for overlay-based services," ICC, pp. 1148–1153, 2007.

[8] I. Stoica, R. Morris, D. R. Karger, M. F. Kaashoek, and H. Balakrishnan, "Chord: A scalable peer-to-peer lookup service for internet applications," in SIGCOMM, 2001, pp. 149–160.

[9] E. K. Lua, "Securing peer-to-peer overlay networks from sybil attack," in ISCIT'07, Sydney, Australia, 2007.

[10] S. Ryu, K. R. B. Butler, P. Traynor, and P. D. McDaniel, "Leveraging identity-based cryptography for node id assignment in structured p2p systems," in AINA Workshops (1), 2007, pp. 519–524.

[11] L. M. Aiello, M. Milanesio, G. Ruffo, and R. Schifanella, "Tempering kademlia with a robust identity based system," in Peer-to-Peer Computing, 2008, pp. 30–39.

[12] A. Shamir, "How to share a secret," Commun. ACM, vol. 22, no. 11, pp. 612–613, 1979.

[13] I. Baumgart and S. Mies, "S/kademlia: A practicable approach towards secure key-based routing," in ICPADS, 2007, pp. 1–8.

[14] R. Chen, W. Guo, L. Tang, J. Hu, and Z. Chen, "Scalable byzantine fault tolerant public key authentication for peer-to peer networks," in Euro-Par, 2008.

[15] M. J. Freedman and R. Morris, "Tarzan: a peer-to-peer anonymizing network layer," in CCS, 2002, pp. 193–206.

[16] B. Parno, D. Wendlandt, E. Shi, A. Perrig, B. M. Maggs, and Y.-C. Hu, "Portcullis: protecting connection setup from denial-of-capability attacks," in SIGCOMM, 2007, pp. 289– 300.

[17] P. Maymounkov and D. Mazi`eres, "Kademlia: A peer-to-peer information system based on the xor metric," in IPTPS, 2002.

[18] Mohammed Azharuddin and Annapurna P Patil, "Design and Implementation of A secure key issuing scheme for peer to peer Networks",PG Dissertation July 2011,Department of Computer Science and Engineering, M.S.Ramiah Institute of Technology Bangalore-560054,India.

[19] An Examination of Asserted PKI Issues and Pro-posed Alternatives -John Linn, RSA Laboratories, Bedford, MA, USA Marc Branchaud, RSA Security Inc., Vancouver, BC,Canada.

5

Carrier Sense Multiple Access Tuning Parameters using Game Theory

Mahdieh Ghazvini[1], Naser Movahedinia[1], Kamal Jamshidi[1]
[1]Department of Computer Engineering, Faculty of Engineering, University of Isfahan, Isfahan, Iran
{ghazvini, naserm, jamshidi}@eng.ui.ac.ir

Abstract

Ad Hoc and Mesh networks are good samples of multi agent systems, where their nodes access the channel through carrier sense multiple access method, while a node channel access influence the access of neighbor nodes to the channel. Hence, game theory is a strong tool for studying this kind of networks. Carrier sense multiple access parameters such as minimum and maximum size of contention window and persistence factor can be modified based on game theoretic methods. In this study different games for tuning the parameters is investigated and different challenges are examined.

Keywords

Carrier Sense Multiple access, Wireless Networks, Game Theory.

1. Introduction

Medium access layer controls the way of access to common media and permission for transmission over them. Medium access models are divided into two general classes of contention-based and contention-free models. Contention-based models include random access methods. Random access methods are divided into carrier sensing models like CSMA and no carrier sensing models like ALOHA. In CSMA, probability of channel access is implemented through a back off algorithm on contention window (CW) or as persistence probability (P). In back off mechanism, each wireless node keeps a contention window, before transmission, it waits for a random period of time that is limited to the size of contention window. In persistence mechanism, each wireless node keeps a persistence probability and upon observing a channel idle, it accesses the channel with this probability. If persistence mechanism is implemented, the probability of channel access is the persistence probability (p) and in case of using back off mechanism, channel access probability is related to the size of contention window (CW) as follows[1-3]:

$$p = \frac{2}{CW_{\min} + 1} \qquad (1)$$

Choosing the contention criteria and the contention resolution algorithms are performance keys of medium access methods. Improper selection of these parameters leads to poor performance. One of common models of CSMA is DCF algorithm that will be referred to briefly. In DCF, each node to transmit a new packet monitors the channel activity. If the channel is idle for a periods of time, called DCF Inter frame Space (DIFS), the node starts its transmission. Otherwise, the node insists on monitoring the channel so as to see it idle to the amount of DIFS. In the first attempt for transmission, contention window is tuned to the size of minimal contention window (CW_{min}). After each unsuccessful transmission, CW will be larger in order to reach its maximal value namely CW_{max}.Once CW reaches CW_{max}, this value remains constant so that the packet is transmitted successfully or the number of retransmissions reaches retry bound.

In this way retransmission is stopped and the packet is dropped [4]. In the original version of DCF, each new transmission starts with CW_{min} value regardless of the contention level in the network. Therefore, there is a need to other models for determining the size of *CW* or persistence probability in order to reach high performance (throughput and low collision) and a better fairness.

As nodes decide independently about their channel access and access of a node to the channel affects the neighbouring nodes, game theory is a powerful tool for studying this kind of network [2, 4-7].

2. Game Theory

Game theory is a branch of applied mathematics which is used in most of sciences. Whenever utility and profit a person is looking for is not only effected by his own attempt and decision but also by attempt and decision-either positive or negative- of another party, it is referred to as a game. Game theory is a tool for analysis of systems as well as a tool for optimizing. Although optimization theory is not able to consider interactions between different players (like users), game theory has turned out to be a good approach for studying and analyzing the behavior of complicated and interacting systems [7,8]. The main feature of decision making in game is that each player before decision making should analyze the reaction of other players toward him and then make a decision that is the best for him and contains the most benefit with respect to reaction of another party. An environment in which exists such effect and mutual reaction among decisions of individuals is called a strategic environment and each decision maker in this environment is called a player[8, 9].

A game consists of a set of players, a set of movements and a set of payoff functions. Usually in every game, a payoff function is defined as the subtraction of the utility function and the user cost. Utility function is a mapping of all behaviours of player into a set of real numbers, its definitions plays an important role in the game. In some researches, however, payoff function and utility have been considered equivalent. Strategy of an agent can be every action from his actions space or a combination of them. If players choose a behaviour clearly, it is called pure strategy but generally players choose their strategy with probability and do not have complete confidence in selection of opposite player; such a behaviour is called mixed strategy. Mathematical representation of a game is $\{N, A_i, \{u_i\}\}$ which represent set of players, action space and set of payoff functions, respectively[10, 11].

Utility is a criterion for measuring a node satisfaction level resulted of its actions. Utility function is a reward (Reward minus cost) that is expected to be obtained by each link from the network after their transmission. Users have different utility functions with respect to the kind of their services. With maximizing the network utility (for example, total utility of all users) social welfare of system is also maximized. On the other hand, utility functions can be interpreted as handles for controlling tradeoff between performance and fairness. In game theory, normally, the goal is finding equilibrium(s) in the game. In equilibrium point, each player chooses a strategy that is the best response to selected strategies of the other players. Several equilibria have been proposed, among which the most well-known is Nash equilibrium. Nash equilibrium is a solution from a game including a number of players in which no user gains benefit through one-way change of her/his strategy [12-16].

Players can be cooperative or non-cooperative. In the first case, players cooperate with each other and the problem is converted to an optimization problem that each player leads the system toward a social equilibrium. In a cooperative game, players through coalition decision making, bargain, dealing, and discussion try to reach agreement so that they may gain maximal benefits. The most standard criterion for expressing equilibrium performance in cooperative games is Pareto performance. Pareto means that a user cannot increase his utility without decreasing the utility of at least one of the other players. Another kind of games is non-cooperative game in which each decision maker chooses his strategy selfishly. In fact, in non-cooperative games each player chooses strategies without coordinating with others. In this case, equilibrium if any will be Nash equilibrium.

3. Game Theory in Wireless Networks

Game theory studies decision making in a shared environment by some decision makers with different goals. Nodes of a network, is a good example of such a condition. Hence, game theory in wireless networks especially in mesh and Ad Hoc networks is applied and such networks can be easily modeled by game theory. For mapping network components to a game, usually network nodes and all resources required by nodes for communication (spectrum, power, bandwidth, time, etc) are considered as players and resources respectively[9, 10]. In wireless networks games the strategy space can include some actions related to network functions for example, decision on whether transmit the packets, forward or no, tuning of power level, selection of modulation method, contention window adjustment or persistence probability adjustment, regulation of hold off parameter, etc. finally Utility or payoff functions can be defined based on some performance criteria like utilization, fairness, throughput, target SNR as listed in Table1[11, 17-20].

Table1. Mapping Between Wireless Networks Elements and Game Theory Elements

Components of games	Components of wireless networks
Players	Network links, Network nodes
Strategy	Transmission power, Bandwidth, transmission time, Transmission probability, Contention window, Forwarding a packet or no.
Payoff	Delay, Channel utilization, Fairness, Throughput, Power consumption, Target SNR, Packet loss ratio

In recent years game theory has been mainly used in some problems like routing and resources allocation in competitive environments. The main advantages of game theory in wireless networks are analysis of distributed systems, cross layer optimization and designing of encouraging methods for avoiding selfish behaviors. Application of game theory, its models and challenges in networks have been well studied in [2, 8-10, 17, 21-26]

4. Carrier Sense Multiple Access Tuning Parameters Based On Game Theory

Medium access control is a distributed approach in access to a shared wireless channel among competitive nodes. Since wireless networks use a shared transmission medium, collision may occur because of simultaneous transmissions by two or several interfering nodes. In order to control the contention between transmissions of different nodes, increasing the performance of system and fairness, it seems necessary to use distributed media access protocols.

In random access games, the strategy of a player is usually transmission probability or equivalently, the size of her/his contention window. Payoff functions include utility portion from channel access and packet collision cost [27, 28]. By studying the previous works it was revealed that strategies of more players were contention window, transmission power and data rate. Following, proposed CSMA games, from two perspectives of cooperative and non-cooperative games has been investigated.

It has shown that in Ad Hoc networks in which all nodes cooperate with each other, game theory can be used. In cooperative games, agents interact with each other and choose their strategies after agreement. Usually implementation of cooperative solutions is difficult and requires interaction of agents with each other and needs to be repeated some times. But generally cooperative solutions are much better than non-cooperative solutions.

Incomplete cooperative game theory is possible through estimating the current state of the game (e.g. the number of competing nodes) and achieves equilibrium strategy of each node based on estimation of game state. Zhao et al. have proposed an incomplete cooperative game for improving the performance in Ad Hoc [6] and mesh networks [7]. In these games each node estimates the number of competing nodes (n) in the network using the method presented in [29, 30]. Then it tunes its equilibrium strategy through adjusting its minimum size of contention window (CW_{min}) as follows:

$$CW_{min} = \begin{cases} \lfloor n \times rand(6,7) \rfloor & n \leq 5 \\ \lfloor n \times rand(7,8) \rfloor & 6 \leq n \end{cases} \quad (2)$$

Since all nodes are able to obtain the number of competing nodes, they form a cooperative game and nodes with fewer numbers of competitors tune a smaller CW_{min}. The best strategy for rest of nodes with more competitors is regulation of a larger CW_{min} for reducing collision probability. One advantage of [6, 7] compared with other games is having no need of information exchange like SNR [5]. With respect to partial swarm optimization, [31] has proposed a game called (G-PSO) for WMNs as follows: in the game, mesh routers operate complicated mechanisms like ARMA and Kalman filters [30] and estimate the game state more exactly and broadcast it across all nodes. Thus, router and nodes form a cooperative game with respect to estimated condition so as to reach global best (g_{best}). In G-PSO, each node tunes its contention window according to the following equation [31]:

$$CW_{min} = \begin{cases} CW_{min}^{p} & \text{a node does not know the game state} \\ CW_{min}^{g} & \text{a node knows the game state} \end{cases} \quad (3)$$

[32] proposed a cooperative game (G-EDCA), in which each node estimates the number of users in each traffic class through listening and counting the data frames with different source addresses from every kinds of traffic and then tunes its equilibrium strategy namely the size of contention window for each traffic class in each stage of transmission as follows. This game is also an incomplete cooperative game [32] :

$$CW_{i,min} = [n_i \times rand(7,8)] \quad i = 3, 2, 1, 0 \quad (4)$$

In [33], it is supposed that each node knows conditions of its channel-SNR, number of existing nodes as well as probability distribution of channel conditions for other nodes. Then based on this assumption a CSMA game has been proposed. The game is an opportunistic CSMA game based on user's channel conditions and it is robust to selfish users. Since the proposed game is a complete game; cost, probability distribution, and utility functions are similar for all users. Although cooperative algorithms are more effective and optimal, they need information exchange like PRICE, SNR, etc, to be updated.

Wireless nodes usually are not exactly aware of number of nodes in network and each node can obtain some limited information about channel state (e.g. collided packets, busy or idle state of channel) through listening to channel. In such conditions the best thing a node can do, is to optimize its personal goals. Therefore, for modelling such situation, non-cooperative game models are the best choice. It is shown that MAC protocols can be reverse engineered and studied within the game theory framework [2, 22, 23]. A vector of channel access probability *p* is equilibrium if for the given contention of network; no node is willing to change its transmission probability. In [34, 35] has been shown that(EB) exponential back off protocol can be modelled with a non-cooperative game in which links try to maximize their utility function in the form of reward for successful transmission. Therefore, for EB protocol, non-cooperative game is more suitable than global optimization model and it has been shown that without coordination between nodes, selfish behaviours reduce network performance.

In forward engineering, a cooperative NUM problem has been formulated according to persistence probability of each node and links for the purpose of improving medium access protocol [15, 34, 35]. With a little exchange of contending prices the best coordination for balancing the nature of non-cooperative EB and presentation of a new protocol that moves toward the best converging point can be obtained. Convexity feature is necessary for designing an optimal distributed algorithm and each node has a global convex utility function [15, 34, 35]. In [2, 21, 23, 36] there are examples of how utility functions are designed and also it includes determination of contention criterion through reverse engineering from existing protocols and from desirable performance points (according to throughput and fairness) and through forward engineering in a heuristic manner as the following.

$$U_i(p_i) = p_i - \frac{e\ \varsigma}{1-\alpha}(1-p_i)^{1-a} \qquad (5)$$

$$U_i(p_i) := \frac{1}{a_i}\left(\frac{(a_i-1)b_i}{a_i}\ln(a_i p_i - b_i) - p_i\right) \qquad (6)$$

In [22] service differentiation has been provided through defining different utility functions according to traffic classes with different weights as a non cooperative game. In most researches including [2, 21, 23, 36] only cellular wireless local networks have been taken into account in which each node hears transmissions of other nodes. Also it is assumed that all nodes are saturated, channel is free from error and packet drop occurs only in a result of collision. In many researches e.g.[21, 37], the payoff functions is defined as below:

$$u_i = U_i(p_i) - p_i q_i(p) \qquad (7)$$

$$q_i(p) = 1 - \prod_{j \in N-\{i\}}(1-p_j) \qquad (8)$$

In fact, the cost functions of these researches are linear functions of multiplication of transmission and collision probabilities. In above equations, $U_i(p_i)$ is utility function, p_i is channel access probability for node i, and q_i is the criterion for measurement of contention like collision probability, idle time within channel access, etc. but $U_i(p_i)$ has been defined with different models. Authors of [37] has used game theory as an optimization tool. The proposed game is an iterative multi-stage game non-cooperative game with selfish users and Each node does not need know the total number of nodes. The utility function of the game is defined as the following;

$$U_i(p_i) := \frac{p_i(\ln w_i - \ln p_i + 1)}{\ln r_i}, \quad r_i = {}^{w_i}/_{v_i}, \quad v_i \le p_i \le w_i \qquad (9)$$

[38]studies behavior of non-cooperative users who tune their access probability through changing their persistence coefficient or the back off exponential control parameter. It has defined the disutility function for each user according to collision rate and power consumption. In [38] like other researches, the effect of physical layer and noise has not been considered and frame loss has been regarded only because of collision.

In studies like [39-42] channel contention problem has been implemented as a non-cooperative power control game (GMAC) in a distributed manner with quasi CSMA/CA access mechanism. GMAC uses a shared channel for data and control. In definition of the utility, GMAC uses a linear pricing factor of power consumption.

$$u_i(p_i, p_{-i}) = \log(1+\gamma_i) - \alpha_i p_i \qquad (10)$$

In the game, γ_i denotes the SINR at the receiver and power control decisions are made for each packet and each receiver uses the knowledge of previous scheduled transmissions among its neighbors in order to join the scheduling program. Simulations have shown that GMAC makes possible more concurrent transmissions and improves power consumption.

In [43, 44], using static non-cooperative game a distributed power-aware MAC algorithm called PAMG has been modeled for Ad Hoc networks. In these games, each active link has been taken as a player. Strategies vector is two-dimensional including transmission probability and power probability. In the game, each node according to its place in the network and received feedback from the channel, tune their transmission probability and power selfishly. Decision making in each link occurs based on optimality of a payoff function that has been defined on two-dimensional strategy space and the value of this function shows the payoff resulted by medium. Each link can improve its data rate through increasing its power and transmission probability. Authors of [45] has proposed a non-cooperative and contention-based medium access game (CAG) with initial frameworks like [22] and with selfish users. In CAG the number of users is pre-determined and constant. The strategy of each user is his transmission probability α_i. Payoff function has been also defined through combination of payoffs resulted by successful transmission u_s, collision u_c, and waiting u_w as follows:

$$u(\alpha_i) = u_s - u_c + u_w = A\alpha_i \prod_{j \in N(i)}(1 - \alpha_j) - B\alpha_i (1 - \prod_{j \in N(i)}(1 - \alpha_j)) + C(1 - \alpha_i) \quad (11)$$

Then CAG has converted into an optimization problem with Constraint condition and the strategy has been updated by gradient for reaching Nash equilibrium. Author of [46] has considered a scenario same as [47] with presence of a specific number of users with different QoS requirements. She modelled the priority of traffic in this CSMA/CA network through game theory. Each node labels its traffic with a specific traffic class and these classes do not change. In this mode, it is expected that nodes with higher priorities take over the wireless resources and access the wireless medium with lower values of contention window. Nodes with low priority should not continue reducing their contention window to eliminate the higher priority nodes action in order to have more medium access. Each time when throughput of a node is estimated, punishment mechanism through jamming of those nodes which have violated the expected throughput, does not allow for confirmation of proper reception of the traffic transmitted by a cheater to a specific destination and thus the transmitter does not receive any acknowledgement [46]. Yang formulated a non-cooperative game [48, 49], in which each link tries selfishly to maximize its payoff function, its throughput, by its local information. Power constraints have been also added to the conditions and price of each link reflects the interference it gets from the price receiver. Although performance of non-cooperative game does not equate cooperative game but in case of using non-cooperative games, message exchange cost reduces. [3] Has proposed a non-cooperative persistence model with purpose of minimizing the exchanged information between nodes. In the non-cooperative game, users access the channel in order to maximize their successful transmissions based on collision cost and power consumption. In [50] a two-stage mathemathical model called TRG/CSMA has been proposed. in the first step, throughput and delay have been selected as optimization goals and a game between (n) nodes is played. In the second step of game, throughput and delay as players play a two-players game with infinite strategy and obtain their weight in the game and contention window is tuned.

Study of cooperative and non-cooperative models proposed so far for the purpose of improving carrier sense multiple access model leads toward this fact that the desirable solution for carrier sense multiple access games should be unique, fair, and Pareto optimal. In cooperative games, using bargaining techniques, etc, these conditions can be obtained. In non-cooperative games, however, these techniques are not suitable and to reach the desirable result, some other techniques like punishment should also be used [17].

5. Conclusions

In recent years application of game theory has been introduced in wireless communications. Applications of game theory in data link layer is related to medium access control problem and in these games, selfish users through unfair sharing in channel access try to maximize their performance. This reduces the ability of other users in accessing the channel. Users have to overcome the collision incurred by limited transmission resources, thus game theory has turned into a powerful tool to analyze and improve the contention-based protocols. Numerous games have been proposed for modelling such environment in which behavior of users is (transmit and wait) and in some others behaviours include transmission probability. Studying the researches showed that cooperative random access can use radio resources more effectively and fairly compared to non-cooperative random game. Thus for achieving an efficient and fair system with low collision, reception of controlling messages from other nodes seems necessary. Of course, it should be noted that message passing between nodes is based on cooperation assumption with purpose of acquiring a system with a global objective.

Acknowledgment

This work is partially supported by Iran Telecommunication Research Canter (ITRC No.18507/500T).

References

[1] Y. Jin, H. Miao, H. Liu, and Y. Xiong, "An improved backoff algorithm for wireless sensor network based on game theory," in *IET International Communication Conference on Wireless Mobile and Computing (CCWMC 2009)*, Shanghai, China 2009, pp. 197-200.

[2] L. Chen, T. Cui, S. H. Low, and J. C. Doyle, "A game-theoretic model for medium access control," in *WICON '07*, Austin, Texas, USA, 2007.

[3] X. Guan, B. Yang, G. Feng, and C. Long, "Random access game in ad hoc networks with cooperative and noncooperative users," *Int. J. Syst. Control Commun*, Vol. 1, No. 1, pp. 13-30, 2008.

[4] "IEEE 802.11, IEEE Standard for Information Technology Telecommunications and Information Exchange Between Systems Local and Metropolitan area Networks Specific Requirements. Part 11: Wireless LAN Medium Access Control (MAC) and Physical Layer (PHY) Specifications," 2007.

[5] Y. Xiao, X. Shan, and Y. Ren, "Game theory models for IEEE 802.11 DCF in wireless ad hoc networks," *IEEE Commun. Mag.*, Vol. 43, No. 3, pp. S22-S26, 2005.

[6] L. Zhao, J. Zhang, K. Yang, and H. Zhang, "Using incompletely cooperative game theory in mobile ad hoc networks," in *IEEE International Conference on Communications,ICC '07*, 2007, pp. 3401-3406.

[7] L. Zhao, J. Zhang, and H. Zhang, "Using incompletely cooperative game theory in wireless mesh networks," *IEEE Network*, Vol. 22, No. 1, pp. 39-44, 2008.

[8] M. Debbah, S. Lasaulce, and H. Tembine, *Game Theory and Learning for Wireless Networks: Fundamentals and Applications*: Academic Press, 2011.

[9] Z. Han, D. Niyato, W. Saad, T. Basar, and A. Hjorungnes, *Game Theory in Wireless and Communication Networks: Theory, Models, and Applications*: Cambridge University Press, 2012.

[10] Y. Zhang and M. Guizani, *Game Theory for Wireless Communications and Networking*: Taylor and Francis, 2011.

[11] W. Huang, "Application of game theory in wireless communication networks," PhD dissertation, Dept. Elec. and Comp. Eng. University of British Columbia, 2012.

[12] M. J. Osborne and A. Rubinstein, *A Course in Game Theory*: MIT press, 1994.

[13] D. E. Charilas and A. D. Panagopoulos, "A survey on game theory applications in wireless networks," *Computer Networks,* Vol. 54, No. 18, pp. 3421-3430, 2010.

[14] V. Georgiev, "Using Game Theory to Analyze Wireless Ad Hoc Networks," RWTH Aachen,Department of Computer Science 2008.

[15] J. W. Lee, M. Chiang, and A. R. Calderbank, "Utility-optimal random-access control," *Wireless Communications, IEEE Transactions on,* Vol. 6, No. 7, pp. 2741-2751, 2007.

[16] A. B. MacKenzie and L. A. DaSilva, *Game Theory For Wireless Engineers*: Morgan & Claypool, 2006.

[17] K. Akkarajitsakul, E. Hossain, D. Niyato, and D. Kim, "Game theoretic approaches for multiple access in Wireless networks: a survey," *IEEE Commun. Surv. Tutorials,* Vol. 13, No. 3, pp. 372-395, 2011.

[18] S. Mehta and K. S. Kwak, "Application of Game Theory to Wireless Networks," in *Convergence and Hybrid Information Technologies*, M. Crisan, Ed.: InTech, 2010.

[19] T. V. P. Sundararajan and A. Shanmugam, "Selfish Avoidance Routing Protocol for Mobile Ad Hoc Network," *International Journal of Wireless & Mobile Networks,* Vol. 2, No. 2, pp. 80-92, 2010.

[20] S. S. Prasad, R. Baruah, and S. Audhinarayanan, "Convergence Performance of SINR and SIR utilities for a Distributed Resource Allocation Game," *International Journal of Wireless & Mobile Networks,* Vol. 3, No. 3, pp. 29-40, 2011.

[21] L. Chen, S. H. Low, and J. C. Doyle, "Contention control: A game-theoretic approach," in *46th IEEE Conference on Decision and Control*, New Orleans, LA, 2007, pp. 3428-3434.

[22] L. Chen, S. H. Low, and J. C. Doyle, "Random access game and medium access control design," *IEEE/ACM Trans. Networking,* Vol. 18, No. 4, pp. 1303-1316, 2010.

[23] T. Cui, L. Chen, and S. Low, "A game-theoretic framework for medium access control," *IEEE J. Sel. Areas Commun.,* Vol. 26, No. 7, pp. 1116-1127, 2008.

[24] Z. Kong and Y. K. Kwok, "Efficient wireless packet scheduling in a non-cooperative environment: Game theoretic analysis and algorithms," *Journal of Parallel and Distributed Computing*, 2010.

[25] D. N. Zhu Han, Walid Saad, Tamer Başar, Are Hjørungnes, *Game Theory in Wireless and Communication Networks: Theory, Models, and Applications*: Cambridge University Press, 2011.

[26] I. Tinnirello, L. Giarre, and G. Neglia, "MAC design for WiFi infrastructure networks: a game-theoretic approach," *IEEE Transactions on Wireless Communications,* 99, pp. 2510 - 2522, 2011.

[27] Z. Fang and B. Bensaou, "Design and implementation of a MAC scheme for wireless ad-hoc networks based on a cooperative game framework," in *IEEE International Conference on Communications*, 2004, pp. 4034-4038.

[28] T. Hisch, "Game Theory and Distributed Systems," 2006.

[29] G. Bianchi, "Performance analysis of the IEEE 802.11 distributed coordination function," *IEEE J. Sel. Areas Commun.,* Vol. 18, No. 3, pp. 535-547, 2000.

[30] G. Bianchi and I. Tinnirello, "Kalman filter estimation of the number of competing terminals in an IEEE 802.11 network," in *INFOCOM '03*, 2003.

[31] L. Zhao, H. Zhang, W. Ding, and J. Zhang, "Game-theoretic particle swarm optimization for WMNs," in *19th International Symposium on Personal, Indoor and Mobile Radio Communications(PIMRC)*, Cannes 2008, pp. 1-5.

[32] L. Zhao, L. Cong, H. Zhang, W. Ding, and J. Zhang, "Game-theoretic EDCA in IEEE 802.11 e WLANs," in *Vehicular Technology Conference, 2008. VTC 2008-Fall. IEEE 68th*, 2008, pp. 1-5.

[33] Y. Cho, C. S. Hwang, and F. A. Tobagi, "Design of robust random access protocols for wireless networks using game theoretic models," in *INFOCOM 2008. The 27th Conference on Computer Communications*, 2008, pp. 1750-1758.

[34] J. W. Lee, A. Tang, J. Huang, M. Chiang, and A. R. Calderbank, "Reverse-engineering MAC: A non-cooperative game model," *IEEE Journal on Selected Areas in Communications,* Vol. 25, No. 6, pp. 1135–1147, 2007.

[35] J. W. Lee, M. Chiang, and A. R. Calderbank, "Utility-optimal medium access control: reverse and forward engineering," in *25th IEEE International Conference on Computer Communications*, 2006, pp. 1 - 13.

[36] T. Cui and T. Ho, "On Multiple Access Random Medium Access Control," in *Proceedings of the 2009 IEEE international conference on Symposium on Information Theory (ISIT'09).* Vol. 3, 2009, pp. 1869-1873.

[37] D. K. Sanyal, M. Chattopadhyay, and S. Chattopadhyay, "Performance improvement of wireless MAC using non-cooperative games," *Adv. Ele. Eng. Comput. Sci.,* Vol. 39, No., pp. 207-218, 2009.

[38] Y. Jin and G. Kesidis, "Distributed contention window control for selfish users in IEEE 802.11 wireless LANs," *IEEE J. Sel. Areas Commun,* Vol. 25, No. 6, pp. 1113-1123, 2007.

[39] F. Wang, O. Younis, and M. Krunz, "Throughput-oriented MAC for mobile ad hoc networks: A game-theoretic approach," *Ad Hoc Networks,* Vol. 7, No. 1, pp. 98-117, 2009.

[40] F. Wang, O. Younis, and M. Krunz, "GMAC: A game-theoretic MAC protocol for mobile ad hoc networks," in *4th International Symposium on Modeling and Optimization in Mobile, Ad Hoc and Wireless Networks*, 2006, pp. 1-9.

[41] F. Wang, O. Younis, and M. Kiunz, "Throughput-Oriented MAC for Mobile Networks with Variable Packet Sizes," in *3rd Annual IEEE Communications Society on Sensor and Ad Hoc Communications and Networks, 2006. SECON '06*, Reston, VA, 2007, pp. 421-430.

[42] C. Long, Q. Chi, X. Guan, and T. Chen, "Joint random access and power control game in ad hoc networks with noncooperative users," *Ad Hoc Networks,* Vol. 9, No. 2, pp. 142-151, 2011.

[43] A. Ghasemi and K. Faez, "A non-cooperative game approach for power-aware MAC in ad hoc wireless networks," *Comput. Commun.,* Vol. 33, No. 12, pp. 1440-1451, 2010.

[44] A. Ghasemi and K. Faez, "Power-aware MAC for multihop wireless networks: a cross layer approach," *Wireless Communications, IEEE Transactions on,* Vol. 7, No. 10, pp. 3917-3929, 2008.

[45] X. He and L. Tan, "Contention Access Game Method in Wireless Ad-Hoc Networks," in *4th International Conference on Wireless Communications, Networking and Mobile Computing, WiCOM '08.* , 2008, pp. 1-4.

[46] L. Galluccio, "A game-theoretic approach to prioritized transmission in wireless CSMA/CA networks," in *69th Vehicular Technology Conference,VTC* Spain, 2009.

[47] M. Cagalj, S. Ganeriwal, I. Aad, and J. P. Hubaux, "On selfish behavior in CSMA/CA networks," in *24th Annual Joint Conference of the IEEE Computer and Communications Societies.* , 2005, pp. 2513-2524.

[48] B. Yang, G. Feng, and X. Guan, "Noncooperative Random Access Game via Pricing in Ad Hoc Networks," in *Decision and Control, 46th IEEE Conference on,* New Orleans, 2007, pp. 5704-5709.

[49] B. Yang, G. Feng, and X. Guan, "Random access in wireless ad hoc networks for throughput maximization," in *9th International Conference on Control, Automation, Robotics and Vision, ICARCV '06.*, 2006, pp. 1-6.

[50] J. Yang and H. Shi, "Two Rounds Game CSMA algorithm of WSNs," in *2nd International Conference on Networking and Digital Society (ICNDS)*, Wenzhou 2010, pp. 270-273.

6

MOBILITY BASED CHECKPOINTING AND TRUST BASED RECOVERY IN MANET

Suparna Biswas[1], Sarmistha Neogy[2] and Priyanka Dey[3]

[1,3]Department of Computer Science & Engineering, West Bengal University of Technology, Salt Lake, Kolkata
mailtosuparna@gmail.com
priyankadey24@yahoo.co.in

[2]Department of Computer Science & Engineering
sarmisthaneogy@gmail.com

ABSTRACT

Proposed work is a mobility based checkpointing and trust based rollback recovery algorithm to provide fault tolerance in Mobile Ad hoc Network (MANET). Here each mobile host maintains a count of number of clusters a mobile host traverses through, during a single checkpoint interval. A mobile host increments 'cluster-change-count' by 1, each time it leaves a cluster and joins another. Each mobile host saves a checkpoint independently if its 'cluster-change-count' exceeds a predefined threshold. This measure is important because each mobile host leaves its last checkpoint and logs at different clusters that it has visited earlier. If the mobile host fails, the time to search and collect its last checkpoint and logs gets added to the recovery time of the mobile host. In MANET, retrieval of checkpoint and logs has to be done through a number of intermediate mobile hosts because each mobile host has short area coverage, hence direct communication among distant mobile hosts is not possible. Now if any of these mobile hosts fail, depending on the nature of failure, the checkpoint or log may be lost or forwarding of them to the failed host may be delayed causing unsuccessful or delayed recovery of the failed host respectively. This can be avoided if it is ensured that the checkpoints and logs are forwarded only through trusted nodes. Trust model proposed here computes trust value of a mobile host based on four factors: failure rate, availability in network, unused energy and recommendations from neighbour mobile hosts. Simulation results show that proposed algorithm achieves low recovery cost and high recovery probability of failed mobile hosts.

KEYWORDS

MANET, checkpoint, mobility, trust, recovery

1. INTRODUCTION

Rapid development of communication technology from wired to wireless network led almost all service oriented systems to "all time everywhere" service from "anywhere anytime" service. This has been possible due to stabilization of portable devices e.g. laptop, smart phones, mobile phones and advancement of wired communication to wireless infrastructure as well as wireless infrastructure less communication. Today's portable devices are equipped with sufficient resources hence can do computing as well as communication while on the move. In this kind of systems, computing devices as well as communication links are failure prone. Mobile hosts(MH) in MANET are not physically protected hence can come easily under intruder's control. Failure prone nodes are security attack prone and vice versa. In failure prone system fault tolerance measures should be there for quick recovery. Checkpoint and rollback recovery is popular technique for fault tolerance in mobile computing [1], [2], [3], [4]. In a checkpoint the state of a process saved contains process structure, register values, segments, actual data and open file information [5]. Starting address of each segment of a process to be checkpointed, size of a segment, actual data of a segment etc. are saved in checkpoint file. The recovering process will recover and resume execution from saved register values in checkpoint file. The checkpoint is generally saved in cluster head. Now the MH moves across different clusters and may fail in other cluster. To recover from saved checkpoint which is saved in another cluster head, the checkpoint needs to be transferred to the cluster head in which the failed host will recover. Checkpoint is transferred through a number of cluster heads and gateway nodes. If any of these nodes fail, then the checkpoint may be lost and the recovery process will be terminated. So, the challenge is to select a set of nodes through which checkpoint can be successfully transferred without any fail. In this paper we propose a mobility and trust based checkpointing algorithm. Here we consider cluster based hierarchical MANET [6]. A MH moves across different clusters, leaves behind checkpoint and logs in different cluster heads. If the MH fails, these dispersed recovery information need to be collected at the time of recovery. To limit this search and retrieval cost of recovery information, MHs save checkpoint based on threshold of cluster_change_count [7]. All the cluster members of a cluster save checkpoints in current cluster head and if any cluster member fails in another cluster, last saved checkpoint of failed host is transferred through intermediate gateway nodes and cluster heads to the cluster head in which the failed host will recover. So, cluster head and gateway nodes must be available in the network without any failure. We compute trust value of a mobile host based on following four factors : failure rate, availability of node, unused battery power and recommendation from neighbor node. A mobile host having low failure rate and high unused battery power, will be highly available in network ensuring successful execution of any assigned task enhancing positive recommendations from neighbor nodes. We define a threshold of trust value. A mobile host is considered as trusted if its trust value is greater than threshold.

The rest of the paper is structured as follows: Section 2 discusses some of the related works, our observation and problem definitions. Section 3 describes system model, preliminary assumptions, data structures and notations. Section 4 explains proposed checkpointing scheme and proposed trust model. Section 5 presents algorithms and its correctness proof. Section 6 illustrates simulation and performance analysis. In Section 7 we conclude our work.

2. RELATED WORKS

The proposed work combining trust model with checkpoint and rollback recovery technique to implement secure checkpointing is unique of its kind. Hence similar related work cannot be presented. Overviews of some of existing works related to our proposed algorithm are described in this section.

In [7] Biswas and Neogy present a movement based independent checkpointing and log based rollback recovery algorithm in cluster based MANET. Mobile hosts move among different clusters and leave saved checkpoint and logs scattered in different previously visited cluster heads. If number of inter-cluster movements of a mobile host exceeds a predefined threshold then the mobile host takes checkpoint. Cluster head is like any other node and does not have stable storage, hence copy of checkpoint of a mobile host is saved in one of the cluster members of same cluster having sufficient resources. Threshold of Cluster_change_count limits recovery cost. In [8] P.K.Jaggi et.al describes a staggering based synchronous checkpointing and recovery scheme adapted for handling the limited storage and bandwidth problems of

Mobile ad hoc network which is cluster based. Here cluster member will save checkpoint and logs in cluster head. Cluster member will save only in channel message in log. Any cluster head can initiate checkpointing scheme. After taking checkpoint cluster head sends a request message to take checkpoint to the other cluster head s in its transmission range and then initiates the checkpoint in its cluster. In [9], Juang and Liu presented an efficient asynchronous recovery algorithm in cluster based MANET. In this work they have studied the fundamental problem of crash recovery in mobile distributed environment. Each processor takes checkpoint independently without any synchronization and coordination overhead. Yi, Heo, CHo and Hong presented an adaptive mobile checkpointing facility for wireless sensor networks in [10]. Each node saves copy of checkpoint in its neighbor nodes. A node saves checkpoint, sends its copy to all its neighbor nodes which send acknowledgement along with a probability. This probability is inversely proportional to the number of neighbor nodes of a node. As a node cannot save all the checkpoints of its neighbor nodes due to limited storage space, this probability helps each node to adaptively decide on saving checkpoints of its neighbor nodes calculated based on each node's neighbor node. In [11] X. Zhao et.al proposed a work to provide security in Mobile ad hoc network based on availability based trust model. They divided availability into three main elements: collaboration, honesty and ability. They build their trust model based on security and safety capacity of node, experience of some certain trusted nodes and ability of some certain nodes which participate in web collaboration and stand alone capacity of data transfer. Their trust model based security mechanism ensures the validity of trust computation under limited resource. In [12] A.fathi et.al proposed an energy efficient topology control protocol for Mobile ad hoc network which ensures minimum connectivity in the network at all times. They defined energy is the main factor of life time of a node. So, dynamic topology depends on energy. So, based on energy cluster head and gateway is elected so that they get maximum lifetime. In [13] J. Luo et.al proposed an efficient trusted node discovery method on recommendation based on fuzzy logic. They decide based on recommendation whether a node is trusted or selfish. In [14] R. Manoharan et.al proposed a trust based Hybrid Gateway selection scheme. According to their proposed scheme gateway is selected based on three metric: load taking capacity, recommendation based on observation from other neighbour node and route selection based on trust.

2.1. Problem Identification

After analyzing the above literature survey we identify the following problems that motivated us to propose the work.

- **Cluster Head has limited memory space:** In [8], checkpoints are always saved in cluster head. A cluster head has limited memory space. Hence, there should be an upper limit of number of cluster members that can be connected to a single cluster head. But they do not specify number of Cluster member in a cluster so that cluster head can save all the cluster members' checkpoint.

- **Random mobility of clusterhead and cluster member nodes:** Above mentioned related works do not consider random mobility of cluster members and cluster heads as a factor of saving checkpoints.

- **Frequent and sudden failures of mobile hosts:** Frequent transient or crash failures may occur in mobile hosts suddenly. Existing works consider that all mobile hosts that participate in checkpoint- recovery process do not fail.

- **Trust value calculation based on a single factor:** In [11], [12], [13] cluster head is selected based on only a single factor. In [11], cluster head is selected based on a mobile host's availability. In [12], mobile node is selected on the basis of only available battery power. In [13], cluster head is selected based on only recommendation from others. The problem of considering a single factor is that other factors that are not considered at all, may cause failure.

In our work we have addressed above problems and tried to give solutions as given below:

- **Limited cluster members in a single cluster:** Maximum number of cluster member nodes of a cluster is defined by dividing allocated memory space of a cluster head to save copy of checkpoints of its cluster member nodes by size of checkpoint.
- **Inter-cluster movement based checkpointing:** Mobile hosts save checkpoints based on their inter-cluster movement.
- **Only trusted nodes evaluated based on multiple factors can participate in checkpoint-recovery process:** Trust value of a mobile host is calculated based on its failure rate, unused battery power, availability in the network and recommendation from others based on previous performance.

3. System Model

MANET considered here consists of a number of clusters each having a cluster head and multiple cluster member nodes. Cluster heads communicate with each other through gateway nodes. In MANET there is no stable storage, fixed access point, topology changes dynamically.

Assumptions: Our proposed algorithm is true based on following assumptions:

- Recovery of a mobile host is possible for both transient and crash failures.
- Each mobile host saves last checkpoint, logs and log_tags of current checkpoint interval at any instant of time.
- The mobile host that saves checkpoint and the cluster head that carries the copy will not fail simultaneously

3.1. Data Structures & Notations:

CH = Cluster head, CM =cluster member, GW = gateway, c3 = cluster_change_count, c3_th = threshold of cluster change count, F_r = failure rate, $\%B_{left}$ = percentage of available battery power, R_{t_chkp}= Rate of checkpoint transfer, A = availability, R= Recommendation from other node, T_r = recovery time, T_{val} = trust value, T_1= Time required to send CH id where checkpoint and log is saved to the currently connected CH, T_2 = Time required to send request to the CHs where checkpoint and log reside, T_3= Time required to transfer checkpoint and log from requested CHs to the CH where failure CM is connected after failure,T_4= Time required to transfer checkpoint and log from currently connected CH to failed CM,T_5= Time required to rollback,T_6= Time required to replay log, mh_dep [ ] = during current interval, list of mhs from which computation messages received, CH_mh_list[ ] = list of mobile hosts connected to a clusterhead currently, CH_mh_list[i] = 1, mh_i connected, CH_mh_list[i] = 0, mh_i disconnected, CH_mh_list[i] = -1, mh_ifailed, CH_traversed_list[ ] = Array of cluster heads through which a mobile host traverses during a checkpoint interval, mhi_s = mobile host that sends computation message, mhi_r = mobile host that receives computation message, m_c= computation message, m_{co} = coordination message, log = $log_{s,\ r,intv,seq}$: s = sender mh, r = receiver mh, intv. = interval, seq = sequence number of computation message, log_tag = $log_{r,s,intv,seq}$, $log_{(1,*,1,*)}$: s = mh_1, intv = 1, r = seq = 'any', UID = $UID_{s,r,intv,seq}$: s=sender mh, r = receiver mh, intv. = interval, seq = sequence number of computation message, e.g. $UID_{(*,1,1,*)}$: r = mh_1, intv = 1, s = seq = 'any'

Checkpoint_ file:	**MH_id, CH_id** **Process status** **Data, intv**	**Log_file:**	**$log_{s,r,intv,seq,}$** **copy of m_c**

4. PROPOSED WORK

4.1 Checkopinting and Recovery Technique

During a checkpoint interval mobile host computes, sends and receives computation messages to and from other mobile hosts. A mobile host saves log of sent computation message and forwards to current cluster head. A cluster head saves logs of its cluster members, coordinates recovery of its failed member hosts besides own computations. A mobile host increments 'cluster-change-count' by 1 each time the mobile host leaves a cluster and joins another. Each mobile host saves a checkpoint independently if its 'cluster-change-count' exceeds a predefined threshold. If a mobile host fails before saving a checkpoint during current checkpoint interval, the cluster head in which it will recover will search for its last checkpoint and logs saved in other cluster heads. The cluster head that saved failed MH's last checkpoint will forward checkpoint to the cluster head in which the failed MH will recover. The cluster head then forwards checkpoint to the MH. In similar way the logs also will reach the MH. Now the MH will roll-back upto last checkpoint, replay logs, sends request message to the MHs from which the MH has received computations messages according to log_tags saved before failure. The recovered mobile host is now same as that before failure.

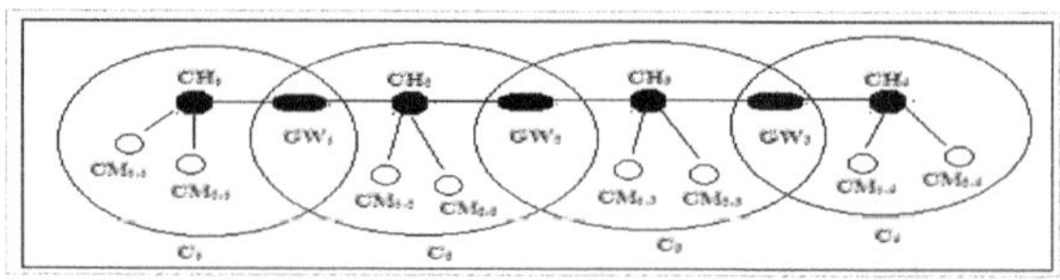

Figure.1. Working example of the network considered in our work

Above figure represents the cluster based mobile ad hoc network considered in our work. In the above example, we consider 4 clusters C_1, C_2, C_3, C_4. Each cluster contains 4 cluster members. Within 4 cluster members, one cluster member will play the role of cluster head, another cluster member will play the role of gateway node, rest of the two are normal cluster members. So clusters C_1, C_2, C_3 and C_4 have cluster heads CH_1, CH_2, CH_3 and CH_4 respectively. Each cluster head of one cluster is connected to cluster head of another cluster through gateway nodes, GW_1, GW_2 and GW_3. Here we consider each cluster member will take checkpoint when c3_th =3. So, for example we consider $CM_{1,1}$ initially take checkpoint at C_1 and save checkpoint and logs in CH_1..So after visiting C_2, C_3 when CM_1 will reach C_4 its c3_th will be 3. So CM_1 will have to save checkpoint and logs in CH_4. After taking checkpoint in C_3 two cases may happen: case1:$CM_{1,1}$ may fail before saving checkpoint CH_4, case2: $CM_{1,1}$ may save checkpoint and logs in CH_4.

Case1: If $CM_{1,1}$ fails before saving checkpoint to CH_4, after reconnect to CH_4 after failure $CM_{1,1}$ will give the id of CH_1 where checkpoint and logs of send messages [15] are saved. Then CH_4 will send requests to CH_1 through the path CH_4-GW_3-CH_3-GW_2-CH_2-GW_1-CH_1.CH_1will send checkpoint and logs through the above discussed path.CH_4 then sends checkpoint and logs to $CM_{1,1}$. $CM_{1,1}$ will rollback to this checkpoint and replay the logs. This mobility based checkpoint and log based rollback recovery is consistent. If mobile hosts save checkpoint independently and do not saved log of send message then if a mobile host fails without saving next checkpoint whereas the receiver host of that message has saved checkpoint then the message has been received without being sent i.e. orphan message which will cause inconsistent recovery. Here our proposed checkpointing and recovery algorithm is consistent.

Case2: $CM_{1,1}$ saves checkpoint and logs to CH_4 and sends message to CH_4 to delete last saved checkpoint and logs which is saved in CH_1. Ch_4 sends message to $CH_1$to delete checkpoint and logs of $CM_{1,1}$. After receiving message CH_1 deletes checkpoint and logs of $CM_{1,1}$ and acknowledges CH_1.

This recovery technique does not consider failure probability of intermediate MHs through which the checkpoint is transferred to the failed mobile host from the cluster head that saved it. So if any checkpoint forwarding nodes fails, checkpoint transfer will be delayed or unsuccessful depending on the type of failure. Any of these are undesirable. Alternatives may be: i) Cluster head will keep a copy of the checkpoint till a successful recovery message comes from failed MH after its recovery. Then the cluster head will delete the checkpoint. Within stipulated time if successful recovery message does not come or failed recovery message come, the cluster head will again resend a copy of checkpoint and wait for reply. Reasons of this could be due to failure of forwarding mobile hosts. In worst case, this may continue repeatedly causing resource consumptions of cluster head, intermediate forwarding mobile hosts.

ii) If it can be ensured that checkpoint forwarding is done through only 'trusted MHs' that does not fail during recovery process of a MH. To ensure that, before forwarding checkpoint to next MH, current MH must be sure about trust status of next MH. For that, trust value of next MH has to be calculated. That can be done using a trust model [6].

The second alternative solution is much more acceptable because checkpoint forwarding through trusted MHs will ensure that checkpoint will reach to the failed mobile host. Here we propose a trust model that evaluates trust of a mobile host based on multiple factors like its unused battery power, life time in the network, failure rate and recommendation from other MHs.

4.2 Proposed Trust Model

Trust of cluster member in a cluster is a belief that all the cluster member will resist different types of attack. A cluster member can be trusted or distrusted. But sometimes it may be impossible to decide whether a cluster member is 'trusted' or 'distrusted' due to lack of communication or improper knowledge. In this case we consider cluster member as 'uncertain' cluster member. According to [16], probability of a mobile host being trusted, distrusted and uncertain can be expressed in following way-

$$b + d + u = 1 \quad \text{.................................} \quad (1)$$

Where b=belief, d=disbelief and u=uncertain. In our paper we term belief as trusted, disbelief as distrusted.

There are various methods of trust value evaluation of a system [11, 13] and how to remove uncertainty [17]. In proposed algorithm trust value is defined based on trust matrix in which values of availability, recommendation from neighbour nodes, failure rate, and unused energy per mobile host are saved.

Availability: This is a statistical data accumulated during a MH's communications with other mobile hosts. Each MH estimate its neighbour's availability based on its accumulated observations using Bayesian inference as discussed in [17]. Bayesian inference is a statistical inference in which evidence or observations are used to update statistics. Beta distribution, Beta (a, b), are used here in the Bayesian inference, since it only needs two parameters that are continuously updated as communications are made. When a new communication is made, if it is successful, then parameter 'a' is updated, otherwise parameter 'b' is updated. Every mobile host always maintains experience statistics record of its neighbour mobile hosts. Every mobile host always maintains availability record of its neighbour mobile host. When availability of a node is being observed or calculated, availability record from its neighbour hosts are taken.

Algorithm to evaluate availability:

```
if ( a >= b && a != 0) availability = 1;
else {  if ( b= = 0 )   availability = 0.5;
        else  availability = 0; }
```

Failure rate: It is an important factor to decide whether a MH is trusted or distrusted. According to our algorithm, a mobile host having low failure rate will be considered as trusted because it will not fail frequently and hence will be recommended by others.

Unused energy: Energy is also an important factor to decide whether a MH is trusted or distrusted. If a MH has high energy then its probability of disconnection from network will be low. So, MH will be more reliable.

Recommendation from neighbour nodes: When we have to decide whether a thing is good or bad, sometimes it is necessary to take recommendation from others. Many trust models are already implemented based on this concept of recommendation. So, we also include this parameter in the trust matrix defined here. Recommendation from others is a very important factor to get to know about the behaviour of a particular MH. In MANET, as no dedicated router is there, forwarding of checkpoint has to be done through normal mobile hosts. To select next node on a path, recommendation helps. A mobile host will give recommendation about other host based on its availability, failure rate and unused energy.

For example, suppose A will have to give recommendation of B. In that case, first B has to share its failure rate, availability and unused battery power to A. Then B checks its availability by using algorithm as already discussed in the above availability section. Then give recommendation given as below:

Table 1. Set recommendation value on different factors

Recommendation	**Failure rate (F_r)**	**Availability(A)**	**Unused battery power(P)**
1	< 0.5	> 0.5	> 0.5
0.5	$= 0.5$	$= 0.5$	$= 0.5$
0	> 0.5	< 0.5	< 0.5

So, a mobile host having failure rate < 0.5, availability > 0.5 and unused battery power > 0.5 will have a recommendation 1 from other mobile hosts. Fifty-fifty recommendation is given for all parameter value of 0.5. Hence, 0.5 is threshold level of all three parameters.

So, Trust_value_threshold (**T_{val_th}**) = (F_{r_th} * r) + ($A_{\_th}$ * r) + ($P_{\_th}$ * r) ---------------------- (2)
= (0.5 * 0.5) + (0.5 * 0.5) + (0.5 * 0.5)
= 0.25 + 0.25 + 0.25 = 0.75

By using following algorithm we consider whether a node is trusted, distrusted or uncertain.

<u>How to decide a mobile host's trust status:</u>

if ((trust_value >= trust_value_threshold) && A >=.5&&F<=.5&&P>=.5)
mobile host is 'trusted' ;
else { if (A = = 0) mobile host is 'uncertain' ;
else mobile host is distrusted ; }

If a node is uncertain, we check failure rate of the node. If failure rate is greater than or equal to 0.5, mobile host will be considered as trusted, otherwise distrusted. For example, suppose the trust matrix of three mobile host of a cluster is as follows:

Table 2. An example of labelling of node after computing trust value

MH	**F_r**	**A**	**P**	**R**	**$T_{\_val}$**	**Label of node**
A	0.2	0.2	0.7	(1,0,1)	= (0.2*1) + (0.2*0) + (0.7*1) = 0.2+0+0.7 = 0.9 > 0 .75	Trusted
B	0.2	0	0.7	(1,0,1)	= (0.2*1) + (0*0) + (0.7*1) = 0.2+0+0.7 = 0.9 but A==0	Uncertain
C	0.6	0.02	0.6	(0,0,1)	= (0.6*0) + (0.02*0) + (0.6*1) = 0+0+0.6 = 0.6 < 0.75	distrusted

In the above table trust value of A is greater than threshold and availability is non-zero. So it is a trusted node. Trust value of node C is less than threshold, so distrusted. Trust value of B is greater than threshold but availability is zero. So, it is uncertain. Justification is that when a mobile host is not available in the network, how its trust status can be defined?

In a cluster, the mobile host that has highest trust value will be selected as cluster head, the mobile host with next higher trust value will be gate way node. A mobile host once trusted may not be trusted forever because its battery power may be reduced, its availability in the network may be poor, its failure rate may be high and if any or all of these are true, then other nodes may not recommend it. Then a trusted node may become distrusted or undefined. If due to reduced trust value, or random mobility or any other reason, cluster head or gateway nodes get changed, these nodes will transfer all the saved checkpoints, logs, data structures to the newly selected cluster head or gateway nodes respectively. The node may change but its ID will remain same like cluster head of cluster 1 is designated as CH_1. Now irrespective of the node, the cluster head ID of cluster 1 will always be CH_1.So trust value evaluation of mobile hosts is a continuous process which must repeat itself after certain time interval which can be set based on specific system or the network being used. As only trusted nodes can be cluster head and gateway nodes, by our proposed trust model it is ensured that checkpoint will be forwarded through only trusted nodes.

5. ALGORITHM

1. Mobile hosts compute, communicate with each other through message passing.
2. Each mobile host sends a beacon message to own cluster head at regular interval to convey that it is connected. Through beacon message continuous authentication of nodes also goes on.
3. If mh_{i_s}wants to send a computation message to mh_{i_r}, mh_{i_s}saves a log, forwards m_c and log to current cluster head. c_ch saves log in its memory, checks own ch_mh [ ], if finds a 1 in corresponding array field then $mh_{i_r}$is in cluster 1, if finds a 0 then mhi_r is in other clusters, if finds a -1, then mh_{i_r}failed.

   case 1: $mh_{i_s}$and mh_{i_r}in same cluster:

   c_ch forwards m_c to mh_{i_r}andmh$_{i_r}$sends ack to mh_{i_s} throughc_ch.

   case 2: $mh_{i_s}$and mh_{i_r}in different cluster:

   c_ch of mhi_s broadcasts a look up (mh_{i_r}) message to all cluster heads through gateway nodes. All cluster heads searches their corresponding ch_mh [ ]. The CH that founds 1 replies to c_chof mh_{i_s}, c_ch forwards m_c and UID of m_c gateway node which saves UID of m_c and forwards m_c to replyingch, ch forwards to mh_{i_r}.

   case3: mh_{i_r}fails:

   mh_{i_s} stops sending m_c temporarily, will try to send later.
4. mh_{i_r} receives computation message,saves updates dependency array, sends UID of m_c to gateway node.
5. mh_{i_s}moves randomly, leaves current cluster and joins another cluster .
6. In cluster based adhoc network each mobile host increments cluster_change_counter (c3) each time a mobile host changes cluster.
7. The mobile host saves ID of cluster heads being traversed in an array, CH_traversed_list[ ].
8. Before the mobile host leaves a cluster head, sends different data structures e.g. dependency array, CH_traversed_list etc. saved being connected to current cluster head along with a leave message to current cluster head.
9. The mobile host saves ID of previous cluster head in CH_traversed_list[ ], e.g. CH_traversed [0] = CH_1, if mh leaves CH_1.
10. The mobile host next joins another cluster head say CH_2
11. repeat steps 2 to 6 .
12. The mobile host saves ID of previous cluster head in CH_traversed_list[ ], e.g. CH_traversed [1] = CH2, if mh leaves CH2.
13. Repeat steps 2 to 6.
14. The mobile host saves ID of previous cluster head in CH_traversed_list[ ], e.g. CH_traversed [2] = CH3, if mh leaves CH3.
15. If c3 > c3_th, mobile host invokes checkpoint procedure ( ).

    15.1.The mh takes a snapshot of current state of computation of running application process.

    15.2 Checkpointingmh searches for an immediate neighbour with highest remaining memory and saves a back up copy of checkpoint in it. The mh ends checkpoint_backup_node = ID of$mh_{neighbour}$ to CH_2.
16. The mobile host sends delete_logof prev_interval message to CH_2 which forwards this message tothe cluster heads that are saved in CH_traversed_list of the mobile host through gateway nodes.

16.1 If any gateway node finds a match between log_ID and UID, sends do_not_delete message to the cluster head.

16.2 The cluster head forwards log to the gateway node.

16.3 Gateway node saves log.

17. The mobile host enters into next checkpoint interval.

18. The mobile host saves current cluster head in its CH_traversed_list[ ]

19. repeat steps 2 to 6 till c3 ≤ c3_th.

20. The mobile host fails, last checkpoint can be recovered from the host itself or cluster head

20.1 **CASE 1:** Last checkpoint can be recovered from the failed host itself, the host rollsback upto last checkpoint.

20.1.1 **Case 1:** The failed host is mhi_s, Replay_ log (UID) message to c_ch, c_ch checks if the log is saved in its memory, if yes replays log to mhi_r, else broadcasts replay_log (UID) message to the chs saved in failed host'sch_traversed_list.

20.1.2 **Case 2:** The failed host is mhi_r, Receive_m_c (UID) message to c_ch, c_ch checks if the log is saved in its, memory, mhi_r receives mc from, c_ch, else c_ch broadcasts receive_$m_{c,}$ (UID) message to the chs saved in mhi_s' ch_traversed_list.

20.1.3 **Case 3:** The failed host is both mhi_s and mhi_r, Do tasks in case 1 and case2.

20.2 **CASE 2:** Last checkpoint needs to be recovered from cluster head, Current cluster head sends recovery_message of the failed mobile host to the cluster head saved in first field of ch_traversed_list. Then that cluster head forwards copy of checkpoint of failed host saved to it following the same path through which the recovery_message has been sent to it.

21. The failed host rolls back upto last checkpoint and recovers following any of the above three cases.

5.1 Correctness Proof

Theorem 1: The algorithm ensures consistent recovery

Proof: With the help of following two lemmas the above statement can be proved.

Lemma 1: There is no orphan message.

Proof: save log of each sent $m_{c,}$ TRUE for $\forall$ mh
if (mh fails without saving checkpoint)
sent m_c is retrieved from saved logs, TRUE for $\forall$ sent m_c

Lemma 2: There is no lost message.

Proof: save UID of each received $m_{c,}$ TRUE for $\forall$ mh
if (mh fails without saving checkpoint)
received m_c is retrieved according to saved UIDs, TRUE for $\forall$ mh

Theorem 2: Proposed algorithm causes minimum checkpoint and log overhead per mobile host per checkpoint interval

Proof: Each mh saves log, $\forall$ sent computation message, interval=current
Each mh saves UID, $\forall$ received computation message, interval=current
Each mh saves checkpoint, C3>C3_th, delete logs and UIDs, interval=next
Each mh saves log, $\forall$ sent computation message, interval=current
Each mh saves UID, $\forall$ received computation message, interval=current

Each mh saves checkpoint, C3>C3_th, delete previous checkpoint, delete logs and UIDs, interval=next

So, per interval each mh saves checkpoint, logs and UIDs of single checkpoint interval.

Theorem 3: Checkpoint traverses only through trusted mobile hosts

Proof: Following two lemmas help to prove this theorem.

Lemma 1: Only Trusted mobile hosts can be selected as cluster heads and gateway nodes.

Proof: If there is n number of mobile hosts in a cluster then the mobile host having highest trust value will be selected as cluster head and mobile hosts having next higher trust value/values will be selected as gateway node/nodes. If trust value of current cluster head deceases due to reduced battery power or unavailability in the network etc., then it will be replaced by another mobile host having highest trust value currently.

Lemma 2: Saved last checkpoint copy of failed mobile host is forwarded through only cluster heads and gateway nodes to the failed host for its recovery

Proof: Last checkpoint copy of any mobile host is saved in cluster head. When it fails after some inter-cluster movements, the cluster head in which it will recover communicates with other cluster heads to find and retrieve its last checkpoint. Then the checkpoint saving cluster head forwards it to the communicating cluster head through intermediate cluster heads and gateway nodes.

6. Performance Analysis

We simulate the whole MANET environment like clusters, cluster heads, gateways, cluster members, movement, cluster change, communication through message passing, logging, checkpointing, failure, recovery all these non-deterministic events randomly using C programming. The mobility model considered here is random way point model [18]. We use rand () function in C to generate join, leave, send and receive function randomly. We have implemented random movement, handoff, trust model computation message sending and receiving, logging, checkpoint, failure and rollback recovery of mobile hosts to get different parameters for the performance analysis of the proposed algorithm. We get all this computation time using clock () function. By using clock() function we get the time to take checkpoint and log, time required to decide whether a node is trusted and distrusted, time required to cluster change and failure time by using clock(). We consider following parameter set to show performance analysis of proposed mobility based checkpointing and trust based rollback recovery technique. The parameter values which we use in our simulation are represented in tabular form in below:

Table 3. Performance Analysis parameters value

Parameter	**value**
Checkpoint Size	2-256 KB
Log Size	50 B
computation message size	50 B
coordination message size	2.5 B
UID size	2.5 B
time to transfer checkpoint per hop through wireless channel	0.08s

time to transfer log or computation message per hop through wireless channel	0.002s
time to transfer coordination message or UID of a log.	0 .0001s
Energy capacity of a node	1800J
Availabity	10-100%
Mobility rate	0.40-0.48
Channel bandwidth	1 MB
Failure rate	0.02-0.08
Cluster change threshold	3-20
Recovery probability	0.4-0.99

c3 vs. failed node's recovery time (assuming mobile host fails when c3 = c3_th) :

A cluster member node moves from one cluster to another cluster, c3 increases by 1, distance between old and new clusterhead increases by 1 hop. Here we assume that a node fails when c3 = (c3_th) i.e. just before c3 exceeding (c3_th), the condition to save checkpoint of current state. At this point recovery information, logs have to be retrieved from (c3_th) number of cluster heads and checkpoint will be transferred from the cluster head which is at a distance of (c3_th) number of hops from recovery clusterhead. Thus log transfer cost and checkpoint transfer cost both will be maximum. So it is justified that if checkpoint is saved based on c3_th then maximum recovery cost can be calculated beforehand. This will help to set the value of (c3_th) as per application.

Recovery time = cost to transmit recovery request message to (c3_th) no. of cluster heads + cost to transfer logs saved in c3_th no. of cluster heads + cost to transfer last checkpoint from the cluster head which is at a distance of c3_th hops

= (Cm_{co}* (c3_th) + Clog_transfer * (c3_th) + Ccheckpoint_ transfer*(c3_th) hops) unit

= ((Cm_{co}+ Clog_transfer + Ccheckpoint_transfer) * (c3_th))unit

= ((0.0001 + 0.002 + 0.08) * (c3_th)) unit

= (0.0821 * (c3_th)) unit -------------------- (3)

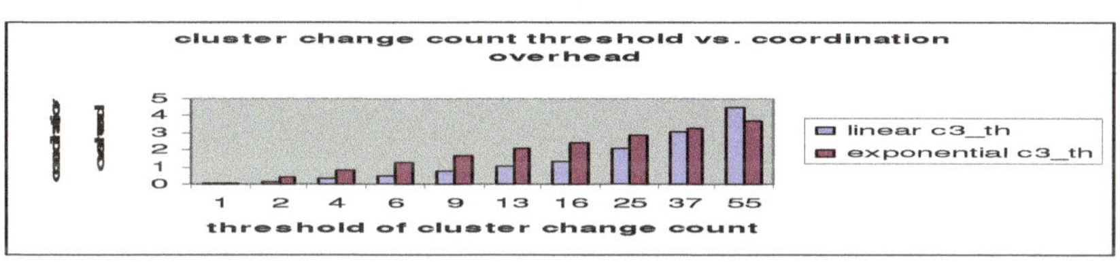

Figure 2. maximum coordination overhead at the time of recovery varies with cluster change count threshold

A mobile node fails when c3= (c3_th), just before saving checkpoint. Hence number of cluster heads where the node's recovery information are scattered is c3_th. For simplication we consider that distance between last cluster head where the ID of the cluster member node that saves backup copy of last checkpoint of failed node is saved and the cluster head where the failed node recovers is (c3_th) hops. Here checkpoint transfer cost is bounded by (c3_th).

Number of cluster member nodes vs. coordination overhead :

Coordination message for Checkpoint-recovery

= recovery $message_{mh\text{-}CCH}$ + $transfer\_checkpoint\_message_{CCH\text{-}CH}$ + $transfer\_checkpoint\_message_{CH\text{-}mh}$
= m_{co}+ CH* m_{co} + m_{co} = $m_{co}(1\text{+}CH\text{+}1) \approx$ CH (for high value) --------------------------(4)

coordination message for Log recovery
=log recovery $message_{mh\text{-}CH}$+ transfer log $message_{CH\text{-}CH}$
$\approx (1\text{+}CH)^{*}m_{co} \approx$ CH, hence total coordination message = CH + CH = 2*CH

Total coordination message ∞ CH ------------------------ (5)

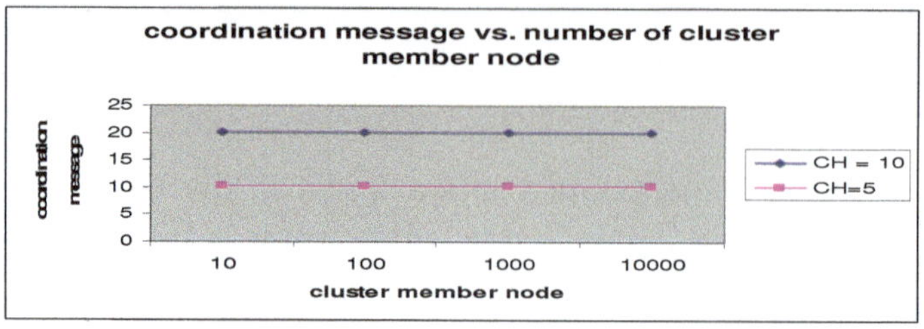

Figure 3. recovery coordination message varies with cluster head irrespective of number of cluster member nodes of a cluster

In figure 3 it is shown that coordination among different nodes at the time of recovery of a failed node is restricted to cluster head level. Coordination message overhead does not depend on number of cluster member nodes.

Next, we compare recovery probability by considering selection of CH and GW based on trust model and considering selection of CH and GW based on individual factor like failure rate, available battery power, percentage of availability and recommendation. We have calculated and plotted recovery probability vs. varying individual factor among the four mentioned above while the other three is constant,

Case1: Selection of CH and GW based on failure rate.
Case2: Selection of CH and GW based on percentage of availability of CM.
Case3: Selection of CH and GW based on available battery power of CM.
Case4: Selection of CH and GW based on recommendation from neighbour CM.
Case5: Selection of CH and GW based on Trust Model.

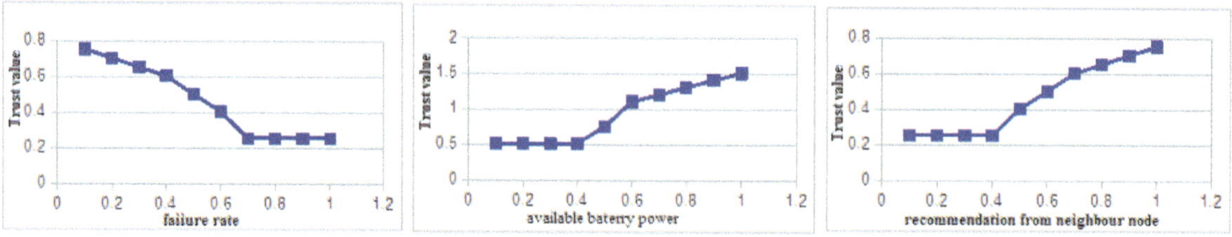

Figure 4.(a) failure rate of node vs. Trust value; (b)available battery power of node vs. Trust value;(c) recommendation from neighbour node vs. Trust value

In figure.4.(a) we see that if failure rate of node increases trust value will decrease.So, failure rate is inversely proportional to trust value. In figure 4.(b) we see that if availability of baterry power increases trust value will increase.So, it is directly proportional to trust value. In figure

4.(c) we see that if recommendation from neighbour node increases trust value will increase.So, it is directly proportional to trust value.

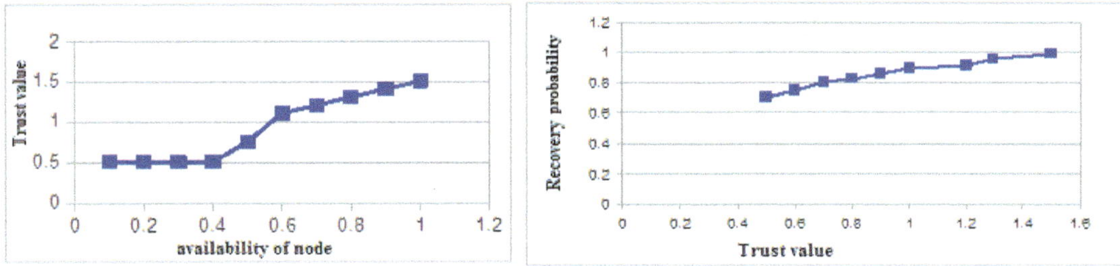

Figure 5.(a) Availablity of node vs. Trust value; (b) Trust value vs. Recovery probability

In figure 5.(a) we see that if availability of node increases trust value will increase. So,availability of node is directly proportional to trust value.In figure 5.(b)we see that recovery probability increases with trust value. we get 0.99 as highest recovery probability. In all the comparison we consider failure rate 0.08. At this failure rate we consider available battery power, percentage of availability and percentage of recommendation will be 100%.

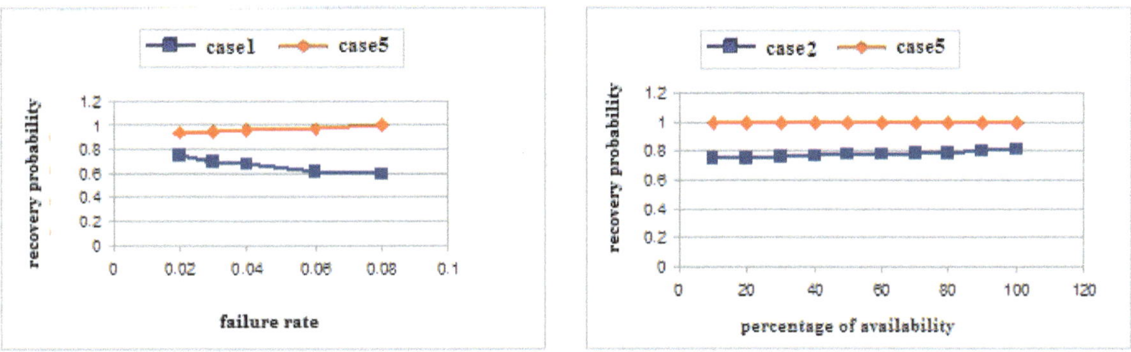

Figure 6.(a) Failure rate vs. recovery probability considering case1 and case5 ; (b) Percentage of availability vs. recovery probability considering case2 and case5

In figure 6(a) we see that if failure rate of CH and GW is increased recovery probability will decrease because if failure rate of CH and GW is high then checkpoint and log may not be transferred all time to the destination cluster. Therefore the recovery probability decreases as the failure rate increases. In the above figure we compare between case5 and case1.We see that recovery probability of case5 is more than case1. In figure 6(b) we see that if percentage of availability of CH and GW is increased recovery probability will increase because if percentage of availability of CH and GW is high then failure rate will be low. In above we have already discussed failure rate is inversely proportional to recovery probability. As percentage of availability is inversely proportional to failure rate, percentage of availability is directly proportional to recovery probability. In the above figure we compare between case5 and case2.We see that recovery probability of case5 is more than case2.

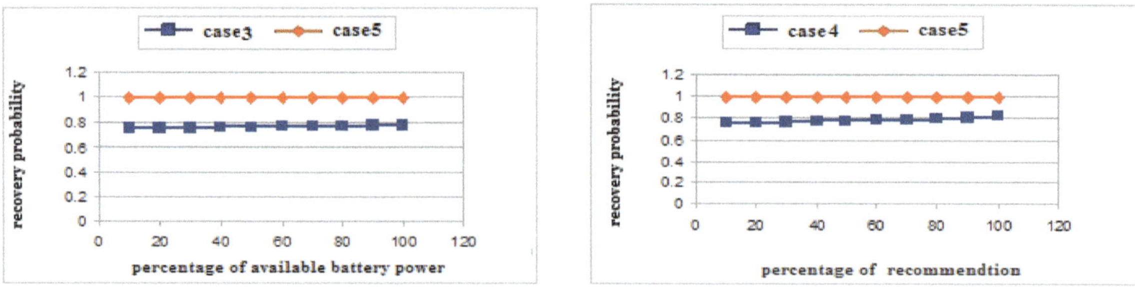

Figure 7.(a) Percentage of available battery power vs. recovery probability considering case3 and case5; (b) Percentage of recommendation vs. recovery probability considering case4 and case5

In figure 7(a) we see that if available battery power of CH and GW is increased, recovery probability will increase because if available battery power of CH and GW is high then failure rate will be low. We have already discussed failure rate is inversely proportional to recovery probability. As available battery power is inversely proportional to failure rate, available battery power is directly proportional to recovery probability. In figure 7.a, we compare between case5 and case3.We see that recovery probability in case5 is more than case3.

In figure 7(b) we see that if percentage of recommendation of CH and GW is increased, recovery probability will increase because if percentage of recommendation of CH and GW is high then failure rate will be low. We have already discussed that failure rate is inversely proportional to recovery probability. As percentage of recommendation is inversely proportional to failure rate, percentage of recommendation is directly proportional to recovery probability. In figure 7.b, we compare between case5 and case4.We see that recovery probability of case5 is more than case4.

After analyzing the above comparisons we get the following results.

Table 4. Performance analysis result

desired recovery probability	**Highest Recovery Probability we get from our experimental result**				
	Case1	Case2	Case3	Case4	Case5
1	.75	.81	.78	.82	.99

After analyzing table 4, we see that we get highest recovery probability in case5. So, selection of CH and GW based on proposed trust model gives high recovery probability which is desired in any fault tolerant system. We get all the above values through our simulation result. In case1, CH and GW get selected based on only failure rate without considering available battery power, availability and recommendation. So, after failure when checkpoint is transferred through GW or CH, there is a possibility of failure of CH and GW due to low available battery power or availability or poor recommendation. So, checkpoint and log may not be successfully transferred all the time to the destination. Hence recovery probability may be very low. This may be true in case2, case3 and case4 mentioned above. But in case5, we select CH and GW considering above four factors so that chances of failure of CH and GW will be less. So we can get maximum 99% recovery probability using proposed trust model.

7. Conclusion

In this work we propose a mobility based checkpointing and trust based rollback recovery algorithm combined with message logging to provide fault tolerance in cluster based mobile ad hoc network. Mobility based checkpointing limits recovery time and trust based recovery increases recovery probability of failed mobile hosts. Each cluster member keeps a count of its change of clusters and if this count exceeds a predefined threshold, then the mobile host will save checkpoint. What makes more challenging the task of checkpointing in MANET is lack of stable storage. In MANET, checkpoint and log placement is also a very trivial issue. Checkpoint traversal only through trusted mobile hosts ensures successful recovery of failed mobile host. Failure prone components are security attack prone also. Moreover insecurity leads to distrust and vice versa. So, security of checkpoints in mobile hosts can be added as a factor to calculate trust value of a mobile host. Proposed trust model will be extended towards that.

REFERENCES

[1] T. Park , N. Woo , H.Y. Yeom,(2003) "An Efficient recovery scheme for fault-tolerant mobile computing systems", Future Generation Computer System, 19(1)pp: 37-53.

[2] C. Men, Z. Xu , D. Wang , (2007)"An Efficient Handoff strategy for Mobile Computing Checkpoint System", EUC 2007, LNCS 4808, pp. 410–421.

[3] F. Quaglia, B. Ciciani , R. Baldoni (2006)"Checkpointing Protocols in Distributed Systems with Mobile Hosts: a Performance analysis",Workshop on Fault-Tolerant Parallel and Distributed Systems, pages 742-755.

[4] R. Prakash, M. Singhal, (1996)"Low Cost Checkpointing and Failure Recovery in Mobile Computing Systems", IEEE Transactions on Parallel and Distributed Systems, Vol. 7, pp:1-38.

[5] G.Hong, S.J.Ahn, S.C.Han, T. Park, H.Y. Yeom, Y.Cho,(2000) " Kckpt: Checkpoint and Recovery Facility on UnixWare Kernel" [J], Computers and Applications, pp. 303-308.

[6] P. Basu, N. Khan, and T.D.C. Little, (2001)"A Mobility Based Metric for Clustering in Mobile Ad Hoc Networks," in Proc. IEEE ICDCSW' 01, Apr., pp. 413–418.

[7] J.H.Cho, A.Swami (2011)"A Survey on Trust Management for Mobile Ad HocNetworks", Communications Surveys & Tutorials, IEEE, Computer & Information Science,Volume: 13, pp. 562 – 583.

[8] S.Biswas,S.Neogy,(2012)"Checkpointing and Recovery using Node Mobility Among Clusters in Mobile Ad hoc Network", , Springer, NECOM pp. 447-456.

[9] P.K. Jaggi, A.K. Singh,(2011) "Staggered Checkpointing and Recovery in Cluster Based Mobile Ad Hoc Networks ", Advances in Parallel Distributed Computing Communications in Computer and Information Science, , Volume 203, pp. 122-134

[10] T. Ying., T. Juang, M.C. Liu, (2002) "An efficient asynchronous recovery algorithm in wireless mobile adhoc networks", Journal of Internet Technology Special Issue on Wireless Internet: Applications and Systems". Vol 3, No.2, pp 147-155

[11] Y. Sangho, J. Heo, Y. Cho, J. Hong,(2006) "Adaptive Mobile Checkpointing Facility for Wireless Sensor Networks", LNCS 3981, pp. 701-709.

[12] X.Zhao, Z.You, Z. Zhao, D.Chen, F.Peng,(2010)"Availability Based Trust Model of Clusters for MANET" Service Systems and Service Management (ICSSSM), 7th International Conference on,pp: 1 – 6.

[13] A. fathi, H. taheri,(2010)" Enhance Topology Control Protocol (ECEC) to Conserve Energy based clustering in Wireless Ad Hoc Networks",ICCSIT, 3rd IEEE International Conference , Volume: 9,pp. 356 – 360.

[14] J. Luo, X.Liu , M.Fan,(2009)"A trust model based on fuzzy recommendation for mobile ad-hoc networks", Journal Computer Networks: The International Journal of Computer and Telecommunications Networking , September,Volume 53,pp. 2396–2407.

[15] R. Manoharan, S. Mohanalakshmie,(2011)"A Trust Based Gateway Selection Scheme for Integration of MANET with Internet" , IEEE-International Conference on Recent Trends in Information Technology,MIT, Anna University, Chennai, ,pp543-548.

[16] E.N. (Mootaz) Elnozahy, L.Alvisi, Y.Wang and D.B. Johnson, (September 2002), "A Survey of Rollback Recovery Protocol in Message Passing System"ACM Comput. Surv., Vol. 34, No. 3. pp. 375-408.

[17] A.Jøsang,(1999)"An Algebra for Assessing Trust in Certification Chains", In J.Kochmar, editor, Proceedings of the Network and Distributed Systems Security (NDSS'99) Symposium, The Internet Society, pp.1-10

[18] F.Li, J.Wu, (2007) "Mobility Reduces Uncertainty in MANETs", INFOCOM.26th IEEE International Conference on Computer Communications. IEEE, pp.1946 – 1954.

[19] T. Camp, J. Boleng, and V. Davies, (2002,)"A survey of mobility models for adhoc network research", Wireless Comm. & Mobile Comp. (WCMC), vol. 2, no. 5,pp: 483-502.

7

REVIEW ON THE SECURITY RELATED ISSUES IN CONTEXT AWARE SYSTEM

Saad Almutairi[1], Hamza Aldabbas[1] and Ala Abu-Samaha[2]

[1]De Montfort University, Software Technology Research Laboratory (STRL)
Leicester, United Kingdom
{Almutairi,hamza}@dmu.ac.uk
[2] Middle East University, Amman, Jordon
asamaha@meu.edu.jo

ABSTRACT

A context aware system is recognized as a "system which uses any context information previous to, or in the duration of, service stipulations", whereas the main goal of this system is to track and identify the users. The mobility feature of most computing and personal assistance devices have made the 'context of the user' an important aspect of the system, affecting its development process in terms of end-user requirements and system behaviour. Context Aware Systems are different from traditional systems as they provide unique features such as heterogeneity, high complexity and artificial intelligence. Since the parameters that constitute a context aware system, such as location and time of the day are rapidly changing. The use and importance of this system is increasing and therefore becoming a major part in prospective wireless systems. However, beside the numerous advantages, context-aware systems have also come up with the numerous issues and challenges related to privacy and security of the user's information in these systems. Therefore, this paper is mainly aimed at a review of the security and privacy issues related with context-aware systems. It also gives a brief introduction regarding the preferred and suitable frameworks required to implement the security requirements. It further focuses on the main question for most researchers about the frameworks used for context-aware systems.

KEYWORDS

Context, context-aware systems, security requirements, frameworks in context-aware systems.

1. INTRODUCTION

"The most profound technologies are those that disappear. They weave them selves into the fabric of everyday life until they are indistinguishable from it" [30]. This statement is quoted from Mark Weiser's visionary article `The Computer for the Twenty-First Century'. This article was published in 1991 and it was considered as a head of its time, Mark Weiser's vision has given birth to a new field in computing namely pervasive computing (aka ubiquitous computing) which is also guided to explore the vital concept of context awareness. Context awareness is the initial element of pervasive computing. Since the first introduced by Schilit and Theimer [31], the term 'context aware system' attracted many researches, for example, Dey and Abowd [1] defined Context as any information which is used in characterising the current condition of any object or entity. In a real time scenario, numerous examples related to the context are given such as location, time, temperature, noise, the bandwidth of communication, connectivity of network etc [2]. In a much broader perspective, a context is being termed as "some information which can be utilized to identify the present condition of any entity". Moreover, Dey and Abowd have further talked about the definition on the basis on which a system can be considered as a context aware system. According to the definition, "A system is considered to be context aware if it utilizes the context in an effort to provide the appropriate information or service to the user where appropriate and significant information depends upon the requirement and need of the user" [1]. Sitou and Spenfelner pointed out that context aware application "seem to be particularly prone to problems related toa discrepancy between user expectation and system behaviour" [25]. In almost the same manner, a context aware system is recognized as a "system which uses any context information previous to, or in the duration of, service stipulations, whereas the main goal of this system is to track and identify the users" [3].

Context Aware Systems are different from traditional systems as they provide unique features such as heterogeneity, high complexity and artificial intelligence. Since the parameters that constitute a context aware system, such as location and time of the day are rapidly changing [32]. Thus the use and importance of context aware system is increasing and therefore becoming a major part in prospective wireless systems. It is being highlighted as important and quite useful due to several reasons such as context can reduce the input cost, it can provide an exciting user experiences without much effort on the user`s part and also users can benefit through context sharing [4,34].

This paper is organized into five sections. In section 2, we will have some discussion about background history along with some related works which have been done in the past by various researchers. Section 3 will review the security issues which related to context aware systems. Section 4 will present some frameworks which have been proposed in the literature review, and finally Section 5 will summarize our conclusion.

2. BACKGROUND AND CONTEXT AWARE APPLICATIONS

Various researchers have highlighted and discussed the basic concept of context and context aware systems in their own way. Most of them are not quite in favour of the most basic concept and have tried to come forward with their own definition in a more precise manner. For instance, Schilit *et al.* [5] have mentioned three points which are required to be considered when discussing a context. Firstly, at what location you are, with whom you are roaming around and what kind of various resources you are utilizing. Furthermore, Chen *et al.*[6] has termed context as a collective combination of environmental states and settings that judges the behaviour of an application or which are important from the user's perspective. Moreover, Dey*et al.*[7] has defined context as information that is used to determine the situation of any entity such as a person, place. These entities are further considered to be very useful for establishing an efficient interaction between the users and any application.

There are many applications which can be considered as context aware or having greater context aware applications to be used. These applications were used in the recent past and many examples can be quoted from the current scenario. According to various researches, many context aware applications were lab based and provided the designers with the facility to work in a proper laboratory environment. A typical example of the most initial kind of context aware application is the Active Badge introduced in the early 90's [8]. Regarding its working, it made use of badges in order to transfer infrared signals to various sensors located in numerous buildings. This case helped to track the user's location and therefore automatically redirect the user's call to that exact tracked location. Another example of context aware application is ParcTab which also serves and provides the user with the same kind of functionality as provided by Active Badge. However this application performs this procedure by making use of much simpler badges [9]. Furthermore, it was also implemented in a form of a PDA (Personal Digital Assistant) in order to introduce much richer context aware applications.

There are many others quite efficient and effective context aware applications which are mobile based and therefore provide some potential facilities and functionalities for the users. These mobile applications come up with various user friendly services such as displaying the relevant information to the user, automatically performing any functionality or providing a service to the user, etc [10].As mentioned above, tourist information systems are a kind of application which makes some major use of context aware softwares nowadays. The major example in this regard is Cyber Guide, a mobile based context aware system used by visitors. It provides the users with information on various places and locations which could be of interest to users by making efficient use of the history of the places visited by users and the direction of movement, etc [11]. Another example in this regard is a GUIDE system which serves as a tour guiding system for the visitors in a city of Lancaster [12]. Moreover, Mobile Location-Aware Handheld Event is another tour guiding system which acts as an event planner and therefore serves as a tour guiding system for the visitor by making GPS location acquisition [13]. Furthermore, Access Sights also acts as a tour guide system and therefore serves people with special needs, like blind users or weak sighted users, with a tour guiding facility [14].

3. REQUIREMENTS & DESIGN CONSIDERATIONS

Before embarking into the discussion especially on some potential security issues regarding the context aware applications, a brief insight will be thrown on to the basic categories of context as a whole. Most of the studies have come forward with various categories of context. However, the majority have agreed collectively upon three - four as the major ones. For instance, in research conducted by Chen and Kotz [14], four basic categories of context have been mentioned. These are: *Computing Context, User Context, Physical Context and Time Context.* A brief description about the information and areas covered by each is given below [37]:

- **Computing context:** covers areas and information related to system connectivity, inter networking, other computing related assets like printers and workstations, CPU, Memory resources etc [14]. Another important aspect and characteristic of context aware applications in terms of computing context is that users sometimes use multiple devices. Therefore they prefer the devices which provide the facility of performing multiple functionalities at the same time [15]. These kinds of devices cover mobile applications and PDAs etc.
- **User context**: contains information related to the user's application usage which includes the user's personal information, preferences, current location, and potential activity, etc [14]. The choice of users can be characterised by the details related to the kind of preferences they have [16]. For instance, some users are willing to go to any specified location by taxi or hiring private transport whereas some other users are willing to travel walking distance only [16]. Therefore, for this reason, a component or facility for storage and efficiently using preference related information is required.
- **Physical context**: covers areas and provides information regarding Location, Time, Destinations, lighting, Physical and environmental conditions, etc [14]. For instance, in most cases the users require access to information such as weather forecasts, route directions, ongoing traffic situations and temperature related updates, etc [16]. Therefore the requirement is always kind of useful information which is going to be updated on a regular basis and should always be made available to the users.
- **Time context**: provides information and access about the time and calendar related information on a daily, weekly or monthly basis [14].

3.1. PRIVACY ISSUES IN CONTEXT AWARE

Privacy concern which related to current context aware systems is a very challenging issue, and is being considered as quite an important and a hot topic to be discussed. However, it has been discussed and mentioned by the researchers that there are many factors which play quite a major role in raising some considerable privacy concerns and potential security issues in the mind of the user who is utilising and accessing the context aware based applications. These factors can be further classified into two categories; small scale and large scale. Small scale factors directly affect the personal, private and sensitive information of the users and therefore users are directly affected by these factors. On the other hand, large factors are the total opposite of the small ones and therefore make much less impact directly to the private or sensitive data of a user.

Furthermore, a brief highlight will be thrown onto some various potential factors that raise some potential privacy concerns among the users of context aware system applications [17].

- *Information Receiver reliability:* One of the factors which raises some privacy concerns among the users of context aware system applications is the reliability and authenticity of the receiver who receives the user's information. Users sometimes have an ambiguity or doubt on the credibility of the person who receives and uses their information. Therefore, the only option left for the user is to show some trust and confidence on the credibility and authenticity of the receiver's information.
- *Possible usage of user's information:* Another factor mentioned by some researchers which raises privacy concerns among the user is the kind of usage which is going to be made related to their private and sensitive data at the receiver's end.The concerns arise when the users think or become doubtful about the proper or positive usage of their sensitive information.
- *Level of sensitivity in terms of users data:* According to studies, some times the level or extent of sensitivity and privacy of user's data also raises a concern in the user's mind whether to share or transmit this kind of data or not . The user thinks twice whether it will be secure and good enough to make the private and sensitive data available and accessible to a third party or not.
- *Environment in which user's information is shared or its privacy is being disclosed:* This is a factor which raises concern in the user's mind whether they should be sharing and giving access to their information in various contexts and environments. The users are sometimes reluctant to share their information in any context or environment due to the concerns about the reliability and authenticity of the context. Moreover, a change or any update in the context also triggers some serious privacy concerns among the users and therefore prevents them further in sharing their sensitive and highly secure information in that very context.

3.2. SECURITY REQUIREMENTS FOR CONTEXT AWARE SYSTEMS

Many researchers have come up with, and collectively agreed upon, some common steps to be undertaken in an effort to implement the most efficient and effective level of computer security and applications working in a context aware system environment. The main purpose is to ensure that the level of risk is also going to be reduced by implementing these security steps. These requirements will be discussed to consider the security steps required to be undertaken in order to provide reliable security to context aware systems and applications. The following are the security requirements which been specified briefly below by International Telecommunications Union (ITU-T) represented in their recommendation X.805 and X.800 [17, 18, 25, 26, 27, 28, 29].

- **Authentication:** Authentication is basically a process of verifying the identity of the entity which is accessing the facilities of context based systems and applications. Authentication is essential to verify the identity of every node in the system and its eligibility to access the network. This means that, nodes in the Context aware systems are required to verify the identities of the communicated entities in the network, to make sure that these nodes are communicating with the correct entity, in order to make sure that the user who is trying to access the context aware system is reliable or eligible to access the system. This will help to validate the user's identity and will assist in ensuring that only the authorized user can access the system. A need of mutual authentication is justified in the case of security infrastructure to validate the user's identity. Furthermore, it has been said that in order to validate and run the authentication process, it is not always required that the user or device should be connected to the application identity. This could be done by authentication of users in terms of their context e.g., nicknames in IRC chat rooms.
- **Access Control:** This step in the security requirement is defined as a process of giving permission or access to the authentic and real users to utilise the system facilities. At the same time, its other purpose is to restrict access to the system utilities against fake or unauthorized users. In addition, users should be capable of restricting each other from accessing their private information. There are many techniques that can be used for access control such as Discretionary Access Control (DAC), Mandatory Access Control (MAC) and Role Based Access Control (RBAC) [25].This requirement is actually executed to verify or to validate the related privileges with each user. If the verified or reliable user matches and fulfills the required privileges then the access or authority to use the system resources will be granted to the user. Otherwise the request for accessing the resources will be rejected due to the failure of providing the relevant permission information [33, 36].
- **Privacy and confidentiality:** This requirement is basically related to protect or restrict the use of some highly sensitive, private or secure information or data from being shared or being made available to anyone else without your permission or willingness to do so. Each node in the context aware system has to secure both the information that is exchanged between each other; and secure the location information and the data stored on these nodes. Privacy means preventing the identity and the location of the nodes from being disclosed to any other entities. While confidentiality means keeping the secrecy of the exchanged data from being revealed to those who have not permission to access it. According to some previous research, it is considered as being someone's capability to protect or restrict your information from going public to others or keeping the sensitive data quite confidential and not disclosing it to unwanted entities in some way. However, implementing this concept in context-aware is considered a very challenging task. Most of the studies have claimed that, it is highly possible that the users who want to access and use the functionalities and facilities of context-aware systems quite increasingly would get somehow drawn in sharing the contextual information. In order to make sure such users would not be sharing or providing access to their private information to other entities, there should be a need to apply some methods to make the information confidential from unwanted users.
- **Data Integrity:** The data transmitted between users in the context aware systems should be received to the intended entities without been tampered with or changed by unauthorized modification. This requirement is essential especially in military, banking and aircraft control systems, where data modification would cause potential damage. Previous studies have showed that the integrity of the user's data can be acquired by providing assurance that the data is not going to be accessed and therefore altered or

modified by the fake, illegal or invalid users. The integrity requirement in context-aware systems is defined as a concept in which the access to the user's resources is not allocated or assigned to any illegal or incorrect user. Moreover, this will create a serious problem in tracking the illegal and unauthorized users accessing and modifying the user's context in an illegal manner.

- **Availability and survivability:** The services and applications in the system should be accessible, when needed, even in the presence of faults or malicious attack such as denial-of service attack (DoS). While survivability means the capability of the network to restore its normal services under such these conditions. These two requirements should be supported in any context aware system. Availability in this kind of systems is related with the concept of making sure that the services, facilities and functionalities required by the users are made available or accessible to them. This requirement is considered as important in the case of wireless systems where there is a total lack or unavailability of required services and functionalities by the user. This availability issue could be quite serious in many other critical applications as well.

4. RELATED FRAMWORKS IN CONTEXT AWARE SYSTEMS

Previous studies have proposed various frameworks for implementing efficient and effective security on user's information in the context and the context-aware system as a whole. The main purpose of implementing these frameworks is to gain and acquire various security related requirements and models which mentioned before. However, each framework has come up with its own purpose of implementing and serving with a unique kind of security requirements and models. For this reason, each separate framework has got its own importance and is therefore definitely required to be implemented and fulfilled. The following are brief details about various frameworks which been used to implement various security requirements along with the kind of unique purpose they fulfil and which are therefore being used in the context-aware systems.

- **Confab Framework:** As discussed before, users should be able to apply some privacy, confidentiality and therefore protect their private or personal information from the illegal access of unauthorized users. Therefore, in an effort to fulfil this security requirement. Jason *et al.* [19] came forward with an architecture called Confab in order to provide privacy to the users for the protection of their sensitive information. Talking specifically regarding its purpose and workings, Confab framework is basically designed for providing protection and reliable security to the information relevant to the user's location in ubiquitous systems. Moreover, its working hierarchy is based on some specific analysis related to basis privacy requirements of end-users and the application developers [20]. The private information of the user is acquired, stored and therefore processed on the user's device instead of taking and storing it on some other device for security purposes.

- **Uniform Access Control (UAC):** Covington *et al.* [21] came up with a uniform access control framework specifically for serving the purpose of environmental roles. Moreover, it has been declared and claimed as a further extension to the Role-Based Access Control (RBAC) model. Regarding the RBAC model, specific privileges or access rights to the users towards the system services are linked with the environmental roles. Talking specifically about the concept behind the roles, a role can be a developer or a manager in a top level domain. Moreover, it is responsible for analysing and determining the security aspects related to Context-aware systems and applications in ubiquitous environments [22, 36].

- **General Role Based Access Control (GRBAC):** Furthermore, these environment roles are based on another framework called the General Role Based Access Control. Ahamad *et al.*[22] have highlighted a main difference between RBAC and GRBAC models. Generally RBAC framework is a basic model which only covers and relates towards subject-oriented approach whereas GRBAC allows the defining of access control policy depending upon not only the subject but also on other essential and important factors like object or environment.

- **Gaia:** It is designed to assist the building of smart space applications [5,35], such as smart homes and meeting rooms. It consists of a set of core services and a framework for building distributed context-aware applications. Gaia's event manager service enables applications to be developed as loosely coupled components, and can provide basic fault tolerance by allowing failed event producers to be automatically replaced. Gaia's remaining four services support various forms of context-awareness, and include the following :

    - **Context service:** This allows applications to find providers for the context information they require.

    - **Presence service***:* This monitors the entities entering and leaving a smart space (including people as well as hardware and software components).

    - **Space repository:** This maintains descriptions of hardware and software components.

    - **Context file system:** This associates files with relevant context information and dynamically constructs virtual directory hierarchies according to the current context [23].

  According to Roman *et al.* [24] defined generic context-based software architecture for physical spaces, such as Gaia. Moreover, they defined physical space as a "geographic region with limited and well defined boundaries, containing physical objects, heterogeneous networked devices, and users performing a range of activities". Based and depending upon this very physical space, active space providers assist the users of context-aware systems and application to directly connect and therefore interact with the physical space. Again, this framework is being used to cover and acquire the step in the security requirement which is called the access control. Its main purpose is therefore giving permission or access to the authentic and real users to utilize the system facilities.

- **Cerberus & Kerberos:** Other frameworks like Cerberus & Kerberos are basically used for achieving the purpose of fulfilling and implementing various security requirements in the context-aware systems and applications such as identification, authentication and access control. Cerberus, however, is a framework for which the centre of attention concentrates on verification and validation of the identity of the user who is requesting access to services and facilities of context-aware systems. This validation and verification can be done by making use of information and data related with the user's context which includes fingerprint, voice and face recognition etc [24].

FRAMEWORKS	COMPATIBLE SECURITY REQUIREMENTS
Confab	Authentication
Uniform Access Control (UAC):	Access Control
General Role Based Access Control (GRBAC):	Access Control
Gaia	Authentication , Access Control
Cerberus	Authentication , Access Control, Privacy
Kerberos	Authentication , Access Control, Privacy

Table 1: Inter relationship between frameworks and security models

5. CONCLUSION

From the current study, we have come to understand the detailed concept of not only the context and context-aware systems them selves but also some constructive awareness regarding the privacy and security concerns along with the potential methods and requirements to be implemented. However, in order to overcome the privacy concerns mentioned by the users of context-aware systems, there are many essential security requirements raised by previous researchers recently which are required to be fulfilled and acquired for efficient and effective security. These are authentication, access control, privacy, availability and integrity of the data of relevant and authorized users of context-aware systems.

Many frameworks have been proposed and discussed for implementing and acquiring these security requirements on the user's context with the required level of efficiency and effectiveness. The most commonly used and preferred are Confab, Gaia, Cerberus, Kerberos, Uniform Access Control (UAC), General Role Based Access Control (GRBAC).Regarding a discussion about the most preferable and suitable one, we can make a conclusion that each of these have their own importance and therefore they are serving and providing assistance in fulfilling the security requirements individually in a much more efficient and effective manner. However, Kerberos along with Cerberos framework is highly preferable compared to the others as they concentrate on verification and validation of the identity of the user who is requesting access of services and facilities of context-aware systems. This validation and verification can be done by making use of information and data related with the user's context, which includes fingerprint, voice and face recognition, etc. Moreover, RBAC framework is a basic model which only covers and relates towards a subject-oriented approach whereas GRBAC allows defining an access control policy depending upon not only the subject but also on other essential and important factors like object or environment. Some further work needs to be done in order to overcome the core issues and challenges which faced by the context based security professionals and to come up with a more suitable and reliable framework. The issues and challenges are discussed briefly in the current study and there is still a need to give them serious thought in future potential research projects and high level studies linked with the implementation of Security and Privacy in Context Aware Systems.

REFERENCES

[1] G. D. Abowd, A. K. Dey, P. J. Brown, N. Davies, M. Smith, P. Steggles, Towards a better understanding of context and context-awareness, in: HUC '99: Proceedings of the 1st international symposium on Handheld and Ubiquitous Computing, Springer-Verlag, London, UK, 1999, pp. 304–307

[2] B. Schilit, M. Theimer, Disseminating active map information to mobile hosts, IEEE Network 8 (5) (1994) 22–32. URL citeseer.ist.psu.edu/schilit94disseminating.html

[3] John, Meaney. Context. Amherst, NY: Pyr, 2005, pp.15-17.

[4] Dan Hong, Dickson K.W. Chiu. Vincent Y. Shen, Requirements Elicitation for the Design of Context-aware Applications in a Ubiquitous Environment, 2005.

[5] B. N. Schilit, N. Adams, and R. Want.Context-aware computing applications. In IEEE *Workshop on Mobile Computing Systems and Applications*, pages 85-90, Santa Cruz, CA, US, 1994. IEEE.

[6] G. Chen and D. Kotz.A survey of context-aware mobile computing research. Technical Report TR2000-381, Department of Computer Science, Dartmouth College, 2000.

[7] A. K. Det and G. D. Abowd, and D. Sabler. A conceptual framework and a toolkit for supporting the rapid prototyping of context-aware applications, *Human-Computer Interaction,* 16:97-166, 2001.

[8] R. Want, A. Hopper, V. Falcao, J. Gibbons, The active badge location system, Tech. Rep. 92.1, ORL, 24a Trumpington Street, Cambridge CB2 1QA (1992). URL citeseer.ist.psu.edu/want92active.html

[9] R. Want, B. Schilit, N. Adams, R. Gold, K. Petersen, J. Ellis, D. Goldberg,M. Weiser, The PARCTAB ubiquitous computing experiment, Tech. Rep. CSL- 95-1, Xerox Palo Alto Research Center (Mar. 1995).

[10] G. D. Abowd, C. G. Atkenson, J. Hong, S. Long, R. Kooper, and M. Pinkerton. Cyberguide: A mobile context-aware tour guide, *Wireless Networks,* 3(5):421-433, Oct. 1997.

[11] K. Cheverst, N. Davies, K. Mitchell, A. Friday, and C. Efstratiou.Developing a context-aware electronic tourist guide: Some issues and experiences. In *CHI 2000,* pages 17-24, The Hague, The Netherlands, April 2000. ACM Press.

[12] R. Fithian, G. Iachello, J. Moghazy, Z. Pousman, and J. Stasko.The design and evaluation of a mobile location aware handheld event planner. In *Mobile hci 2003,* pages 145-160, Udine, Italy, September 2003, Springer-Verlag.

[13] P. Klante, J. Krosche, and S. Boll.Accesssighs – a multimodal location-aware mobile tourist information system. In *International Conference on Computers Helping People with Special Needs (ICCHP)*2004, pages 187-294, Paris, France, 2004. Springer-Verlag.

[14] Chen and Kotz, "A survey of Context-aware Mobile computing research," Dartmouth Computer Science Technical Report TR2000381.

[15] Y. Sumi, T. Etani, S. Felsyand, N. Simonetz, K. Kobayashix, and K. Mase.Cyberguide: A mobile context-aware tour guide. In *Community Computing and Support Systems*, pages 137-154, Kyoto, Japan, 1998. Springer-Verlag.

[16]B. Shneiderman and C. Plainsant. Designing the User Interface: Strategies for Effective Human-Computer Interaction, Addison-Wesley, San Francisco, USA. Fourth edition, 2005.

[17] Tariq Bashir Ahmad, Security and Privacy in Context-AwarComputing inside a Hospital, 2008.

[18] Douglas McIlwraith, Security and Privacy for Context AwareComputing, 2006.

[19] J. I. Hong and J. A. Landay, An architecture for privacy-sensitive ubiquitous computing, in MobiSYS '04: Proceedings of the 2nd international conference on Mobile systems, applications, and services. ACM Press, 2004,pp. 177–189.

[20] N. Shankar and D. Bafanz, Enabling secure ad-hoc communication using context-aware security services, in UNBICOMP 02: Workshop on Security in Ubiquitous Computing, 2002. http://www.teco.edu/philip/ubicomp2002ws/organize/palo.pdf

[21] Z. Z. Michael J. Covington PrahladFogla and M. Ahamad, A context-aware security architecture for emerging applications, in Proceedings of the Annual Computer Security Applications Conference (ACSAC), Las Vegas Nevada USA, 2002.

[22] M. J. C. M. J. Moyer and M. Ahamad., Generalized role-based access control for securing future applications, 2000.

[23] M. R. C. K. H. R. C. A. R. R. H. Campbell and K. Nahrstedt, Gaia: A middleware infrastructure to enable active spaces, 2002. http://gaia.cs.uiuc.edu/papers/GaiaSubmitted3.pdf

[24] M. R. Bussard L., Roudier Y., Untraceable secret credentials: Trust establishment with privacy, in PERCOMMW'04. Second IEEE Annual Conference on Pervasive Computing and Communications Workshops, 2004.

[25] Aldabbas, H. Alwada'n, T. Janicke, H. andA.Al-Bayatti. Data Confidentiality in Mobile Ad hoc Networks, in International Journal of Wireless & Mobile Networks (IJWMN), Vol. 4, No. 1, February 2012.

[26] W. Li and A. Joshi. Security Issues in Mobile Ad Hoc Networks-A Survey. Department of Computer Science and Electrical Engineering, University of Maryland, 2008.

[27] A.J. Menezes, P.C. Van Oorschot, and S.A. Vanstone.Handbook of applied cryptography.CRC, 1997.

[28] A. Toninelli, R. Montanari, L. Kagal and O. Lassila, "A semantic context-aware access control framework for secure collaborations in pervasive computing environments," The Semantic Web-ISWC 2006, pp. 473-486, 2006.

[29] K. Wrona and L. Gomez, "Context-aware security and secure context-awareness in ubiquitous computing environments," in f Polish Information Processing Society Conference Proceedings Conference, pp.255-265, 2005.

[30] M. Weiser. The Computer for the Twenty-First Century.Scientific American,265(3):94{104, 1991.

[31] Schilit, B.N., Theimer, M.M.: Disseminating active map information to mobile hosts. IEEE Network, 22–32, (1994).

[32] Choi, J. "Context-driven requirement analysis". In Proceeding of Computational Science and its Application [ICCSA 2007], Lecture Notes in Computer Science, vol. 4707, pp.739-748, 2007.

[33]Almuatiri, A. and Siewe, F. "CA-UCON: A context-aware usage control model". In Proceeding of the 5th ACM International Workshop on Context-Awareness for Self-Managing Systems [CASEMANS '11], pp. 38-34, 2011.

[34]Alkhaldi, W, Almutairi, S. Almutairi, A. Aldrawiesh, K.: Toward Development Context Aware Advertisement system (Case Study). In: IEEE, International Conference on Computer Applications and Network Security (ICCANS 2011). IEEE Press, CFP1182M-PRT/ISBN: 978-1-4244-9764-5. Maldives. (2011)

[35] M, Baldauf ,S.Dustdar and F.Rosenberg: A survey on context-aware systems. Int. J. Ad Hoc and Ubiquitous Computing, Vol. 2, No. 4, 2007.

[36] J. Bardram, R. E. Kjær, and M.ØPedersen, "Context-aware user authentication– Supporting proximity-based login in pervasive computing," in Proc. UBICOMP, 2003, pp. 107–123.

[37] Li Han; Jyri, S.; Jian Ma; Kuifei Yu; , "Research on Context-Aware Mobile Computing," *Advanced Information Networking and Applications - Workshops, 2008. AINAW 2008. 22nd International Conference on*, pp.24-30, 25-28 March 2008 doi: 10.1109/WAINA.2008.115

8

DETERMINATION OF MINIMAL MATRICES OF SIMPLE CYCLES IN LDPC CODES

Zhang Nan[1] and Gao Xiao[2]

[1, 2] Wuhan Maritime Communication Research Institute, Wuhan, 430079, China
[1]nan_zhang313@sina.com , [2]gaoxiao1113@sina.com

ABSTRACT

Simple cycles are the easiest cycles to reveal in Low-Density Parity Check (LDPC) codes. All minimum matrices of simple cycles with the same length will be equivalent after row or column permutations .In this paper, we analysis the structure of simple cycles and show all figures of minimal matrices of simple cycles. Firstly, we introduce a more general definition of cycle and investigate simple cycles in LDPC codes. Secondly, we present the number of simple cycles of arbitrary length and all the minimal matrices of simple cycles. Finally, we have proved that the number of all minimal matrices of 2k-simple cycles is $\frac{(k-1)!}{2}.P_k^k$*, for* $k \geq 3$.

KEYWORDS

Minimal matrices, cycles &LDPC codes

1. INTRODUCTION

The performance of Low-density parity-check (LDPC) codes of finite length may be strongly affected by their cycle property such as girth and stopping sets [1], etc. Here the girth is the minimum length of cycles in the Tanner graph of a given parity-check matrix. In most cases, it is difficult to analyze explicitly these factors of randomly constructed LDPC codes and predict their performance. One advantage of quasi-cyclic LDPC (QC-LDPC) codes based on circulant permutation matrices is that it is easier to analyze their code properties than in the case of random LDPC codes. Recently, several coding theorists proposed some classes of QC-LDPC codes with algebraically strong restriction on the structure and analyzed their properties more explicitly [5], [6], [7], [8]. Gao Xiao and Zhang Nan had analyzed balanced cycle properties of QC-LDPC codes and propose a method to determine the B-girth of a QC-LDPC code in its mother matrix and presented cycle relationships of the mother matrix and the protograph LDPC code [16],[17].

While decoding cycle-free LDPC codes with the sum-product algorithm, the performance converges to the optimal solution. If the Tanner graph of an LDPC code has cycles, very little is known regarding the convergence of iterative decoding methods.

On the other hand, it is known that cycle-free Tanner graphs do not support good codes, as the code minimum distance is asymptotically upper bounded by [3]

$$d_{\min} \leq 2\left\lfloor \frac{1}{R} \right\rfloor$$

where R is the code rate. Therefore, the engineering practice is to construct codes with limited number of short cycles and live with long cycles. In this paper, we will give a more general definition of cycles which is a little different from the definition of cycles in graph theory.

The outline of the paper is as follows. In Section 2, reviews on the quasi-cyclic LDPC codes studies followed by description of QC-LDPC and the definition of the path and cyclic. In Section 3, we analyze Cycle Effects on the error Performance of LDPC Code. In Section 4, we determine the Minimal Matrices of Simple Cycles and all RN minimal matrices of $2k - simple$ cycles and we give concluding in Section 5.

2. QUASI-CYCLIC LDPC CODES

A QC-LDPC code is characterized by the parity-check matrix which consists of small square blocks which are the zero matrix or circulant permutation matrices. Let p be the $L \times L$ permutation matrix given by

$$p = \begin{bmatrix} 0 & 1 & 0 & \cdots & 0 \\ 0 & 0 & 1 & \cdots & 0 \\ \vdots & \vdots & \vdots & \vdots & \vdots \\ 0 & 0 & 0 & \cdots & 1 \\ 1 & 0 & 0 & \cdots & 0 \end{bmatrix} \quad (1)$$

Note that p^i is just the circulant permutation matrix which shifts the identity matrix I to the right by i times for any integer i, $0 \le i < L$. For simple notation, we denote the zero matrix by p^∞. Let H be the $mL \times nL$ matrix defined by

$$H = \begin{bmatrix} P^{a_{11}} & P^{a_{12}} & \cdots & P^{a_{1n}} \\ P^{a_{21}} & P^{a_{22}} & \cdots & P^{a_{2n}} \\ \vdots & \vdots & \ddots & \vdots \\ P^{a_{m1}} & P^{a_{m2}} & \cdots & P^{a_{mn}} \end{bmatrix} \quad (2)$$

where $a_{ij} \in \{0, 1, ..., L-1, \infty\}$. From now on, the code C with parity-check matrix H will be referred to as a QC-LDPC code. When H has full rank, then its code rate is given by $R = 1 - m/n$ regardless of its code length $N = nL$. If the locations of 1's in the first row of the i th row block are fixed, then those of the other 1's in the block are uniquely determined. Therefore, the required memory for storing the parity-check matrix of a QC-LDPC code can be reduced by a factor $1/L$, as compared with random LDPC codes.

The QC-LDPC code defined in (2) may be regular or irregular depending on the choice of a_{ij}'s of H. When H has no blocks corresponding to the zero matrix, it is a regular LDPC code with column weight m and row weight n. In this case, its code rate is larger than $1 - m/n$ since there are at least $m-1$ linearly dependent rows.

For our presentation we introduce the following Lemmas.

2.1. Paths

Suppose that M is a binary matrix. Let $T(M)$ denote the Tanner graph of M. The set of edges of $T(M)$ is denoted by $E(M)$. For $e \in E(M)$, let $d_0(e)$ and $d_1(e)$ denote the check and bit nodes connected by e, respectively. Let $\sigma(e) = \{d_0(e), d_1(e)\}$. Clearly, two edges e, e' are equal if and only if $\sigma(e) = \sigma(e')$. Here, we will mainly work on the paths in $T(M)$. However, differing from in literature, we will despise the nodes on the paths to some extent. Let Ω be a nonempty subset of $E(M)$. For $\tau \in \{0,1\}$, let $\Gamma_\tau(\Omega)$ denote the set of sequences $\gamma = e_1 e_2 \cdots e_k$ with $e_1, e_2, \cdots, e_k \in \Omega$ and

$$d_{1-\tau}(e_{2i-1}) = d_{1-\tau}(e_{2i}), \quad \text{for } 1 \le i \le \lfloor k/2 \rfloor, \tag{3}$$

$$d_\tau(e_{2i}) = d_\tau(e_{2i+1}), \quad \text{for } 1 \le i \le \lfloor (k-1)/2 \rfloor, \tag{4}$$

$$e_i \ne e_{i+1}, \quad \text{for } 1 \le i \le k, \tag{5}$$

where k, e_1, e_k are called the length, origin, terminal of γ and also denoted by $|\gamma|$, $o(\gamma)$,

$t(\gamma)$, respectively. The edges $o(\gamma)$ and $t(\gamma)$ are also called ends of γ. By convention, $\Gamma_\tau(\Omega)$ contains the null sequence ϕ. Clearly, $\Gamma_0(\Omega) \cap \Gamma_1(\Omega) = \Omega \cup \{\phi\}$. We write $\Gamma(\Omega) = \Gamma_0(\Omega) \cup \Gamma_1(\Omega)$. A sequence of edges is called a path over Ω if it is in $\Gamma(\Omega)$. The null sequence is also called a path of length 0. $\Gamma_0(E(M))$, $\Gamma_1(E(M))$ and $\Gamma(E(M))$ are also abbreviated as $\Gamma_0(M)$, $\Gamma_1(M)$ and $\Gamma(M)$, respectively. For any path γ, let $\Gamma(\gamma)$ denote the set of paths consisting of edges on γ. For any sequence $e_1, e_2, \cdots, e_k$ of edges, let $(e_1, e_2, \cdots, e_k)^{-1} = (e_k e_{k-1} \cdots e_1)$. Clearly, for any $\gamma \in E(M) \cup \{\phi\}$, we have $\gamma^{-1} = \gamma$. According to the definitions, we can show the following five lemmas easily.

Lemma 1. A sequence γ of edges is a path if and only if γ^{-1} is a path. For $\tau \in \{0,1\}$ and $\gamma \in \Gamma_\tau(M)$ with $|\gamma| > 1$, the path γ^{-1} is also in $\Gamma_\tau(M)$ if and only if $|\gamma|$ is even.

Lemma 2. For $e_1, e_2 \in E(M)$, $\tau \in \{0,1\}$ and $\gamma \in \Gamma_\tau(M)$ with $|\gamma| > 1$, if $e_1\gamma$, $e_2\gamma$ are paths, then we have $d_\tau(e_1) = d_\tau(e_2) = d_\tau(o(\gamma))$.

Lemma 3. For $\tau \in \{0,1\}$ and $\gamma_1\gamma_2\gamma_3 \in \Gamma_\tau(M)$, we have $\gamma_2 \in \Gamma_\tau(M)$ if $|\gamma_1|$ is even, and $\gamma_2 \in \Gamma_{1-\tau}(M)$ otherwise.

Lemma 4. For $\gamma_1, \gamma_2, \gamma_3 \in \Gamma(M)$ with $|\gamma_2| \ge 2$, the sequence $\gamma_1\gamma_2\gamma_3$ is a path if and only if $\gamma_1, \gamma_2, \gamma_3$ are paths.

Lemma 5. For $e \in E(M)$ and $\gamma_1, \gamma_2 \in \Gamma(M) \setminus \{\phi\}$, the sequence $\gamma_1 e \gamma_2$ is a path if and only if there is a $\tau \in \{0,1\}$ such that $e\gamma_1^{-1} \in \Gamma_\tau(M)$ and $e\gamma_2^{-1} \in \Gamma_{1-\tau}(M)$.

The following lemma can also serve as another definition for paths.

Lemma 6. A sequence $e_1, e_2, \cdots, e_k$ of edges is a path if and only if

$$|\sigma(e_i) \cap \sigma(e_{i+1})| = 1, \quad \text{for } 1 \le i < k, \tag{6}$$

$$\sigma(e_{j-1}) \cap \sigma(e_{j+1}) = \phi, \quad \text{for } 1 < j < k. \tag{7}$$

Proof. Only-if-part: Assume that $e_1, e_2, \cdots, e_k$ is a path of length k. Clearly, (6) follows from (3) to (5). To show (7), assume in contrast that $\sigma(e_{j-1}) \cap \sigma(e_{j+1}) \ne \phi$ for some j with $1 < j < k$ Let τ be the integer in $\{0,1\}$ such that $d_\tau(e_{j-1}) = d_\tau(e_j)$ and $d_{1-\tau}(e_{j+1}) = d_{1-\tau}(e_j)$. If $d_\tau(e_{j-1}) = d_\tau(e_{j+1})$, then we have $e_j = e_{j+1}$, which contradicts (5). If $d_{1-\tau}(e_{j-1}) = d_{1-\tau}(e_{j+1})$, then we have $e_j = e_{j-1}$, which contradicts (5) too. Hence, (7) is valid.

If-part: Assume that (6) and (7) are valid. Clearly, (5) follows from (6). For any given j with $1 < j < k$, let τ' and τ'' be integers in $\{0,1\}$ such that $d_{\tau'}(e_{j-1}) = d_{\tau'}(e_{j+1})$ and $d_{\tau''}(e_{j+1}) = d_{\tau''}(e_j)$. Then, from (7) we see $\tau' = 1 - \tau''$. Hence, (3) and (4) are valid for some integer $\tau \in \{0,1\}$.

According to Lemma 6, we can show the following two lemmas.

Lemma 7. For $e_0, e_1, e, e' \in E(M)$ with $e_0 e e_1 \in \Gamma(M)$ and $|\sigma(e) \cap \sigma(e')| = 1$, there are integers v and τ in $\{0,1\}$ such that $d_\tau(e_v) = d_\tau(e) = d_\tau(e')$ and $\sigma(e_{1-v}) \cap \sigma(e') = \phi$. In particular, $e_{1-v} e e'$ is a path.

Proof. Let $\tau \in \{0,1\}$ be the integer such that $d_\tau(e) = d_\tau(e')$. Clearly, $d_{1-\tau}(e) \ne d_{1-\tau}(e')$. Since $e_0 e e_1$ is a path, there is an integer $\gamma \in \{0,1\}$ such that $d_\tau(e_v) = d_\tau(e)$ and $d_{1-\tau}(e_{1-v}) = d_{1-\tau}(e)$. Hence, we have $d_\tau(e') = d_\tau(e) = d_\tau(e_v)$ and $d_{1-\tau}(e_{1-v}) \ne d_{1-\tau}(e')$. From Lemma 6, we see $d_\tau(e_{1-v}) \ne d_\tau(e_v) = d_\tau(e')$. Then, we have $\sigma(e_{1-v}) \cap \sigma(e') = \phi$.

Lemma 8. For $\gamma, \gamma_0, \gamma_1 \in \Gamma(M) \setminus \{\phi\}$ with $o(\gamma_0) \ne o(\gamma_1)$ and $|\gamma| > 1$, if $\gamma\gamma_0$, $\gamma\gamma_1$ are paths, then $\gamma_0^{-1}\gamma_1$ is a path.

Proof. According to Lemma 2, we see $|\sigma(o(\gamma_0)) \cap \sigma(o(\gamma_1))| = 1$. For $v \in \{0,1\}$, let γ_v' denote the path with $\gamma_v = o(\gamma_v)\gamma_v'$. If $\gamma_v' \ne \phi$, from $t(\gamma)o(\gamma_v)o(\gamma_v') \in \Gamma(M)$, $|\sigma(o(\gamma_{1-v})) \cap \sigma(o(\gamma_v))| = 1$, $|\sigma(o(\gamma_{1-v})) \cap \sigma(t(\gamma))| = 1$ and Lemma 7, we see $\sigma(o(\gamma_{1-v})) \cap \sigma(o(\gamma_v')) \ne \phi$. Then, from Lemma 6, we see that

$$\gamma_0^{-1}\gamma_1 = (\gamma_0')^{-1} o(\gamma_0) o(\gamma_1) \gamma_1'$$

is a path.

Lemma 9. Assume that Ω is a nonempty subset of $E(M)$. If there is at least one path in $\Gamma(M)\setminus\Gamma(\Omega)$ with ends in Ω, let $e_1e_2\cdots e_k$ be a such path with the least length, then

$$e_i \notin \Omega, \text{ for } 1<i<k, \tag{8}$$

$$\sigma(e_i)\cap\sigma(e)=\phi, \text{ for } 3\le i\le k-2, e\in\Omega. \tag{9}$$

Proof. (8) is obvious. If there is an edge $e\notin\Omega$ and integer i with $3\le i\le k-2$ such that $\sigma(e_i)\cap\sigma(e)\neq\phi$, according to Lemma 4 and 7, we see that either $e_1e_2\cdots e_ie$ or $ee_ie_{i+1}\cdots e_k$ is a path connecting Ω_1 and Ω_2, contradicts our assumption. Hence, we have (9).

2.2. Cycle

For $\tau\in\{0,1\}$, a path $\gamma\in\Gamma_\tau(M)\setminus\{\phi\}$ of even length is called a cycle if $o(\gamma)\neq t(\gamma)$ and

$$d_\tau(o(\gamma))=d_\tau(t(\gamma)). \tag{10}$$

Clearly, the length of any cycle is even and at least 4. Let $\Theta(M)$ denote the set of cycles. Clearly, $C\in\Theta(M)$ if and only if $C^{-1}\in\Theta(M)$. If $\Theta(M)\neq\phi$, the length of the shortest cycles in $\Theta(M)$ is called the girth of M, and denoted by $g(M)$. If $\Theta(M)=\phi$, we say that the girth of M is $g(M)=\infty$. The following lemma can also serve as another definition for cycles.

Lemma 10. A path γ is a cycle if and only if $|\gamma|>0$ and $\gamma\gamma\in\Gamma(M)$.

Proof. If-part: Assume that $|\gamma|$ is positive and $\gamma\gamma$ is a path in $\Gamma_\tau(M)$ for some $\tau\in\{0,1\}$. Clearly, we have $\gamma\in\Gamma_\tau(M)$, $|\gamma|>1$ and $t(\gamma)\neq o(\gamma)$. Furthermore, from Lemma 3 and $\Gamma_0(M)\cap\Gamma_1(M)=E(M)\cap\{\phi\}$, we see that $|\gamma|$ is even and therefore $d_\tau(t(\gamma))=d_\tau(o(\gamma))$.. Hence, γ is a cycle.

Only-if part: Assume that γ is a cycle of length $2k$. For $1\le i\le 2k$, let e_i denote the $i-th$ edge on γ. Then, $e_1\neq e_{2k}$ and $e_i\neq e_{i+1}$ for $1\le i<2k$. Furthermore, for the integer $\tau\in\{0,1\}$ with $\gamma\in\Gamma_\tau(M)$, we have $d_\tau(e_{2k})=d_\tau(e_1)$ and

$$d_{1-\tau}(e_{2i-1})=d_{1-\tau}(e_{2i}), \text{ for } 1\le i\le k, \tag{11}$$

$$d_\tau(e_{2i})=d_\tau(e_{2i+1}), \text{ for } 1\le i<k. \tag{12}$$

Hence, $\gamma\gamma=e_1e_2\cdots e_{2k}e_1e_2\cdots e_{2k}$ is a path by definition.

Lemma 11. For paths γ, γ' of positive lengths, the sequence $\gamma\gamma'$ is a cycle if and only if $|\gamma|+|\gamma'|$ is even and $\gamma\gamma'\gamma$ is a path.

Proof. If-part: Assume that $|\gamma|+|\gamma'|$ is even and $\gamma\gamma'\gamma$ is a path in $\Gamma_\tau(M)$ for some $\tau \in \{0,1\}$. Clearly, we have $o(\gamma) \neq t(\gamma')$, $d_\tau(o(\gamma)) = d_\tau(t(\gamma'))$ and $\gamma\gamma' \in \Gamma_\tau(M)$. Hence, $\gamma\gamma'$ is a cycle.

Only-if-part: Assume that $\gamma\gamma'$ is a cycle. According to Lemma 10, $\gamma\gamma'\gamma\gamma'$ is a path. Thus, we see that $\gamma\gamma'\gamma$ is a path.

Remark: We note that the condition "$|\gamma|+|\gamma'|$ is even" in Lemma 11 can be dropped if the length of γ is at least 2.

The following lemma is a simple corollary of Lemmas 10 and 11.

Lemma 12. For any two paths $\gamma, \gamma' \in \Gamma_\tau(M) \setminus \{\phi\}$, $\gamma\gamma'$ is a cycle if and only if $\gamma\gamma'$ is a cycle.

Proof. Assume that $\gamma\gamma'$ is a cycle. Clearly, we have $2 \mid (|\gamma|+|\gamma'|)$. From Lemma 10, we see $\gamma\gamma'\gamma\gamma' \in \Gamma(M)$. Then, we have $\gamma'\gamma\gamma' \in \Gamma(M)$ and thus, from Lemma 11, $\gamma\gamma'$ is a cycle.

A path $\gamma \in \Gamma_\tau(M) \setminus \{\phi\}$ is said acyclic if $\Gamma(\gamma)$ has no cycle.

Lemma 13. Let $\gamma = e_1 e_2 \cdots e_k$ be a path in $\Gamma(M)$. Then, γ is a acyclic if and only if $k \geq 1$ and $\sigma(e_i) \cap \sigma(e_j) = \phi$ for all i, j with $|i-j| \geq 3$.

Proof. Only-if-part: Assume γ is acyclic. Clearly, we have $k \geq 1$. If $\sigma(e_i) \cap \sigma(e_j) \neq \phi$

for some i, j with $|i-j| \geq 3$, according to $\sigma(e_i) \cap \sigma(e_{i+2}) = \phi$ and $\sigma(e_j) \cap \sigma(e_{j-2}) = \phi$, we can assume further, without loss of generality, that $\sigma(e_i) \cap \sigma(e_{j-1}) = \phi$ and $\sigma(e_j) \cap \sigma(e_{i+1}) = \phi$. From $|\sigma(e_i) \cap \sigma(e_{i+1})| = 1$ and $\sigma(e_j) \cap \sigma(e_{i+1}) = \phi$, we see $e_i \neq e_j$. Hence, $e_i e_{i+1} \cdots e_j$ is a cycle in $\Gamma(\gamma)$, contradicts our assumption. Thus, $\sigma(e_i) \cap \sigma(e_j) = \phi$ for all i, j with $|i-j| \geq 3$.

If-part: Assume $k \geq 1$ and $\sigma(e_i) \cap \sigma(e_j) = \phi$ for all i, j with $|i-j| \geq 3$. Let γ' be

an arbitrary path in $\Gamma(\gamma)$. One can show easily that there are some integers i and j with $1 \leq i \leq j \leq k$ such that

$\gamma' = e_i e_{i+1} \cdots e_j$ or $e_j e_{j-1} \cdots e_i$.

Clearly, we have $\gamma' \notin \Theta(M)$. Hence, γ is a cyclic.

According to Lemma 6 and 13 we can get the following corollary easily.

Corollary 1. A sequence $e_1 e_2 \cdots e_k$ of edges is an acyclic path if and only if it satisfies (6) and

$$\sigma(e_{i-1}) \cap \sigma(e_{j+1}) = \phi, \quad \text{for } 1 < i \le j \le k. \tag{13}$$

Two cycles C and C' are said equivalent if there are paths γ_1, γ_2 such that $C = \gamma_1\gamma_2$ and $C' = \gamma_2\gamma_1$ or $\gamma_1^{-1}\gamma_2^{-1}$. For cycle C and path γ, we write $\gamma \subseteq C$ if there is a path γ' such that $\gamma\gamma'$ is a cycle equivalent to C, such path γ' is called the complementary of γ in C. A cycle C is said simple if any path $\gamma \subseteq C$ with $0 < |\gamma| < |C|$ is acyclic.

Lemma 14. For $\gamma \in \Gamma(M)$ and $e_0 \in E(M)$, the sequence $e_0\gamma$ is a simple cycle if and only if γ is acyclic and $e_0\gamma e_0 \in \Gamma(M)$.

Proof. The only-if-part is obvious. Now we consider the if-part. Assume $\gamma = e_1 e_2 \cdots e_k$ is acyclic and $e_0\gamma e_0 \in \Gamma_\tau(M)$ for some τ in $\tau \in \{0,1\}$. From $e_0\gamma e_0 \in \Gamma(M)$, we see $k \ge 3$, $|\sigma(e_0) \cap \sigma(e_1)| = |\sigma(e_0) \cap \sigma(e_k)| = 1$ and

$$\sigma(e_0) \cap \sigma(e_2) = \sigma(e_0) \cap \sigma(e_{k-1}) = \phi$$

Since γ is acyclic, we see $\sigma(e_1) \cap \sigma(e_k) = \phi$. Then, $e_0\gamma e_0\gamma$ is a path and thus, from Lemma 10, $e_0\gamma$ is a cycle. From $\sigma(e_0) = \{d_{1-\tau}(e_1), d_\tau(e_k)\}$ and (13), we see $\sigma(e_0) \cap \sigma(e_i) = \phi$ for any i with $3 \le i \le k-2$. Hence, for any j with $1 \le j \le k$, $e_{j+1} \cdots e_k e_0 e_1 \cdots e_{j-1}$ is an acyclic path. Thus, $e_0\gamma$ is a simple cycle.

According to Corollary 1 and Lemma 14, one can show the following corollary easily.

Corollary 2. A necessary and sufficient condition for a sequence $e_1 e_2 \cdots e_k$ of different edges to be a simple cycle is

$$|\sigma(e_i) \cap \sigma(e_j)| = 1 \text{ if and only if } i - j \equiv \pm 1 \bmod k. \tag{14}$$

A path in $\Gamma(M) \setminus \{\phi\}$ is said cyclic if it is not acyclic. We note that the null sequence ϕ is neither acyclic nor cyclic according to our definitions.

Lemma 15. Suppose that γ is a cyclic path. Then, there are a simple cycle C_0 and paths γ_1, γ_2 such that $\gamma = \gamma_1 C_0 \gamma_2$.

Proof. Suppose $\gamma = e_1 e_2 \cdots e_k$ is a cyclic path of length k. According to Lemma 13, there are integers i and j with $|i - j| \ge 3$ such that $\sigma(e_i) \cap \sigma(e_j) \ne \phi$. Without loss of generality, we assume further $i < j$ and $\sigma(e_{i'}) \cap \sigma(e_{j'}) = \phi$ for any i', j' with $i+3 \le i'+3 \le j' \le j$ and $(i', j') \ne (i, j)$. From $\sigma(e_j) \cap \sigma(e_{j-1}) \ne \phi$ and $\sigma(e_i) \cap \sigma(e_{j-1}) = \phi$, we have $e_i \neq e_j$. Hence, from Corollaries 1 and 2, $e_i e_{i+1} \cdots e_j$ is a simple cycle and $\gamma = \gamma_1 e_i e_{i+1} \cdots e_j \gamma_2$ for $\gamma_1 = e_1 e_2 \cdots e_{i-1}$ and $\gamma_2 = e_{j+1} e_{j+2} \cdots e_k$.

A cycle C is called a sub-cycle of another cycle C' if $C \subseteq C'$. Clearly, two cycles are equivalent if and only if they are sub-cycles of each other. Furthermore, according to Lemma 15, the shortest sub-cycles of a given cycle C must be simple. We note that, even if C_1 is a sub-cycle of C_2 and C_2 is a sub-cycle of C_3, C_1 is not necessary to be a sub-cycle of C_3. However, if C and C' are equivalent cycles, then any sub-cycle of C is a sub-cycle of C'.

If C and C's are equivalent cycles, for any given integer t, the cycle

$$C^t = \underbrace{CC\ldots C}_{t}$$

is called a multiple of C'.

Lemma 16. Let C be a simple cycle. Then any cycle in $\Gamma(C)$ is a multiple of C. In particular, any sub-cycle of C is equivalent to C.

Proof. Let $C = e_1 e_2 \cdots e_{2k}$ be a simple cycle and $C' = f_1 f_2 \cdots f_{2l}$ a cycle in $\Gamma(C)$. According to Corollary 2, $|\sigma(e_i) \cap \sigma(e_j)| = 1$ if and only if $i - j \equiv \pm 1 \bmod 2k$. Without loss of generality, we assume that $f_1 = e_1$. From $|\sigma(f_1) \cap \sigma(f_2)| = 1$, we have

$$f_2 \in \{e_2, e_{2k}\}.$$

If $f_2 = e_2$, according to $|\sigma(f_2) \cap \sigma(f_3)| = 1$, we have $f_3 \in \{e_1, e_3\}$. From $\sigma(f_1) \cap \sigma(f_3) = \phi$ and $f_1 = e_1$, we have $f_3 = e_3$. By induction, one can show easily that $f_i \bmod 2l = e_i \bmod 2k$ for any positive integer i. By taking i as the least common multiple of $2l$ and $2k$, we get $f_{2l} = e_{2k}$. Since $e_1, e_2, \cdots, e_{2k}$ are distinct, k must divide l. Hence $C' = C^t$ for some positive integer t.

If $f_2 = e_{2k}$, one can show similarly that $f_i \bmod 2l = e_{2k+2-i} \bmod 2k$ for any positive integer i. Hence, k must divide l and there is some positive integer t such that $C' = C_0^t$, where $C_0 = e_1 e_{2k} e_{2k-1} \cdots e_2$ is a cycle equivalent to C.

3. Cycle Effects on the Performance of LDPC Code

The error performance of an LDPC code with iterative decoding such as SPA depends on a number of structural properties besides its minimum distance. One such structural property is the girth of the code.

By proving the convergence of the sum-product algorithm for codes whose graphs are free of cycles, Tanner was the first to formally recognize the importance of cycle free graphs in the context of iterative decoding. The effect of cycles on the practical performance of LDPC codes was demonstrated by simulation experiments by MacKay and Neal [1] in the mid-1990s, and the beneficial effects of using graphs free of short cycles were shown [2]. Given the detrimental effects of cycles on the convergence of iterative decoders, it is natural to seek strong codes whose Tanner graphs are free of cycles. An important negative result in this direction was established by Etzion et al. [3], who showed that for linear codes of rate $k/n \geq 0.5$ which can be represented by a Tanner graph without cycles, the minimum distance is at most 2. Therefore, we should construct codes with long cycles. The shortest possible cycle in a Tanner graph is a cycle of length 4. A length 4 cycle results if two columns of parity check matrix H have more than one check positions in common. We are interested in knowing and controlling short cycles, since they have a negative impact on the decoding algorithm. Short cycles in LDPC codes cause successive iterations of the sum-product decoding algorithm to be highly correlated after only small number of iterations, thus prevent the iterative sum-product decoding from converging to the optimal solution. One commonly applied constraint in constructing parity check matrix is that the maximum overlap between any two columns in the matrix should be one, which ensures that the code has no 4-cycles.

In LDPC decoding, the sum-product algorithm assumes that the reliability information gathered by the variable or check nodes at each iteration is statistically independent. This assumption is valid for the initial number of iterations equal to $\lfloor (g-1)/4 \rfloor$, as the information has not been propagated around the shortest cycle in the graph, and the messages arriving at any node are indeed statistically independent. Therefore, it is desirable for a Tanner graph to have a large girth. In deriving the bounds on the maximum number of decoding iterations for regular LDPC codes. Gallager [6] established a simple lower bound on the code length of an LDPC code give the girth of the associated Tanner graph. It shows it is difficult to construct high-rate regular LDPC codes with moderate block length (i.e. N = 5000) and large girth (i.e. g = 10).

4. Determination of Minimal Matrices of Simple Cycles

It is obviously the most ordinary cycle in LDPC codes is the simple cycle. We will analysis the property of simple cycles in this section.

For two matrices W and R, we say that W covers R if R can be obtained from some sub-matrix of W by changing some nonzero elements to zero. Clearly, R is covered by W if and only if the Tanner graph of R can be obtained from that of W by deleting some nodes and edges.

A matrix M is called a minimal matrix of a simple cycle C if the following two conditions are valid:

- any sub-matrix of M does not contain the simple cycle C,
- any covered matrix of M does not contain the simple cycle C.

Two matrices M and M' are said equivalent, and denoted $M \equiv M'$, if M itself or its transpose is equal to M' after some permutations of rows and columns.

When we construct LDPC codes with large girth, it is obviously that the shortest cycles are simple cycles. We will present all the minimal matrix of 2k-simple cycles ($k \geq 2$) in the following.

From the definitions of simple cycle and minimal matrix of a simple cycle, we get the following lemma.

Lemma 17. The minimal matrix of a $2k-simple$ cycle is equivalent to a $k\times k$ matrix expressed as follows:

$$\begin{bmatrix} 1 & 1 & 0 & 0 & \dots & 0 & 0 \\ 0 & 1 & 1 & 0 & \dots & 0 & 0 \\ 0 & 0 & 1 & 1 & \dots & 0 & 0 \\ \vdots & \vdots & \vdots & \vdots & \ddots & \vdots & \vdots \\ 0 & 0 & 0 & 0 & \dots & 1 & 0 \\ 0 & 0 & 0 & 0 & \dots & 1 & 1 \\ 1 & 0 & 0 & 0 & \dots & 0 & 1 \end{bmatrix}_{k\times k} \tag{15}$$

It is obviously every minimal matrix of a $2k-simple$ cycle is a $k\times k$ matrix with row and column weights 2. We can get all minimal matrices of a $2k-simple$ cycle after some row or column permutations of (15).

We say a matrix M is row non-equivalent (RN for short) to another matrix M' , if M can not be obtained only by row permutations of M'.

Let (i, j) denote the position which is in the $i-th$ row and $j-th$ column of a matrix.

We will find all figures of $2k-simple$ cycles into two steps. First, find all RN minimal matrices of $2k-simple$ cycle. Then, make exhausted row permutations to each RN minimal matrices. All these RN minimal matrices and their row permutation matrices are matrices we look for.

4.1. Determine all RN minimal matrices of $2k-simple$ cycles

For $k=2$, it is obviously the minimal matrix of a $4-simple$ cycle is $\begin{bmatrix} 1 & 1 \\ 1 & 1 \end{bmatrix}$.

Now, for $k\geq 3$ determine all RN minimal matrices of $2k-simple$ cycles.

Let M be a minimal matrix of $2k-simple$ cycle. Then M is a $k\times k$ matrix, each column and each row of M have weights 2, and M does not contain cycles shorter than $2k$.

Assume two 1's in the first column of M are in the i_1-th and i_2-th rows, where $i_1\neq i_2$. The positions of two 1's in the first column are $(i_1,1)$, (i_1, j_1). Assume the 1 in the same row of $(i_2,1)$ is in position (i_1, j_1), and the 1 in the same row of $(i_2,1)$ is in position (i_2, j_2), where $j_1\in\{2,3,\cdots,k\}$, $j_2\in\{2,3,\cdots,k\}\setminus\{j_1\}$.

Since a matrix, whose 1's in the i_1-th and i_2-th rows are in positions $(i_1,1)$, (i_1, j_1), $(i_2,1)$, (i_2, j_2), are row equivalent to matrix whose 1's in the i_1-th and i_2-th rows are in positions $(i_1,1)$, (i_1, j_2), $(i_2,1)$, (i_2, j_1). We only reserve one of them to obtain all RN minimal matrices.

Assume the position of the 1, which is in the same column of the 1 in position (i_1, j_1), is (i_3, j_1), where $i_3 \neq i_1$, $i_3 \neq i_2$. Assume the position of the 1, which is in the same row of the 1 in position (i3, j1), is (i3, j3), where $j_3 \in \{2,3,\cdots,k\} \backslash \{j_1, j_2\}$.

Iteratively, assume the position of the 1, which is in the same column of the 1 in position (i_l, j_{l-2}), is (i_l, j_l), where $3 \leq l < k$, $j_l \in \{2,3,\cdots,k\} \backslash \{j_1, j_2, \cdots, j_{l-1}\}$, and $i_l \neq i_1, \cdots i_l \neq i_{l-1}$.

Assume the position of the 1, which is in the same column of the 1 in position (i_{k-2}, j_{k-2}), is (i_k, j_{k-2}), where $i_k \neq i_1, \cdots i_k \neq i_{k-1}$. Then the position of the 1, which is in the same row of the 1 in position (i_k, j_{k-2}), is (i_k, j_{k-1}).

Then, all the 1's in M are in positions:

$$\{(i_1,1),(i_1, j_1),(i_2,1),(i_2, j_2),(i_3, j_1),(i_3, j_3),\cdots,$$
$$(i_l, j_{l-2}),(i_l, j_l),\cdots(i_k, j_{k-2}),(i_k, j_{k-1})\},$$

where $(i_1, i_2, \cdots, i_k)$ is a permutation of $\{1,2,\cdots,k\}$, $j_l \in \{2,3,\cdots,k\} \backslash \{j_1, j_2, \cdots, j_{l-1}\}, 1 \leq l < k$.

Take $(i_1, i_2, \cdots, i_k)$ one permutations of $\{1,2,\cdots,k\}$. Take all the possible values of $j_1, j_2, \cdots, j_{k-1}$ described above. We can get all RN minimal matrices of $2k - simple$ cycles.

For each RN minimal matrix, make row permutation of it, we will have p_k^k different minimal matrices.

The number of all minimal matrices of $2k - simple$ cycles can be easily obtained.

Theorem 1. Let N_{2k} be the number of all minimal matrices of $2k - simple$ cycles. Then,

$$N_{2k} = \begin{cases} 1, & \mathrm{k} = 2 \\ \dfrac{(k-1)!}{2} \cdot k!, & \mathrm{k} \geq 3. \end{cases}$$

Proof. It is obviously there is only one minimal matrix of $4 - simple$ cycle.

For simple cycles of length $2\mathrm{k}$, $\mathrm{k} \geq 3$, we prove the number of all figures of minimal matrices is $\dfrac{(k-1)!}{2} \cdot k!$. The number of all minimal matrices of $2k - simple$ cycles is equal to the number of all RN minimal matrices multiple the number of all row permutations. For a $k \times k$ matrix, the number of all row permutations is p_k^k, which is equal to the number of all permutations of $\{1,2,\cdots,k\}$. Now, we calculate the number of all RN minimal matrices. Since j_1 can choose from $k-1$ possible values, j_2 can choose from $k-2$ possible values, All possible chosen of $j_1, j_2, \cdots, j_{k-1}$ is $(k-1)!$. Since exchanging the values of j_1 and j_2 results in row equivalent minimal matrices, the number of all RN minimal matrices is $\frac{(k-1)!}{2}$. Then, the number of all minimal matrices is obtained.

Example All row non-equivalent matrices of 8-simple cycles are as follows:

$$\begin{bmatrix} 1 & 1 & 0 & 0 \\ 1 & 0 & 1 & 0 \\ 0 & 1 & 0 & 1 \\ 0 & 0 & 1 & 1 \end{bmatrix}, \begin{bmatrix} 1 & 1 & 0 & 0 \\ 1 & 0 & 0 & 1 \\ 0 & 1 & 1 & 0 \\ 0 & 0 & 1 & 1 \end{bmatrix}, \begin{bmatrix} 1 & 0 & 1 & 0 \\ 1 & 0 & 0 & 1 \\ 0 & 1 & 1 & 0 \\ 0 & 1 & 0 & 1 \end{bmatrix}.$$

For each row non-equivalent matrix, we can obtain all figures of $8-simple$ cycles after all rows permutations. There are $4!=24$ row permutations, hence the number of all minimal matrices of $8-simple$ cycles are $3\times 24=72$

5. Conclusions

Simple cycles are the easiest cycles to reveal in LDPC codes. All minimum matrices of simple cycles with the same length will be equivalent after row or column permutations. We know the number of all minimal matrices of 2k-simple cycles is $\frac{(k-1)!}{2}.P_k^k$, $for\ k \geq 3$ we introduce a more general definition of cycle and investigate simple cycles in LDPC codes. We present a number of simple cycles of arbitrary length and all the minimal matrices of simple cycles. Future studies include research other cycles different from simple cycles and balanced cycles. For example, a cycle formed by a simple cycle twice will induce a cycle in the expanded matrix for some circulant permutation matrices.

Acknowledgements

This work was supported by the National Defense Pre-Research Foundation of Chinese Shipbuilding industry under the supervision of the Wuhan Maritime Communication Research Institute

References

[1] D.J.C. MacKay and R.M. Neal. Near shannon limit performance of low density parity check codes. Electronics Letters,32(18):1645–,Aug.1996.

[2] D.J.C. MacKay. Good error-correcting codes based on very sparse matrices. Information Theory, IEEE Transactions on,45(2):399–431,Mar.1999.

[3] T. Etzion, A. Trachtenberg, and A. Vardy. Which codes have cycle-free tanner graphs Information Theory, IEEE Transactions on,45(6):2173–2181,Sep.1999.

[4] C. Di, D. Proietti, I. E. Telatar, T. J. Richardson, and R. Urbanke, "Finitelength analysis of low-density parity-check codes on the binary erasure channel," IEEE Trans. Inform. Theory, vol. 48, pp. 1570-1579, June 2002.

[5] J. L. Fan, "Array codes as low-density parity-check codes," in Proc. 2nd Int. Symp. Turbo Codes, Brest, France, Sept. 4-7, 2000, pp. 543-546.

[6] M. P. C. Fossorier, "Quasi-cyclic low density parity check codes from circulant permutation matrices," IEEE Trans. Inform. Theory, vol. 50, pp.1788-1794, Aug. 2004.

[7] J.-L Kim, U. N. Peled, I. Perepelitsa, V. Pless, and S. Friedland, "Explicit construction of families of LDPC codes with no 4-cycles," IEEE Trans.Inform. Theory, vol. 50, pp. 2378-2388, Oct. 2004.

[8] R. M. Tanner, D. Sridhara, T. Fuja, and D. J. Costello, Jr., "LDPC block and convolutional codes based on circulant matrices," IEEE Trans. Inform.

[9] S.Myung,K.Yang,and J.Kim.Quasi-cyclic ldpc codes for fast encoding. Information Theory,IEEE Transactions on,51(8):2894–2901,Aug.2005.

[10] S.Myung and K.Yang.Extension of quasi-cyclic ldpc codes by lifting.In Information Theory,2005.ISIT 2005.Proceedings.International Symposium on, pages 2305–2309,Sep.2005.

[11] S.Kim,J.S.No,H.Chung,and D.J.Shin.Quasi-cyclic low-density parity- check codes with girth larger than 12.Information Theory,IEEE Transactions on,53(8):2885–2891,Aug.2007.

[12] Z.Li,L.Chen,L.Zeng,S.Lin,and W.H.Fong.E?cient encoding of quasi- cyclic low-density parity-check codes.Communications,IEEE Transactions on, 54(1):71–81,Jan.2006.

[13] M.E.O'Sullivan.Algebraic construction of sparse matrices with large girth.Information Theory,IEEE Transactions on,52(2):718–727,Feb.2006.

[14] S.Myung and K.Yang.Extension of quasi-cyclic ldpc codes by lifting.In Information Theory,2005.ISIT 2005.Proceedings.International Symposium on, pages 2305–2309,Sep.2005

[15] .C.A.Kelley and J.L.Walker.Ldpc codes from voltage graphs.In Information Theory,2008.ISIT 2008.IEEE International Symposium on, pages 792–796,Jul. 2008

[16] Zhang Nan, Gao Xiao, "Jointly Iterative Decoding of Low-Density Parity Check codes (LDPC) coded Continues Phase Modulation (CPM)" Multidisciplinary Journals in Science and Technology. JSAT, Vol. 2, No. 3, pp. 25-31, March 2011

[17] Gao Xiao, Zhang Nan, "Determination of the shortest balanced cycles in QC-LDPC codes Matrix" Multidisciplinary Journals in Science and Technology. JSAT, Vol.2, No. 4 , pp. 15-22, April 2011

[18] Najoua Achoura, Riaha Bouallegue, "Impact of feedback on MIMO-OFDM system using joint beamforming " International Journal of Wireless & Mobile Networks (IJWMN) Vol. 3, No. 2, pp. 39-45 April 2011

9

A novel and Efficient Secured Routing Algorithm in Mobile Ad hoc Network

Himadri Nath Saha [#1], Dr. Debika Bhattacharyya [#2], Dr. P. K.Banerjee [*3]

Assistant Professor [#1], Professor [#2], Professor [*3]

Department of Computer Science and Engineering, Institute of Engineering & Management, West Bengal, India [#1, #2]

Department of Electronics and Tele Communication Engineering, Jadavpur university, West Bengal, India [*3]

him_shree_2004@yahoo.com [#1], bdebika@yahoo.com[#2]

Abstract: *Currently the mobile wireless technology is experiencing rapid growth. However the major challenge for deployment of this technology with its special characteristics is securing the existing and future vulnerabilities. The lack of static infrastructure causes several issues in mobile Ad hoc network (MANET) environment, such as node authentication and secure routing. In this paper we propose a new approach for secure routing of data packets in MANET. This approach will reduce the computational overhead to a lot extent. The scheme is based on a specific criterion of the nodes called "fidelity Index" .Data packets are routed based on this fidelity index. Simulation results show our proposed scheme is simple yet robust under some scenarios.*

Keywords: *fidelity; delay; fidelity index; sequence number; hop destination; flooding attack; black hole attack; co-operative black hole attack;routing.*

I. INTRODUCTION

Mobile Ad-hoc Network (MANET) is a collection of wireless mobile hosts without fixed network [1] infrastructure and centralized administration (Figure-1). Communication in MANET [2] is done via multi-hop paths. MANET contains diverse resources; Nodes operate in shared wireless medium; [3] Network topology changes unpredictably and very dynamically; Radio link [4] fidelity is necessary; connection breaks are pretty frequent. Moreover, density of nodes, number of nodes and mobility of these hosts may vary in different applications. There is no stationary infrastructure. Each node in MANET [5] acts a router that forwards data packets to other nodes. Therefore selection of effective, suitable, adaptive and robust routing scheme is of utmost importance .Our proposed scheme that is fidelity based on demand (FBOD)routing is a secure routing scheme based on fidelity index. FBOD is a robust, effective, suitable, adaptive and cost effective scheme.

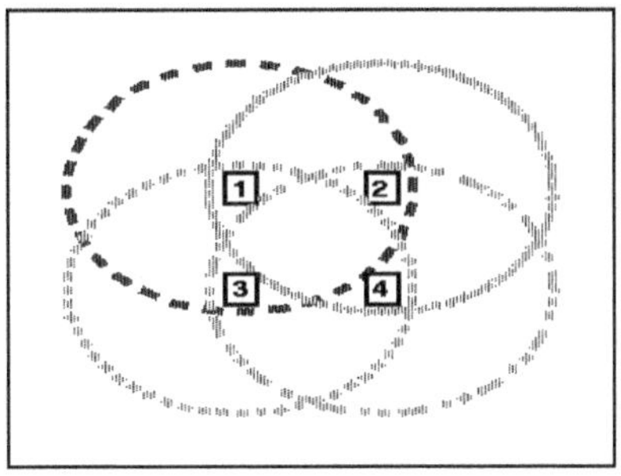

Figure 1: An ad-hoc mobile network with four nodes.

Section II describes some previous work in related field. Section III discusses fidelity. Delay matrix is discussed in section IV. Section V elaborates fidelity index. Section VI presents a description of the scheme Section VII presents FBOD algorithm Section VIII consists of simulation results. Section IX is a treatment on security aspects. Lastly we will discuss performance analysis of the scheme in section X and conclusions in section XI.

II. Related Work

S. Matri [6] proposed to trace malicious nodes by using watchdog/pathrater. In watchdog when a node forwards a packet, the node's watchdog verifies that the next node in the path also forwards the packet by promiscuously listening to the next node's transmissions. If the watchdog finds the next node does not forward the packet during a predefined threshold time, the watchdog will accuse the next node as a malicious node to the source node; The proposal has two shortcomings: 1) to monitor the behavior of nodes two or more hops away, one node has to trust the information from other nodes, which introduces the vulnerability that good nodes may be bypassed by malicious accusation; 2) The *watchdog* cannot differentiate the misbehavior from the ambiguous collisions, receiver collisions, controlled transmission power, collusion, false misbehavior and partial dropping. In *pathrater* algorithm each node uses the *watchdog's* monitored results to rate its one-hop neighbors. Further the nodes exchange their ratings, so that the *pathrater* can rate the paths and choose a path with highest rating for routing. Shortcoming of this algorithm is that the idea of exchanging ratings genuinely opens door for blackmail attack.

SCAN [7] exploits two ideas to protect the mobile Ad Hoc networks [8]: 1) *local collaboration*: the neighboring nodes collectively monitor each other and sustain each other; and 2) *information cross-validation*: each node monitors its neighbors by cross-checking the overheard transmissions, and the monitoring results from different nodes are further cross validated. As a result, the security solution is self-organized, distributed, and fully localized. In SCAN once a malicious node is convicted by its neighbors, the network reacts by depriving its right to access the network by revoking its token. A powerful collusion among the attackers will break SCAN as it violates the assumption of the polynomial secret sharing scheme.

Gonzalez [9] presents a methodology, for detecting packet forwarding misbehavior, which is based on the principle of flow conservation in a network. That states that if all neighbors of a node v_j are queried for i) the amount of packets sent to vj to forward and ii) the amount of packets forwarded by vj to them, the total amount of packets sent to and received from v_j must be equal. They assume a threshold value for non malicious packet drop. A node v_i maintains a table with two metrics T_{ij} and R_{ij}, which contains an entry for each node vj to which vi has respectively transmitted packets to or received packets from. Node v_i increments T_{ij} on successful transmission of a packet to v_j for v_j to forward to another node, and increments R_{ij} on successful receipt of a packet forwarded by v_j that did not originate at v_j. All nodes in the network continuously monitor their neighbors and update the list of those they have heard recently. This algorithm does not require many nodes to overhear each others' received and transmitted packets, but instead it uses statistics accumulated by each node as it transmits to and receives data from its neighbors. Since there is no collaborative consensus mechanism, such an algorithm may lead to false accusations against correctly behaving nodes.

III. Fidelity

Fidelity is the most important concept of this routing scheme. Fidelity is an integer number that is associated with each node. This fidelity of a node denotes many things about the node itself and also deciphers other information regarding the topology of the entire network. It also helps to maintain security [10] to some extent.

To make it understandable in one sentence, "fidelity is a counter that is associated with a node, which is increased whenever it forwards a data packet successfully." Whenever a node comes in a network its fidelity is zero and whenever it goes permanently off from the network its value is again refreshed to zero. Otherwise whenever a node will forward any data packet it will always increase a counter value and that counter value is its fidelity. Note whenever a source node sends a data packet to a destination node, all the intermediate nodes helping to transmit its data packet will increase their counter but the source and the destination node do not increase their fidelity value.

Fidelity is a measure of these two factors:-

A. *How reliable a node is for forwarding a data packet*

Whenever we observe that the fidelity value of a particular node is greater that of another node then we can conclude that the one having the greater value is a more durable node than the other from who's its value is greater. It is quite logical because a node with greater value indicates that it is an experienced node in the network and it has transmitted packets most dutifully than other nodes.

B. *Network topology*

If we can find some nodes with higher fidelity in a region of the network, we conclude that the network activity is higher in that region. More precisely we can also infer that the node density is also higher in that region for it is impossible to have one node having very high fidelity [11] surrounded by nodes with low fidelity because a high fidelity [12] node must send packets to someone in its vicinity which will make that other node's fidelity value also high. Thus a high fidelity value accounts for high network activity as well as high density of nodes in its surroundings.

IV. DELAY MATRIX

Delay is one of the most important factors in this scheme. Delay signifies the time delay between two nodes. Delay matrix is basically a square matrix having dimension n*n, where n=no of nodes. Each $(i, j)^{th}$ entry in the matrix indicates delay between i^{th} and j^{th} node in a network. For example, we are taking a 2*2 matrix which is a delay matrix of a network having 2 nodes 0, 1. The delay between those two nodes is 5 units.

(i,j)	0	1
0	0	5
1	5	0

V. FIDELITY INDEX

Fidelity Index (FI) is basically a real no associated with each node. This index is the main factor in order to choose any neighbor to forward the message to destination. Each node sorts its neighbors in decreasing order of FI. If we get the time delay for a particular node as T and the fidelity value is F, then we can calculate the fidelity index (FI) = X * F + (1-X)/T, where X<1 & its value depends on the practical behavior of the network. The value of the X can be user given input. For our simulation we are considering x=.6 to give more stress on fidelity (security issue).

VI. DESCRIPTION OF THE SCHEME

The term "friends of a node" used in this paper, indicates actually the nodes that fall in the physical range of a particular node. When a node is a message to send, the node will check which nodes are in its neighborhood and what are their fidelity value and the delay between those nodes. Sender will broadcast a request along with an echo message. After getting reply they will make their friend list. More precisely the friend list consists of a table that contains two attributes. The first one is the address [13] of the nodes which are within its range and other is the fidelity index of that particular node. When each node is updated then they will sort that table according to the decreasing order of the fidelity index (FI). Before we enter into the detailed discussion of our scheme (FBOD) there are some concepts that need to be understood. These are as follows-

There will be a sequence counter in every node. If a message is generated in a node then it will be increased by one. This sequence no. will be forwarded as a part of the message. Every node will maintain a buffer where (source, sequence no.) will be stored for last n no. of received messages. After getting a message a node will verify the tuple [9] (source, sequence no) of that message with those tuples in its buffer [14]. If anyone of them matches with that message then that node will reject that message silently. It will prevent flooding attack.

The timeout period of every node through which message is traversed, will be gradually decreased by a critical factor [15] i.e. if timeout period of sender node is x then timeout period of receiver node will be (x-m), where m will be critical factor. This factor [16] helps us to control the max no of hops a message can traverse to reach destination.

Now the scheme is as follows-A node can do either of three activities - message generate, message forward, message receive. If it is not doing any of the three then it is idle. Now if a message is generated in a node and it needs to be sent then the node will remain busy until an acknowledgement is received for this message. It is to be noted that a busy node can accept & process an acknowledgement and can send a fail message.

Now if destination is directly reachable from generator node then it will send message to destination node and will wait for acknowledgement, and remain busy until acknowledgement is received. If the destination node is busy it will send a fail message to generator node. After getting fail message or if timeout period exceeds, generator node will keep on sending the message after a certain time periodically until acknowledgement is received.

If destination is not directly reachable then generator node will send message to the node in its range that has highest fidelity index. If generator node get a fail message from that node or if timeout period exceeds then it will send the message to the node having second highest fidelity index and it will continue like this. If the whole list is exhausted in this way then the process will again continue from the node having highest fidelity index. Only generator node will follow this process. Other nodes will send a fail message to its predecessor if the whole list is exhausted.

When a node receives a message, if it is busy then it will send a fail message to sender, otherwise it will check whether it itself is a destination or not. If it is destination, it will accept the message and send acknowledgement to sender otherwise this node will send message to the node in its range that have highest fidelity index and that process will continue. In that acknowledgement message the sequence no. will be same as received message but source will be substituted by destination.

VII. FBOD Algorithms

Update friend list

STEP 1: Send Hello packet, Echo packet and a special broadcast request to the friends for knowing the identity, delay and fidelity of the friends
STEP 2: Receive replies from friends
STEP 3: Calculate FI=0.6*F+0.4/T, where F=fidelity value of a neighbour node & T=time delay to reach that neighbour. We have taken here the value of X as 0.6.
STEP 4: Update my friend list
STEP 5: Sort friend list in a decreasing order of FI

Generated data

STEP 1: Set my status=busy
STEP 2: If destination directly reachable from here

- Send packet to destination
- Wait for ACK
- If ACK received consider success
- Else if timeout occurs or FAIL received, arrange for resending

Else

- Send data packet to the friend having highest FI
- Wait for ACK
- If ACK received consider success and go to step 3
- Else if timeout occurs or FAIL received, arrange for resending to the friend with next highest FI
- Continue above three steps until ACK received
- If list is exhausted without getting an ACK then again start from the friend with the highest FI and try each node in friend list in the same manner as above.
- While trying to send if the list is exhausted thrice abort

STEP 3: Set my status=free

Received data

STEP 1: If my status=busy send FAIL to sender
STEP 2: Else

- Make my status=busy
- Process received data
- Make my status=free

Process received data

STEP 1: If message destination=my address

- Accept data
- Generate ACK
- Send the ACK to the node from which it directly received the message
- If the received packet is found duplicate then discard the received packet.

STEP 2: Else

- Forward data packet
- Check if forward operation is successful
- If successful increase my fidelity value by 1 and send ACK to the node from which it directly received the message
- Else send FAIL to the node from which it directly received the message

Forward data packet

STEP 1: If message destination is directly reachable from here

- Send packet to destination
- Wait for ACK
- If ACK received consider success
- Else if timeout occurs or FAIL received, arrange for resending to destination.
- If resending fails 3 times consider failure.

STEP 2: Else

- Send data packet to the friend having highest FI
- Wait for ACK
- If ACK received consider success
- Else if timeout occurs or FAIL received, arrange for resending to the friend with next highest FI
- Continue above three steps until ACK received
- If list is exhausted without getting an ACK then consider failure.

VIII. SIMULATION RESULT

We have simulated this scheme with JAVA. We need to know something to make out these simulations. These are

1. Small circle signifies node in the network.
2. Blue circle around node signifies range of that node.
3. Red color indicates that the node is free.
4. Black color indicates that the node is busy.
5. Yellow line between two nodes indicates sending of request & echo message to probe fidelity value & delay.
6. Pink line between two nodes indicates reply of probing with information.
7. Red line between two nodes indicates sending of message.
8. Green line between two nodes indicates sending of acknowledgement.
9. Blue line between two nodes indicates sending of fail message.
10. Any node inside the range of a node is its neighbor node.

Now we will describe one test case simulations.

This is a network having six nodes. Their corresponding fidelity values are written beside the nodes. Here we are trying to send a message from node 0 to node 5. Following figures depict the simulation results

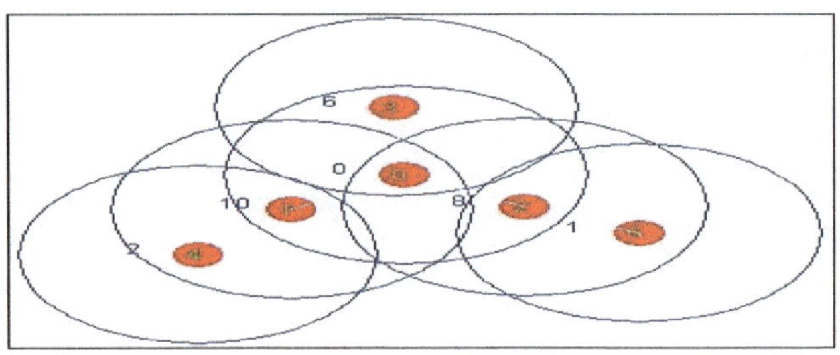

Figure 2: Design of network.

The result we get after network designing is given below-

Output of netdesign.jar:-

6<*Number of nodes*>

0	10	7	6	2	1
-1	0	0	0	-1	-1
0	-1	-1	-1	0	-1
0	-1	-1	-1	-1	0
0	-1	-1	-1	-1	-1
-1	0	-1	-1	-1	-1
-1	-1	0	-1	-1	-1

We got the adjacency list.txt as above and input **delay matrix**:-

0	3	2	1	-1	-1
3	0	-1	-1	2	-1
2	-1	0	-1	-1	4
1	-1	-1	0	-1	-1
-1	2	-1	-1	0	-1
-1	-1	4	-1	-1	0

0<*time instance-0*>
0<source> 5<destination> hello<*msg*>

Then we run the simulation and see the results.

The steps of the visual simulation are given below-

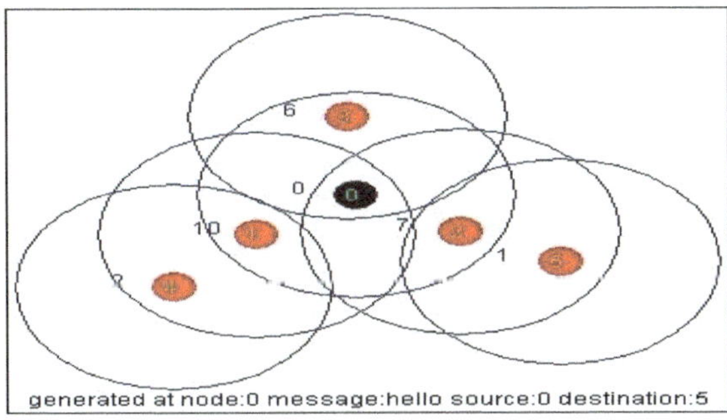

Figure 3: Message generated at node 0.

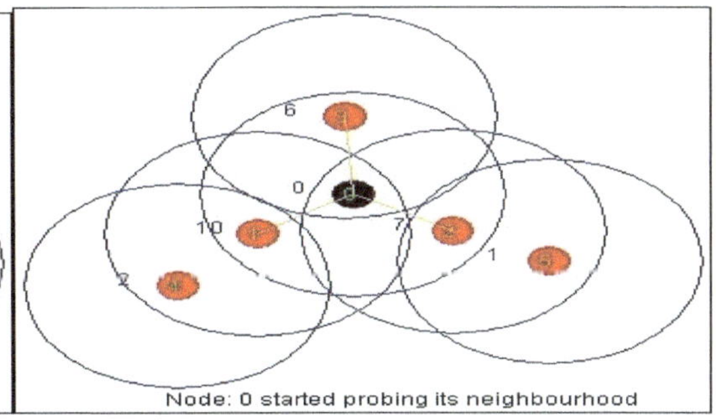

Figure 4: Node 0 started probing.

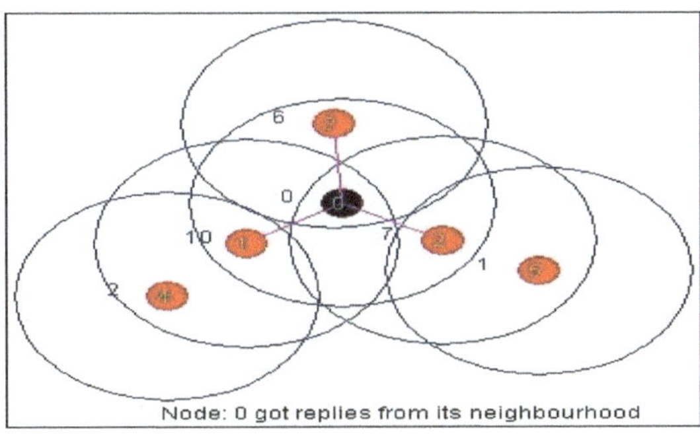

Figure 5: Node 0 got replies from neighbour nodes.

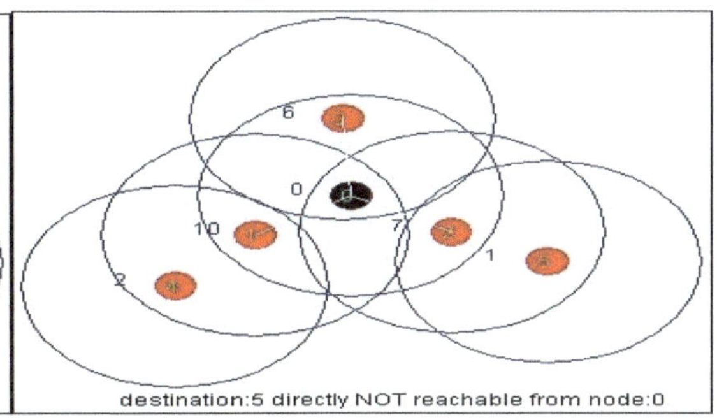

Figure 6: Destination is not directly reachable from source node.

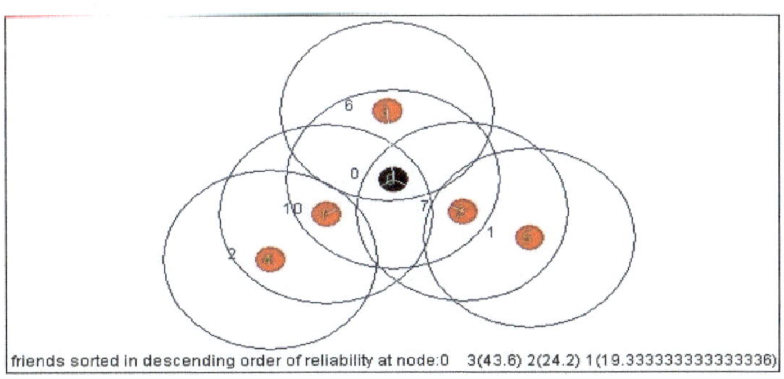

Figure 7: Friend nodes are sorted in descending order of Fidelity Index written beside neighbour nodes in brackets.

For node 1, FI=.6*10+.4*(100/3) =19.333333333

For node 2, FI=.6*7+.4*(100/2) =24.2

For node 3, FI=.6*6+.4*(100/1) =43.6

NOTE: We have taken (100/T) instead of (1/T) for the sake of calculation.

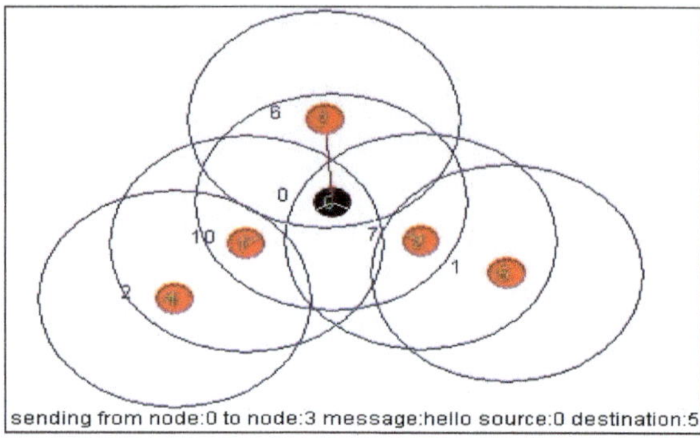

Figure 8: Node 0 is sending message to node 3.

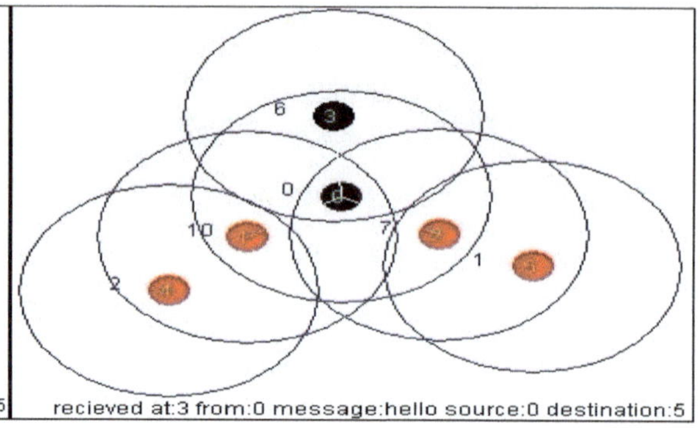

Figure 9: Message is received at node 3.

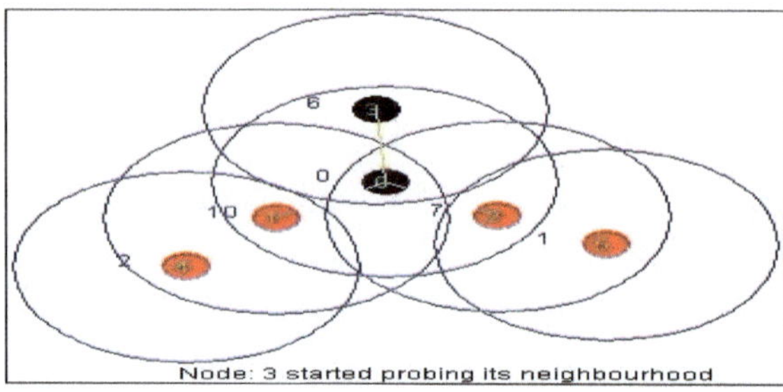

Figure 10: Node 3 is starts probing neighbour nodes.

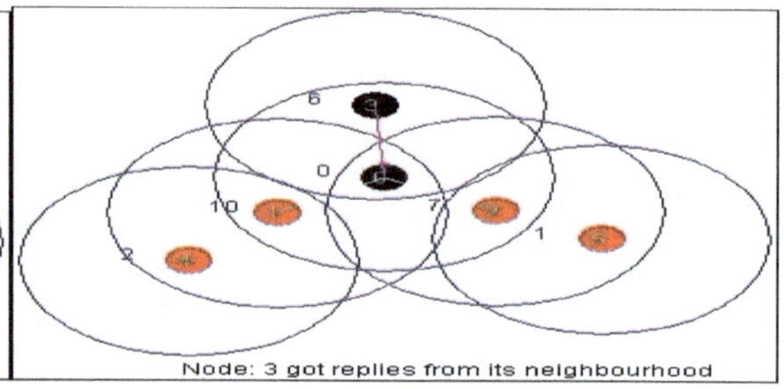

Figure 11: Node 3 receives reply from neighbours.

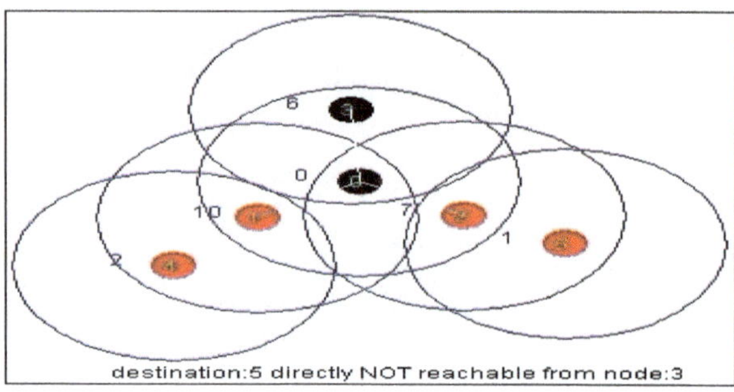

Figure 12: Destination is not reachable from node 3.

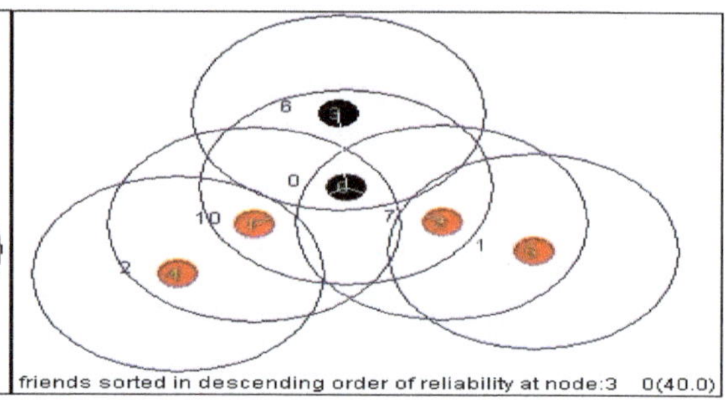

Figure 13: Friend nodes are sorted in descending order of Fidelity Index written beside neighbor nodes in bracket.

For node 0, FI=.6*0+.4*(100/1) =40.0

NOTE: We have taken (100/T) instead of (1/T) for the sake of calculation.

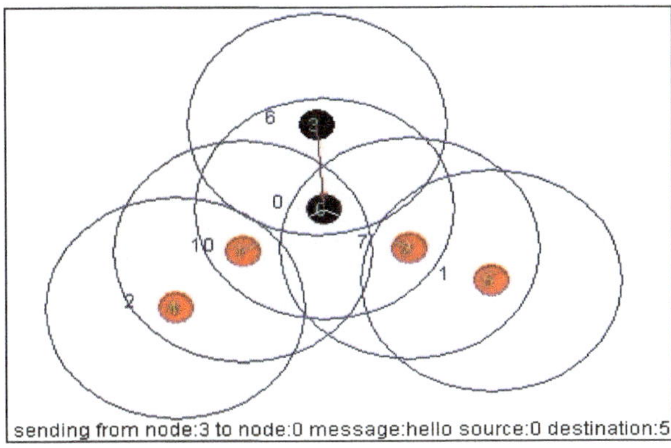

Figure 14: Node 3 is sending message to node 0.

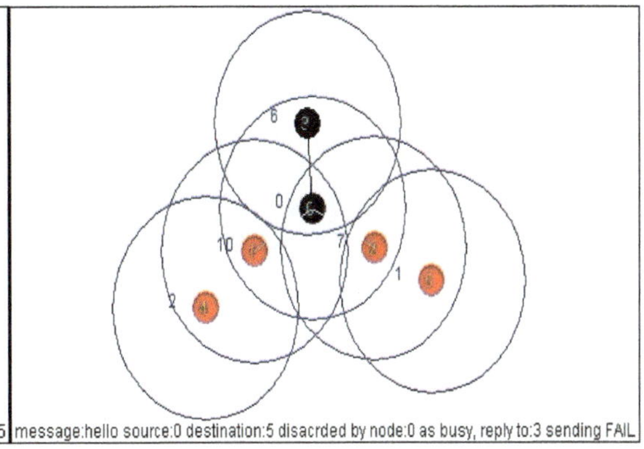

Figure 15: Node 0 discarded the massage.

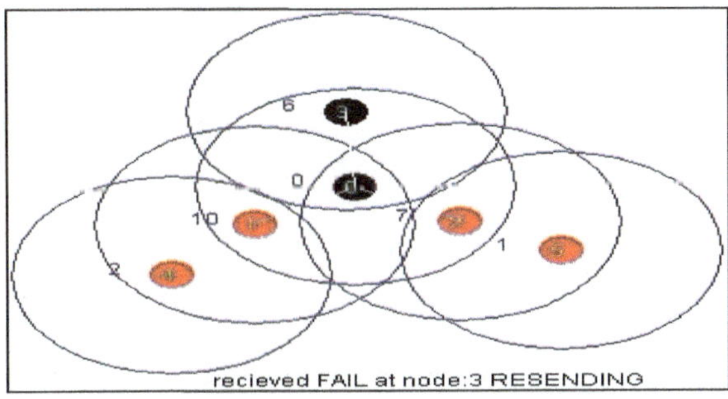

Figure 16: Node 3 receives a FAIL message.

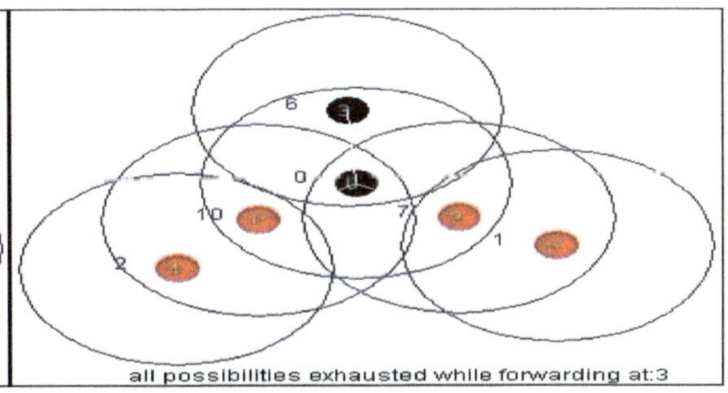

Figure 17: Message cannot be forwarded from node 3.

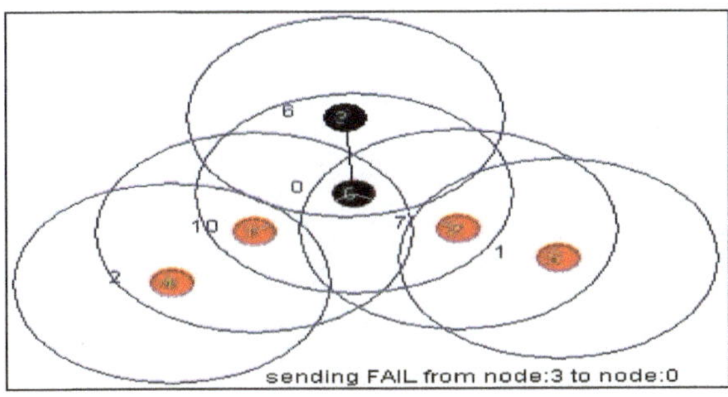

Figure 18: Node 3 is sending FAIL message to node 0.

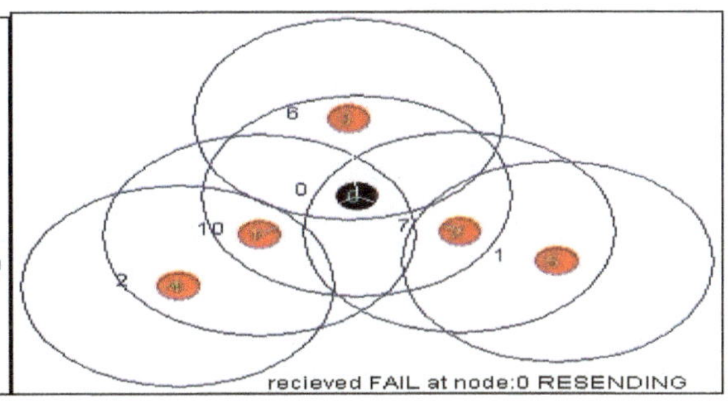

Figure 19: Node 0 receives FAIL message.

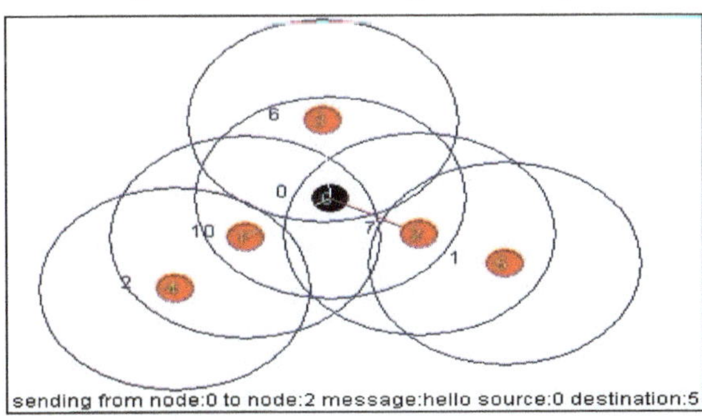

Figure 20: Node 0 forwarding message to node 2.

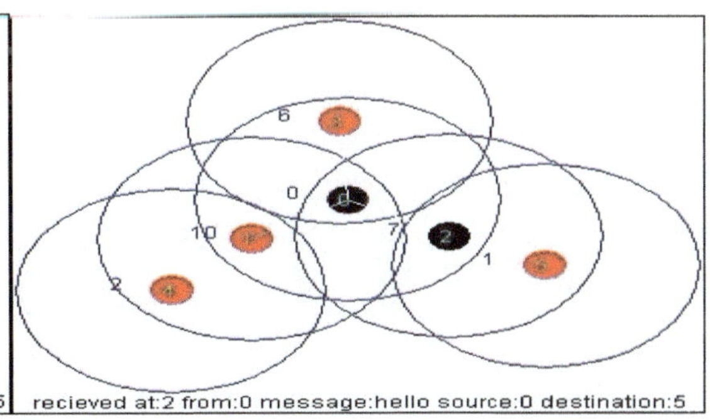

Figure 21: Node 2 receives message.

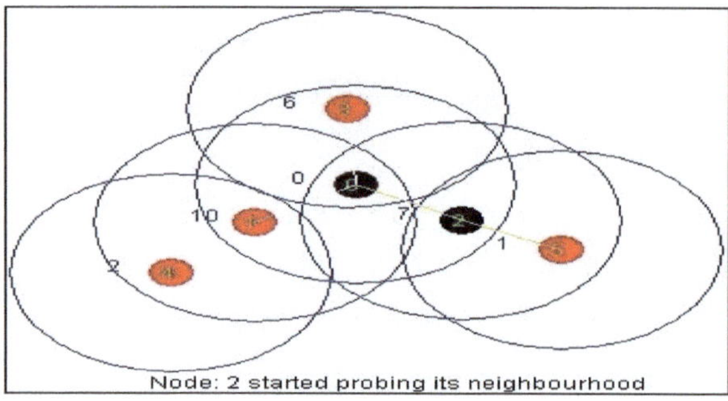

Figure 22: Node 2 started probing neighbours.

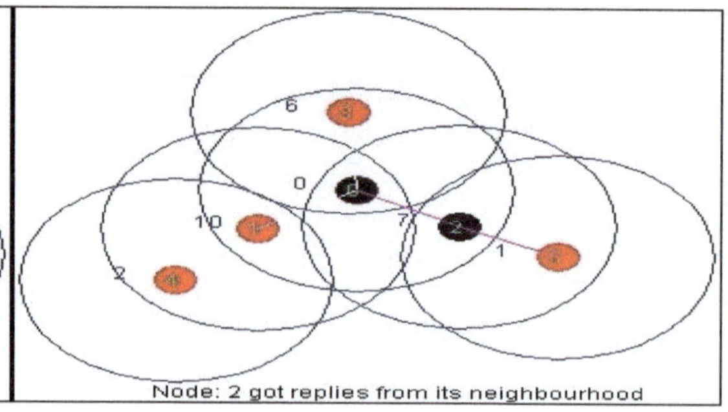

Figure 23: Node 2 receives reply from neighbours.

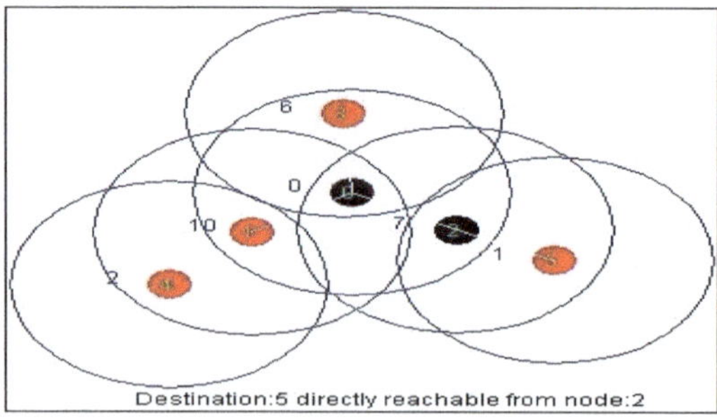

Figure 24: Destination node 5 is reachable from node 2.

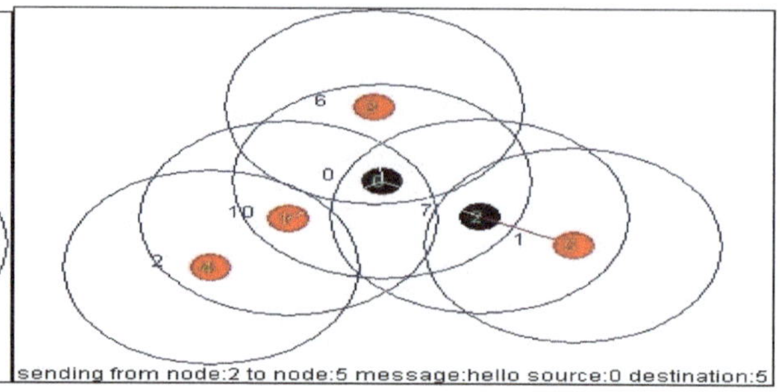

Figure 25: Node 2 sending message to node 5.

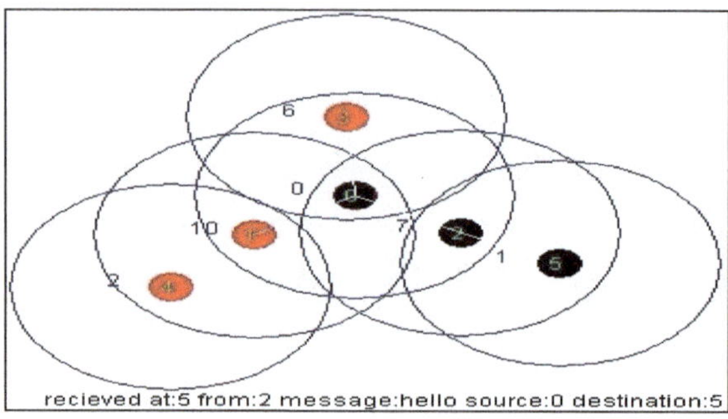

Figure 26: Node 5 receives message from node 2.

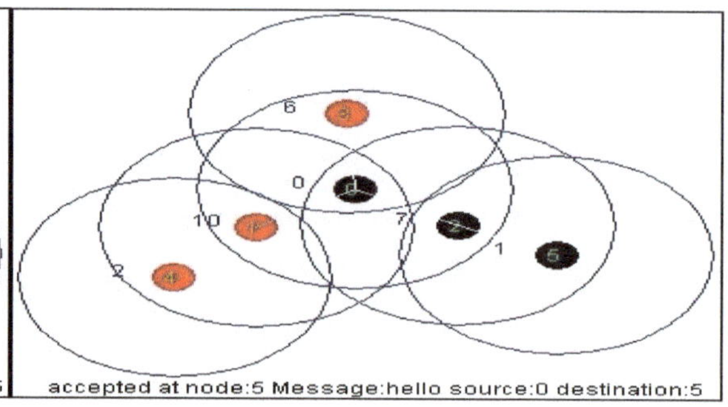

Figure 27: Node 5 accepts the message and generates an ACK.

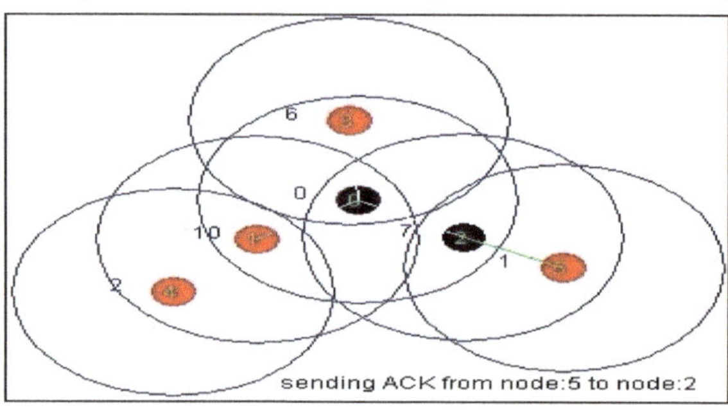

Figure 28: Node 5 sending ACK to node 2.

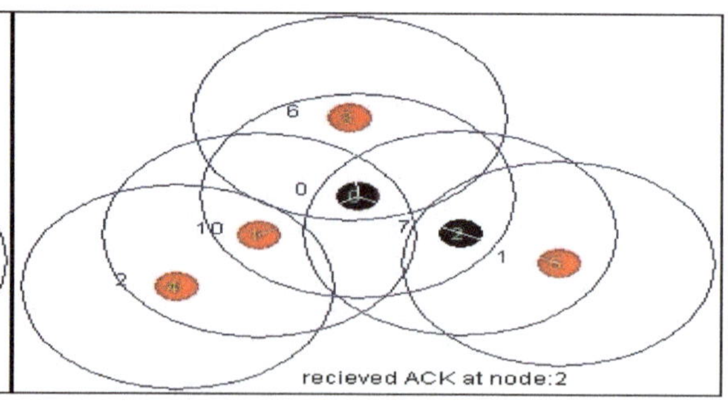

Figure 29: Node 2 receives ACK.

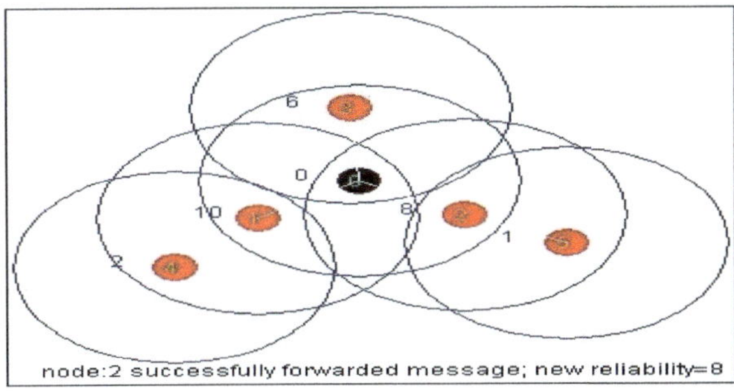

Figure 30: Fidelity of node 2 increases.

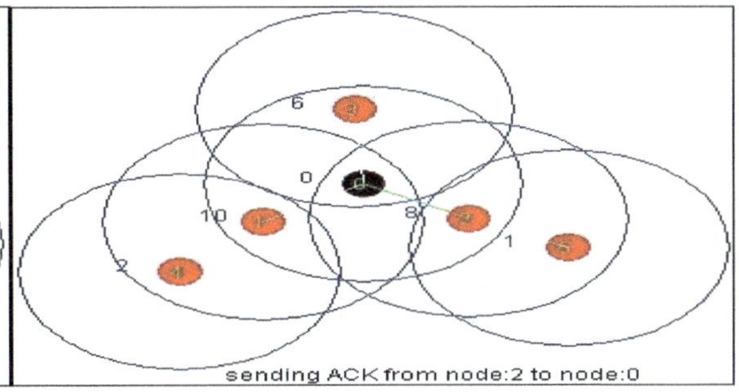

Figure 31: Node 2 forwarding ACK to node 0.

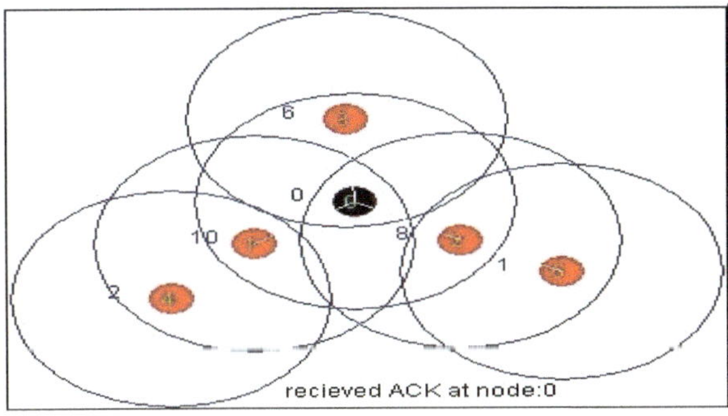

Figure 32: Node 0 received ACK from node 2.

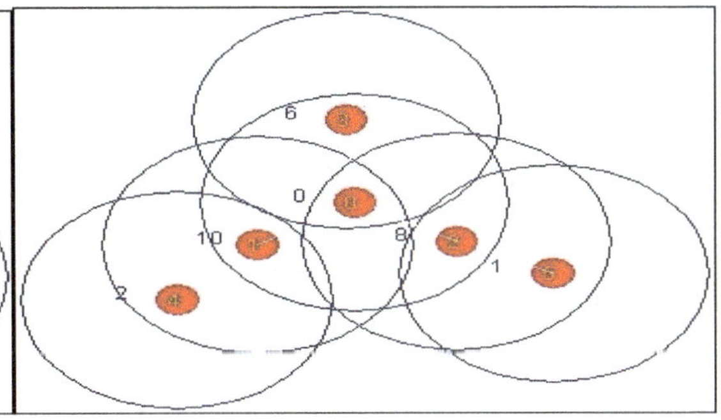

Figure 33: Messages transferred successfully.

Message transfer is completed.

IX. Security Aspects

A. *Flooding Attack*

Flooding attack [17] is an attack that attempts to cause a failure in a computer system or other data processing entity by providing more input than the entity can process properly. Flooding attack occurs when a network or service becomes so weighed down with duplicate packets that it can no longer process genuine connection requests.

In our scheme, we have two fields in the data packet *a)* Sequence number. *b)* Source node number. These two fields together uniquely identifies [18] a message packet. Each data packet generated by a single node in this scheme is guaranteed to have different sequence number. Here if a node encounters a data packet that was recently 'seen' by it, it discards the packet. This prevents propagation of duplicate packet through the network, mitigating flooding attack

B. *Black Hole Attack*

Black holes [19] refer to places or nodes in the network where incoming traffic is silently discarded (or "dropped"), without informing the source that the data did not reach its intended recipient. The black hole problem is one of the security [20] attacks that occur in MANET.

There are two possible solutions. The first is to exploit the packet sequence number included in any packet header. Computer simulation shows that compared to the original *Ad Hoc on-*

demand distance vector (AODV) routing scheme [21] [22] [23]; this solution can verify the route to the destination depending on the pause times at a minimum cost of the delay in the networks. The second is to find more than one route to the destination. Here we are going to discuss about how our scheme implements the second solution vividly.

Let us assume that there are four nodes A, B, C and D. A wants to send a data packet to node C. Node A cannot communicate with destination node directly. Data packet can be sent from A to C in two ways –

i) A→ D→ C

ii) A→ B→ C.

A will pick first path to send the data packet as fidelity of D is greater than B. A will wait for acknowledgement after sending the data packet to node D. But node D is a black hole node. Data packet can't be reached to destination node by this path.

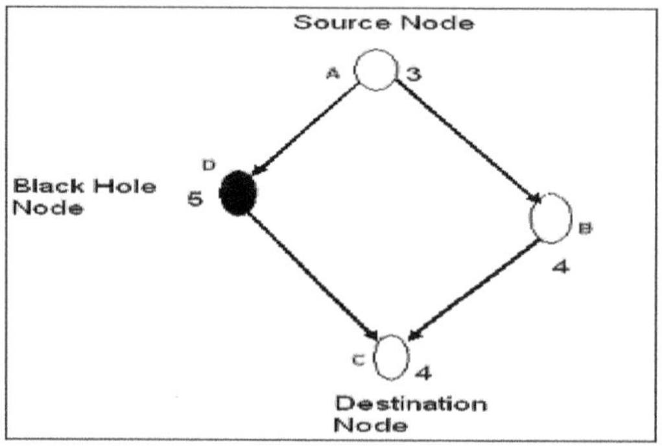

Figure no. 34.1

Node A will pick the second path after timeout period shown in *figure no. 34.1*. Node A will send the data packet to node B and node B will relay the data packet to the destination node C. Node C will send an acknowledgement to node B. Node B will forward the acknowledgement to the source node A. After the successful transmission of data packet, the fidelity of node will be increased and updated. Thus, the fidelity of node B will increase to 5 from 4 as shown in *figure no. 34.2*.

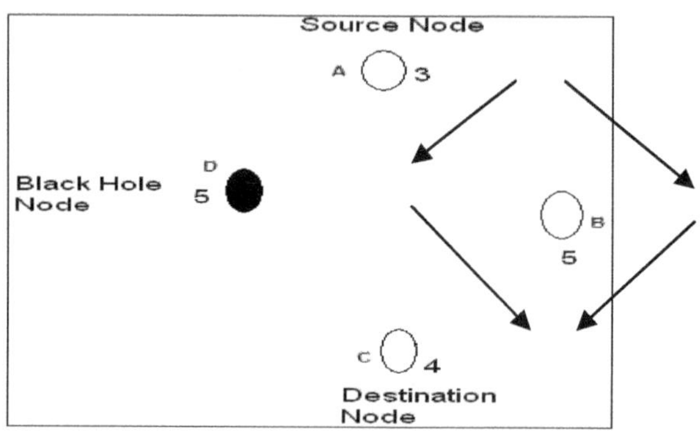

Figure no. 34.2

After every successful transmission of data packets the fidelity and the fidelity of friend nodes will increase. This way fidelity of node B will become 6 shown in *figure no. 34.3* after third transaction.

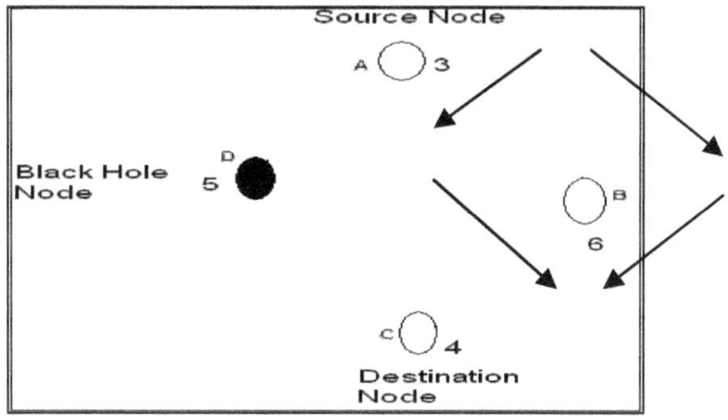

Figure no. 34.3

Afterwards node A will always pick node B to send data packet to node C as the fidelity of B is higher than node D. This way the black hole node will be terminated as shown in *figure no. 34.4.*

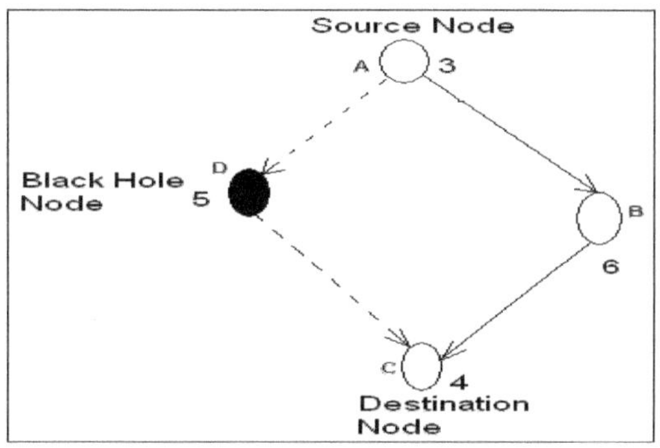

Figure no. 34.4

C. Co-operative Black Hole Attack

In co-operative black hole attack [19], there remains a group of malicious nodes. One of them takes the data packet and keeps on forwarding it among them so that the TTL of that data packet finishes off and the data packet is automatically dropped. We understand that the aim of these co-operative black hole attackers is to form a closed loop among themselves and keep on forwarding the message within this loop..

In our scheme a node will remain busy after sending a packet until it gets the acknowledgement for this message. So whenever any node try to send the message to a node already visited, it will simply reject the message and send a fail message to the sender as it is busy. As this scheme do not allow messages to transfer through the same node more than once there is no possibility for these malicious nodes to form the loop and hence co-operative black hole attack is mitigated.

D. Gray Hole Attack

A gray hole attack [17] is a variation of black hole attack, where an adversary first behave as an honest node during the route discovery process, and then silently drops some or all of the data packets sent to it for further forwarding even when no congestion occurs. Detection [24] of gray hole attack is harder because nodes can drop packets partially not only due to its malicious nature but also due to overload, congestion or selfish nature. A selfish node is unwilling to spend its battery life, CPU cycles or available network bandwidth to forward packets not of direct interest to it, even though it expects others to forward packets on its behalf.

If the probability of packet dropping of the malicious nodes is more than ½ then it is possible to mitigate gray hole attack by this scheme, otherwise it may or may not be possible. Whenever a node drops most of the packets sent to it then it behaves almost like a black hole attack and in that case this scheme can mitigate gray hole attack.

E. Black Mail Attack

Blackmail attack [17] causes false identification of a good node as malicious node. In Ad Hoc [17] [19] wireless networks [25], nodes usually keep information of perceived malicious nodes in a blacklist. An attacker may blackmail a good node and tell other nodes in the network to add that node to their blacklists as well, thus avoiding the victim node in future routes.

In our scheme there is no scope of broadcasting any details of a particular node to the nodes of the network by any other node. Here one node can only broadcast its own details to its neighbor nodes. So no attacker node can convince other nodes that a particular node is malicious. Thus there is no chance of black mail attack.

F. Rushing Attack

An offensive that can be carried out against on-demand routing schemes [23] is the rushing attack. Typically, on-demand routing schemes [26] state that nodes must forward only the first received Route Request from each route discovery; all further received Route requests are ignored. This is done in order to reduce cluttering. The attack consists, for the adversary, in quickly forwarding its Route Request messages when a route discovery is initiated. If the Route Requests that first reach the target's neighbors are those of the attacker, then any discovered route includes the attacker.

In our scheme there is no concept of route discovery which shows the path from sender to receiver. Here each node only has the information of its neighbor nodes and based on this information (fidelity value) sender selects a neighbor and sends the packet to that selected node. So rushing attack cannot be happened in the network following this scheme.

G. Worm Hole Attack

When the source node broadcast the RREQ packet, a malicious node which is at one part of the network receives the RREQ packet. It tunnels that packet to a second colluding party which is at a distant location near the destination, it then rebroadcasts the RREQ. The neighbors of the second colluding party receive the RREQ and drop any further legitimate requests that may arrive later on legitimate multi hop paths [27]. The result is that the routes between the source and the destination go through the two colluding nodes that will be said to have formed a wormhole [17] between them. This prevents nodes from discovering legitimate paths that are more than two hops away.

In our scheme there is no concept of route discovery which shows the path from sender to receiver. Here each node only has the information of its neighbor nodes and based on this information(fidelity value) sender selects a neighbor and sends the packet to that selected node So here is no need of RREQ packet broadcasted from the source node. This type of worm hole attack can not hamper the network following this scheme.

X. Simulation Analysis And Performance Metrics

In order to evaluate the performance of Ad Hoc network routing schemes, the following parameters were considered:

A. *Packet Delivery Fraction(PDF)*

PDF is defined as the ratio between no. of packets originated by application layer [28] in the source node to the no of packets received by the destination node. It will describe the loss rate that will be seen by the transport schemes, which in turn affect the maximum throughput that the network supports. In terms of packet delivery fraction, our scheme FBOD performs well. As the no of nodes getting increased the number of packets generated is higher so it may not transfer some of the packets, but the no of these packets are very small. When the no. of nodes is small then in ideal case PDF value is 1. But in case of DSR [29] the PDF is very fluctuating it is lesser in some of the points with respect to the other schemes but it is very higher in some of the points which are not tolerable. DSDV [30] is better in more no. of nodes but AODV [31] [32] is better in smaller no. of nodes region.

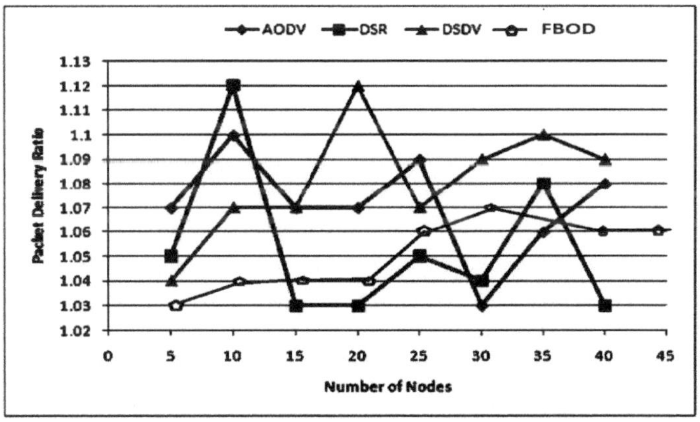

Figure 35.1: Packet Delivery Ratio for AODV, DSR, DSDV, FBOD

B. End to End Delay

The delay is affected by high rate of CBR Packets as well as the buffers become full much quicker, so packets have to stay in the buffer for a longer period of time before they are sent. This can be seen in DSR [2] when it reaches around 2300 packets in 0- mobility. For average end to end delay, the performance of DSR [33] decreases and varies with the number of nodes. In our scheme that is in FBOD the delay is getting increased with the increased no of nodes as the congestion is getting increased. But the rate of this increment is lesser as compared to other because of packets have to stay lesser time in buffer and lass computational overhead. The performance of DSDV [33] id degrading due to increase in the number of nodes the load of exchange of routing tables becomes high and the frequency of exchange also increased. Due to the mobility of nodes the performance of AODV [26] decreases and remains constant as the no of nodes increases.

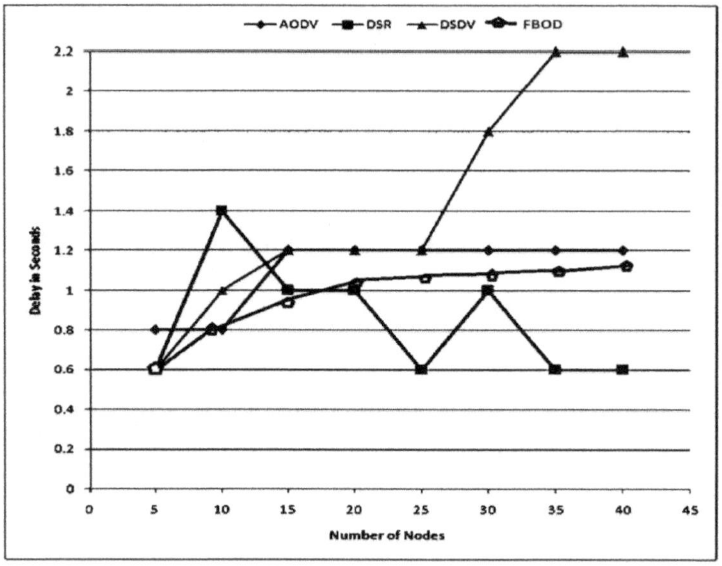

Figure 35.2: Average End to End Delay Ratio for AODV, DSR, DSDV, FBOD

C. Number of Packets Dropped

The number of data packets that are not successfully sent to the destination is the no of packets being dropped. In terms of dropped packets AODV's [2] performance is the worst. The performance decreases with the increase in the number of packets. As the no of nodes increases the no. of packets dropped increases which means that the no of packets not successfully reaches to the destination increases. DSDV [2] [33] performs consistently well with increase in the no. of nodes. DSR [29] [33] performs well when no of nodes is less but fails slightly when no of nodes is increased. In our scheme also in ideal case there is no drop of packets with the increase in no of nodes. It performs consistently well.

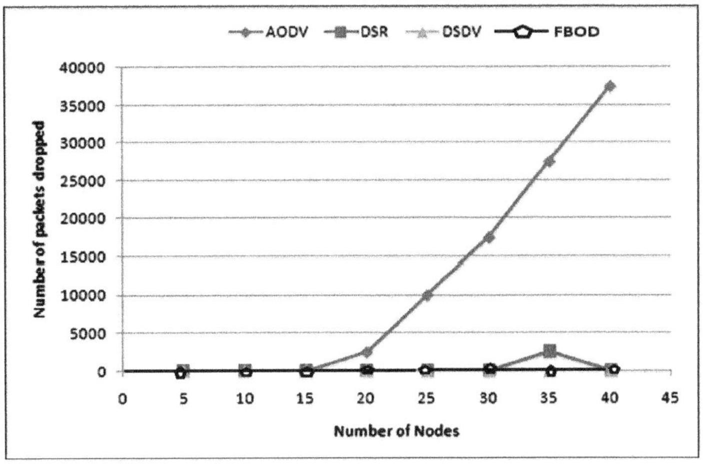

Figure 35.3: Dropped Packets for AODV, DSR, DSDV, FBOD

XI. CONCLUSION

This is a very light weight scheme with minimum computational overheads. In DSDV, we need to maintain a routing table. AODV has a lot of overhead while discovering routes, which clogs the network for sending data packets to desired destination. Not only does no such complicacy exist in our scheme, but it also has some of their benefits. Like AODV it is an on-demand routing scheme and the physical hardware support needed to implement it is substantially low which increases its scalability. This scheme also has added features so as to nullify some of the security threats which cause faults in the MANET networks.

REFERENCES

[1] Hongmei Deng, Wei Li, and Dharma P. Agrawal, "Routing Security in Wireless Ad Hoc Networks", IEEE Communications Magazine October 2002.

[2] V. Ramesh, Dr. P. Subbaiah, N. Koteswar Rao, M. Janardhana Raju, "Performance Comparison and Analysis of DSDV and AODV for MANET", V. Ramesh et al. / (IJCSE) International Journal on Computer Science and Engineering, Vol. 02, No. 02, 2010, 183-188.

[3] Huaizhi Li Zhenliu Chen Xiangyang Qin, "Secure Routing in Wired Networks and Wireless Ad Hoc Networks" IEEE, 2004.

[4] Panagiotis Papadimitratos , Zygmunt J. Haas , "Secure message transmission in mobile ad hoc networks, Ad Hoc Networks" IEEE 2003, 193–209.

[5] Kejun Liu, Jing Deng, Member, IEEE, Pramod K. Varshney, Fellow, IEEE, and Kashyap Balakrishnan, Member, IEEE, "An Acknowledgment-Based Approach for the Detection of Routing Misbehavior in MANETs" IEEE Transaction on Mobile Computing, VOL. 6, NO. 5, May 2007.

[6] S. Matri, T. J. Giuli, K. Lai and M. Baker , Mittigating Routing misbehaviour in Mobile Ad Hoc Networks. *Proceedings of the 6th annual internaional conference on Mobile Computing and Networking (MOBICOM), Boston, Massachusetts, United States, 2000, 255-265.*

[7] H. Yang, J. Shu, X. Meng, and S. Lu, "SCAN: Self-organized network-layer security in mobile ad hoc networks," IEEE Journal on Selected Areas in Communications, vol. 24, issue 2, pp. 261-273, February 2006.

[8] F. Anjum, Anup K. Ghosh, Nada Golmie, Paul Kolodzy, Radha Poovendran, Rajeev Shorey, D. Lee, J-Sac, "Security in Wireless Ad hoc Networks", IEEE journal on selected areas in communications, vol. 24, no. 2, February 2006.

[9] Oscar F. Gonzalez, Michael Howarth, and George Pavlou, Detection of Packet Forwarding Misbehavior in Mobile Ad-Hoc Networks. Center for Communications Systems Research, University of Surrey, Guildford, UK. Integrated Network Management, 2007. IM '07. 10th IFIP/IEEE International Symposium on May 21, 2007.

[10] Nik os Komninos, Dimitris Vergados, Christos Douligeris, "Layered security design for mobile ad hoc networks" journal computers & security 25, 2006 , pp. 121 – 130.

[11] Haiyun Luo, Petros Zerfos, Jiejun Kong, Songwu Lu, Lixia Zhang, "Self-securing Ad Hoc Wireless Networks", 7th IEEE Symp. on Comp. and Communications (ISCC), Taormina, 2002.

[12] H. A. Wen, C. L. Lin, and T. Hwang, "Provably Secure Authenticated Key Exchange Protocols for Low Power Computing Clients," Computers and Security, vol. 25, pp. 106-113, 2006.

[13] Anand Patwardhan, Jim Parker, Michaela Iorga. Anupam Joshi, "Tom Karygiannis, Secure Routing and Intrusion Detection in Ad Hoc Networks" 3rd International Conference on Perv asive Computing and Communications (PerCom 2005), Kauai Island, Hawaii.

[14] Angel R. Otero, Carlos E. Otero and Abrar Qureshi, "A Multi-Criteria Evaluation of Information Security Controls Using Boolean Features", International Journal of etwprk Security & its application (IJNSA), Vol. 2, No. 4, October 2010.

[15] Bing Wua, Jie Wua, Eduardo B. Fernandeza, Mohammad Ilyasa, Spyros Magliveras, "Secure and efficient key management in mobile ad hoc networks" Journal of Network and Computer Applications 30 (2007) 937–954.

[16] J. Nam, S. Cho, S. Kim, and D. Won, "Simple and Efficient Group Key Agreement Based on Factoring" Proc. Int'l Conf. Computational Science and Its Applications (ICCSA '04), pp. 645-654, 2004.

[17] Rashid Hafeez Khokhar, Md Asri Ngadi and Satira Mandala, "A Review of Current Routing Attacks in Mobile Ad Hoc Networks", International Journal of Computer Science and Security, pp. 18-29, Volume-2 Issue-3.

[18] Jung-San Lee, Chin-Chen Chang, "Secure communications for cluster-based ad hoc networks using node identities" Journal of Network and Computer Applications 22 October 2006 International Journal of Computer Science and Security, Volume (1): Issue (1) 67.

[19] "Avoiding Black Hole and Cooperative Black Hole Attacks in Wireless Ad hoc Networks" http://www.scribd.com/doc/26788447/Avoiding-Black-Hole-and-Cooperative-Black-Hole-Attacks-in-Wireless-Ad-hoc-Networks.

[20] P. Papadimitratos and Z.J. Haas, "Secure Link State Routing for Mobile Ad Hoc Networks" Proc . IEEE Workshop on Security and Assurance in Ad Hoc Networks, IEEE Press, 2003, pp. 27–31.

[21] [Perkins94] C.E. Perkins and P. Bhagwat, "Highly Dynamic Destination-Sequenced Distance-Vector Routing (DSDV) for Mobile Computers" , Comp. Comm. Rev., Oct.1994, pp.234-244.

[22] Komala CR, Srinivas Shetty, Padmashree S., Elevarasi E., "Wireless Ad hoc Mobile Networks", National Conference on Computing Communication and Technology, pp. 168-174, 2010.

[23] C. E. Perkins, E. M. Royer, and S. R. Das, "Ad Hoc On Demand Distance Vector (AODV) Routing," IETF Mobile Ad Hoc Networks Working Group, Internet Draft, work in progress, 17 February 2003.

[24] J. Parker, J. L. Undercoffer, J. Pinkston, and A. Joshi., "On Intrusion Detection in Mobile Ad Hoc Network s". In 23rd IEEE International Performance Computing and Communications Conference Workshop on Information Ass urance. IEEE, April 2004.

[25] Huaizhi Li, Mukesh Singha, "Trust Management in Distributed Systems" IEEE Computer Society February 2007.

[26] C. Perkins, E. Belding-Royer and S. Das, "Ad hoc On-Demand Distance Vector (AODV) Routing", 2003.

[27] S. Holeman, G. Manimaran, J. Dav, and A. Chakrabarti, "Differentially secure multicasting and its implementation methods", Computers & Security, Vol 21, No. 8, pp 736-749, 2002.

[28] Jeremy J. Blum, Member, IEEE, and Azim Es kandarian, Member, IEEE, "A Reliable Link-Layer Protocol for Robust and Scalable Intervehicle Communications" IEEE Transactions On Intelligent Transportation Systems, vol. 8, no. 1, March 2007.

[29] Anuj K. Gupta, Dr. Harsh Sadawarti, Dr. Anil K. Verma, "Performance Analysius of AODV, DSR & TORA Routing Protocols", IACSIT International Journal of Engineering & Technology, Vol. 2, No. 2, April 2010, ISSN: 1793 - 8236.

[30] R. Balakrishan, S. Jayabalan, Dr. U. Rajeswar Rao, Dr. T. K. Basak. Dr. V. Cyrilraj, "Performance Issues on AODV and DSDV for MNAETS", Jounnal Theoritical and Applied Information Technoilogy.

[31] Sapna S. Kaushik & P. R. Deshmukh. "Comparison of effectiveness of AODV, DSDV and DSR routing protocols is mobile Ad hoc networks", International Journal of Information Technology and Knowledge Management, July – December 2009, volume 2, No. 2, pp. 499-502.

[32] Luke Klein-Berndt, "A Quick Guide to AODV Routing"

[33] Nor Surayati Mohamad Usop, Azizol Abdullah, Ahmad Faisal Amri Abidin, "Performance Evaluation of AODV, DSDV & DSR Routing Protocol in Grid Environment", IJCSNS International Journal of Computer Science and Netwprk Security, Vol. 9, No. 7, July 2009.

A SWARM INSPIRED MULTIPATH DATA TRANSMISSION WITH CONGESTION CONTROL IN MANETS USING PROBABILISTIC APPROACH

Subhankar joardar[1] , Vandana Bhattacherjee[2] and Debasis Giri[1]

[1]Department of Computer Science and Engineering, Haldia Institute of Technology ,Haldia, West Bengal, India
subhankarranchi@yahoo.co.in, debasis_giri@hotmail.com
[2]Department of Computer Science and Engineering, Birla Institute of Technology ,Mesra, Jharkhand, India
vbhattacharya@bitmesra.ac.in

ABSTRACT

The major causes of network congestion are the lack of network resources and the irrational allocation of network resources. The traditional congestion control methods such as rate control, window mechanism, queue control and others can solve the congestion problem, but for the congestion due to irrational allocation fundamental solution is to make more effectively use of the network resources by adjusting the traffic routing depending on choice may be probabilistic, when congestion occurs. In mobile ad hoc networks congestion creates delay in transmission and also loss of the packet that causes wastage of time and energy on recovery. The Wireless Networks have to play an important role to adopt and execute a large no of innovative application. New challenges have come considering the major limitations of the ad hoc network like node's limited processing power, balance the load of network (to maintain the computation of the node). To overcome the above problem some algorithm is invoked and there may be huge amount of packet loss and this leads to decrease the lifetime of the network. Based on the concept of evolutionary cooperation in swarm Ant, we use the ants swarm intelligence to reinforce good quality routes. The objective of our proposed algorithm is to identify the congestion areas between source and its neighboring nodes to the destination and thus it will help to avoid the congestion of the network in the intermediate links and also minimize the packet loss in the network. Using a new mathematical model considering the swarm-based ant intelligence concept, we found an efficient congestion control mechanism (Ant's probabilistic transition rule).

KEYWORDS

Ad-Hoc Networks, Conditional Probability, Congestion, Distance Vector, Swarm Intelligence, Transmission Queue Length.

1. INTRODUCTION

Mobile Ad hoc Networks are movable and also able to communicate within themselves using a wireless physical medium where there is no need of pre existing network infrastructure. These networks can form stand-alone groups of wireless terminals, which can be again a part of any other cellular system or a fixed network. The main characteristic which draws the attention is that they are able to configure themselves without a centralized administration.

This paper is based on "A Swarm inspired Probabilistic Path Selection with Congestion Control in MANETs " by S.joardar, V.Bhattacherjee, D.Giri which appeared in the proceedings of second international conference on Computer Science, Engineering And Application, Springer-Verlag, Vol.2, pp 617- 626, May 25-27, 2012

The Mobile ad hoc network has several drawbacks that are not found in fixed networks. Unlike in the traditional wireless networks, communication in such a decentralized network is typically multi-hop, with the nodes using each other as relay routers without any fixed infrastructure.

This kind of network is very flexible and suitable for applications such as temporary information sharing in a conference, disaster rescue etc.
However, multi-hop routing, random movement of mobile nodes and other features in MANETs lead to The frequent change in network topology due to the mobility of the nodes causes a great deal of control information onto the network. The small capacity of batteries and the bandwidth limitation of wireless channels are other factors. The scalability is also a major factor because the network performance degrades quickly as the no of nodes increases.

1.1. Routing in Mobile Ad-hoc Wireless Networks

Specially configured routing protocols are engaged in order to establish routes between nodes which are more than a single hop. The ability to trace routes in spite of a dynamic topology is the unique feature of these protocols. These protocols can be categorized into two main types: Reactive (On-demand) and Proactive (Table-driven). Evaluating the routes continuously within the network is done by proactive protocols, so when a packet needs to be forwarded the route is already known and can be immediately used. Reactive protocols appeal to a route determination procedure on demand only.

1.2. Congestion in Mobile Ad-hoc Wireless Networks

The network where the resources are shared, to minimize the network overload the data rate must be adjusted used by the sender. The packets are dropped due to some routers are not able to forward the packets, so due to the bottleneck condition leads to many packet drops. May be the dropped packets already travelled a long way in the network and thus consumed significant amount of resources. So congestion leads to decrease the network throughput. This type of situation occurred on the early Internet, the development of the TCP congestion control mechanism.
In mobile ad hoc networks congestion occurs due to the shared wireless channel and dynamic topology, packet transmissions suffer from interference and fading. The network load is burdened through the transmission errors. There is an increasing demand of multimedia communications in MANETs so large amount of real-time traffic involves high bandwidth and it is liable to congestion. Congestion leads to packet losses and bandwidth degradation and also wastes time and energy on congestion recovery.
Mobile Ad Hoc Networks have some unique properties that affect the design of appropriate protocols and congestion control mechanism in particular. As it turned out, the vast different environment in a mobile ad-hoc network is highly problematic for TCP.
The properties of MANETs are the node mobility and a shared, wireless multi-hop channel. Adapting new routes due to node mobility as well as unreliable medium result in unsteady packet delay and packet loses. Also these are not congestion loses.
The wireless multi-hop channel allows only one data transmission at a time within the transmission range of one node. So close links as per the distance are not independent from each other. Internet routers typically are dedicated hosts connected by high bandwidth links. When congestion occurs on the Internet, it is usually concentrated on one single router, but in MANETs congestion affects a whole area because of the shared medium. Not only nodes are overloaded, but regions of the networks are also overloaded.

2. RELATED WORK

The continuous research on congestion minimization for MANET still needs more new techniques. This section will illustrate the research related to congestion control.

A distributed multi-path DSR protocol (MPDSR) was developed to enhance the quality of service [1]. This protocol forwards the outgoing packets using multiple no of paths and with point to point reliability. Split Multi-path Routing protocol [2] uses a per-packet allocation technique to distribute the data packets into multiple paths which prevents nodes from being congested in heavily loaded traffic conditions.

A novel technique Congestion Adaptive Routing in Mobile Ad Hoc Networks (CRP) [3] was proposed. In CRP each node on a route in the network sends a warn message to its previous node when it is to be congested. Then the previous node uses a new route to bypass the congestion and get the first non congested node on the route. Traffic will be distributed probabilistically over these routes and there will be less chance of congestion occurrence.

A dynamic load aware based load-balanced routing for ad hoc networks (DLBL) algorithm was developed. The DLBL [4] uses intermediate node routing load metric that helps the protocol to discover a route with less network congestion. When a link fails because of the node movement the algorithm uses the path maintenance to re join the broken links to get a route from the source to the destination.

Another novel algorithm proposed is a Congestion-aware routing metric (CARM) [5] which considers Mac-overhead, data-rate and buffer queue delay to select low congestion with high throughput routes. Another mechanism that CARM applied was the use of link data- rate categories to prevent the mismatched link data-rate. CARM performed locally.

An ant based DSR [6] will inform the source node about QoS available to destination node such as acceptable delay, jitter and energy in the case of multimedia and real time application. They proposed DSR using ACO and called AntDSR(ADSR).

Another novel algorithm proposed a QoS-aware routing protocol [7] which consider two scheme admission control and feedback for the Qos requirements. Their major works on the approximation of bandwidth estimation to calculate the rest amount of bandwidth available at each node.

A biologically inspired routing algorithm [8] for multi-hop networks considers the concept of stigmergy and evaluates the optimal and sub-optimal routes without the whole system-wide connectivity information. They also incorporate MAC layer information into the routing decision can prevent the forward data traffic into the congested areas under situation like various nodal mobility, network density and data loads.

The success of data transmission in Multipath routing [9] considers the concept of traffic distribution strategy that defines how concurrently available paths are utilized and data to the same destination as split and distributed over multiple paths. This paper shows when the no of path is single the entire data packet is transmitted through it and maximum probability of success as the no of paths increases the probability of success of data transmission will decrease.

3. Proposed Work

In this section we will discuss Swarm intelligence, Distance metric and the transmission Queue length.

3.1. Swarm Behaviour

This algorithm is inspired by the foraging behaviour of biological ant colony [10], when they find paths to food sources. Each ant deposits a chemical called pheromone when they move from source to the destination and the foragers follow such pheromone trails. By that more ants are attracted by these pheromone trails and in turn reinforce them even more. This natural phenomenon (stigmergy) plays a role in developing and manipulating local information.

3.2. Congestion Metric

The major parameters to control congestion areas probabilistically are the two metrics in a node.

3.2.1 Distance Metric

The first metric is the distance between the single hop by using the Time of arrival [11] estimating the time of signal travelling between nodes (TOA) provides the distance between two nodes. We can use ultrasound sensor for measuring range between nodes. For this purpose very accurate rubidium-based oscillators must be used. In order to calculate the TOA parameter the nodes must have a common clock, or exchange timing information by certain protocols such as two-way ranging protocol.

The distance metric d_{ij} where *j* is the neighbor node of *i.*

$$d_{ij} = C^*(d_{remote} - d_{local})/2\,, \quad C \approx 3^*10^8\ m/s \qquad (1)$$

The time between starting the transmission of a data packet and receiving the corresponding immediate acknowledge will refer as remote delay(d_{remote}) and receiving one data packet and sending out the immediate acknowledgement will be the duration as local delay(d_{local}). C is measured as the speed of light.

3.2.1 Transmission Queue Length Metric

The second metric is a heuristic value transmission queue length in the i_{th} node towards the neighbouring nodes. The transmission queue length **[8]** can affect the packet latency and packet drop because of the size of the packet length.

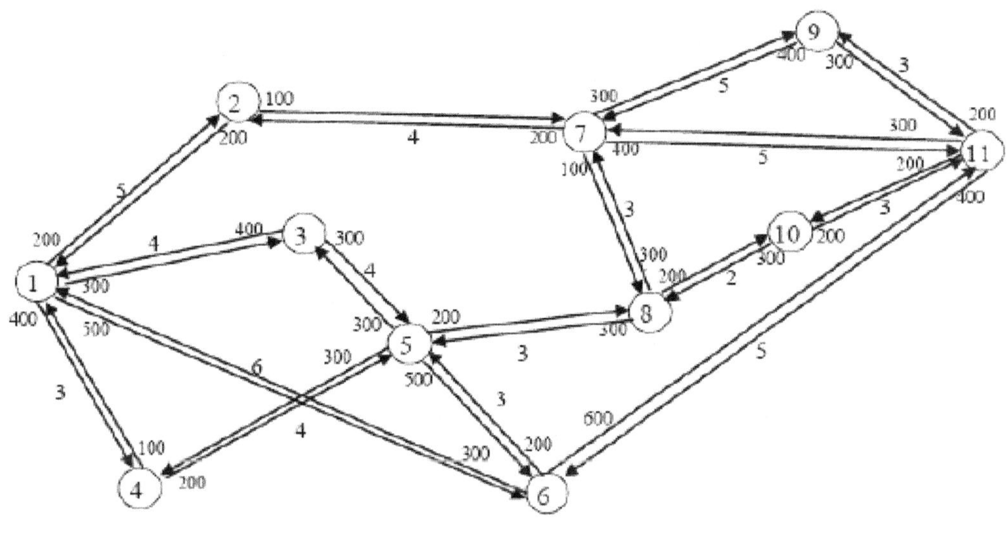

Figure 1 : Mobile nodes with distance and out going transmission queue

$$T_{i,j} = 1 - \frac{q_{i,j}}{\sum_{j \in N_b^i} q_{i,j}} \qquad (2)$$

where $q_{i,j}$ is the outgoing queue length in terms of bits waiting to be sent to the link between i and j and N is all the nodes in the network and $N_b^{[i]}$ is the neighbouring nodes of the node *i.* We have considered some random values as transmission queue in terms of bits. Using (2), we find the goodness of heuristic values means small queue data length will have higher heuristic value. From the figure 1, all the nodes have symmetric link between them. We consider some hypothetical value as distance and transmission queue length, using (1), we found the congestion metric $T_{i,j}$ in the table 1 **[8].**

Table 1: Goodness of congestion metric $T_{i,j}$

$j \Rightarrow$ $i \Downarrow$	1	2	3	4	5	6	7	8	9	10	11
1		.85	.78	.71		.64					
2	.33						.66				
3	.42				.57						
4	.66				.33						
5			.76	.76		.61		.84			
6	.66				.77						.33
7		.80						.90	.70		.66
8					.62		.62			.75	
9							.42				.57
10								.40			.60
11						.63	.72		.81	.81	

3.3. Ant based probabilistic path selection

When a source node s, needs to route a destination d without knowing the global topology, using the initial pheromone information available the ant will chose its next hop with probability $P_{i,j}$. We have proposed an algorithm to find the distance of neighbour node from the i_{th} node and also we calculate the pheromone deposition, pheromone evaporation and residual pheromone on the link $l_{i,j}$

$$\Delta\tau_{ij}(t) = mvl_{i,j} * \frac{1}{d_{i,j}}, j \in N_b^{[i]} \quad (3)$$

The distance $d_{i,j}$ from the node i to its neighbouring nodes can be calculated by using the equation 1, we assume some random distance metric for our algorithm. $mvl_{i,j}$ is the minimum distance selected from the neighbour nodes of *i*. The $\Delta\tau_{ij}(t)$ is the reciprocal of the distance $d_{i,j}$ on the link. We found that the less distance have more reciprocal value of rate of pheromone deposition and vice versa.

$$\tau^d_{ij}(t) = \tau_{ij}(t) + \Delta\tau_{ij}(t)\rho \quad (4)$$

$$\tau^{e}_{ij}(t)=\tau^{d}_{ij}(t)(1-\rho) \tag{5}$$

$\tau_{ij}(t)=0.1$ is the initial pheromone deposition and the ρ [12] specifies the rate at which pheromone evaporates means ant "forget" previous decision for value $\rho=1$ the pheromone evaporates rapidly and search is random, when $\rho=0$ results in slower evaporation rates, $\rho\in[0,1]$. We take $\rho=0.6$. The final pheromone deposition is $\tau^{d}_{ij}(t)$. The final pheromone evaporates from the link is $\tau^{e}_{ij}(t)$. The residual pheromone on the link is.

$$\tau^{r}_{ij}(t)=\tau^{d}_{ij}(t)-\tau^{e}_{ij}(t) \tag{6}$$

Using (3) : $\Delta\tau_{1,2}(t)=3*1/5=.6$, $\Delta\tau_{1,3}(t)=3*1/4=.75$, $\Delta\tau_{1,4}(t)=3*1/3=1$,
$\Delta\tau_{1,6}(t)=3*1/6=.5$

Using (4) : $\tau^{d}_{1,2}(t)=.1+(.6*.6)=.46$, $\tau^{d}_{1,3}(t)=.1+(.75*.6)=.55$, $\tau^{d}_{1,4}(t)=.1+(1*.6)=.7$,
$\tau^{d}_{1,6}(t)=.1+(.5*.6)=.4$

Using (5) : $\tau^{e}_{1,2}(t)=.46*(1-.6)=.18$, $\tau^{e}_{1,3}(t)=.55*(1-.6)=.22$, $\tau^{e}_{1,4}(t)=.7*(1-.6)=.28$,
$\tau^{e}_{1,6}(t)=.4*(1-.6)=.16$

Using (6) : $l_{1,2}(t)=.46-.18=.28$, $l_{1,3}(t)=.55-.22=.33$, $l_{1,4}(t)=.7-.28=.42$,
$l_{1,6}(t)=.4-.16=.24$.

The link $l_{i,j}$ from node i shows when the distance is less compared to other neighbor nodes, more pheromone deposition occurs. The nodes residual pheromone deposition metric in the table 2.

Table 2: Residual pheromone metric $\tau_{ij}(t)$

$j \Rightarrow$ $i \Downarrow$	1	2	3	4	5	6	7	8	9	10	11
1		.28	.33	.42		.24					
2	.42						.35				
3	.42				.42						
4	.42				.33						
5			.33	.33		.42		.42			
6	.33				.42						.20
7		.33						.42	.28		.28
8					.30		.30			.42	
9							.28				.42
10								.42			.30
11						.20	.28		.42	.42	

When the pheromone information are available the ant will chose its next hop with probability $P_{i,j}$. We adopt this concept by using Baye's probability theorem

$$P_{i,j}(t) = \frac{\left[\tau_{ij}(t)\right]^{\alpha}\left[T_{ij}\right]^{\beta}}{\sum_{j \in N_b^{[i]}} \left[\tau_{ij}(t)\right]^{\alpha}\left[T_{ij}\right]^{\beta}}, \quad \text{for i = 1 to N,} \tag{7}$$

where $\alpha \geq 0$ is the relative importance of the pheromone trail. $\beta \geq 0$ is the relative importance of the congestion. α and β are the two tuneable parameter will control relative weight of pheromone trail and heuristic value T_{ij}.

In the following, we calculate the probability corresponding to the node i using j.

$P_{1,2} = (.28^3*.85^1) / (.28^3*.85^1) + (.33^3*.78^1) + (.42^3*.71^1) + (.24^3*.64^1) = .17$

$P_{1,3} = (.33^3*.78^1) / (.28^3*.85^1) + (.33^3*.78^1) + (.42^3*.71^1) + (.24^3*.64^1) = .25$

$P_{1,4} = (.42^3*.71^1) / (.28^3*.85^1) + (.33^3*.78^1) + (.42^3*.71^1) + (.24^3*.64^1) = .48$

$P_{1,6} = (.24^3*.64^1) / (.28^3*.85^1) + (.33^3*.78^1) + (.42^3*.71^1) + (.24^3*.64^1) = .08$

Analogously, we can calculate the probability corresponding to other nodes which are shown in table 3.

Table 3: Probability metric on $l_{i,j}$

i, j	$P_{i,j}(t)$ $\alpha > \beta$	$P_{i,j}(t)$ $\alpha < \beta$
1,2	.17	.31
1,3	.25	.28
1,4	.48	.27
1,6	.08	.11
2,1	.46	.13
2,7	.53	.86
3,1	.42	.28
3,5	.57	.71
4,1	.80	.91
4,5	.19	.08
5,3	.16	.22
5,4	.16	.22
5,6	.27	.15
5,8	.38	.39
6,1	.28	.32
6,5	.68	.65

i,j	$P_{i,j}(t)$ $\alpha > \beta$	$P_{i,j}(t)$ $\alpha < \beta$
6,11	.03	.02
7,2	.23	.26
7,8	.53	.48
7,9	.12	.15
7,11	.10	.09
8,5	.18	.22
8,7	.18	.22
8,10	.62	.55
9,7	.17	.21
9,11	.82	.78
10,8	.64	.29
10,11	.35	.70
11,6	.03	.08
11,7	.11	.17
11,9	.42	.37
11,10	.42	.37

3.4. Algorithm P^2sdt : Node i Processing for distribution of traffic

---

Algorithm P^2sdt( )

Input:

d_{ij} : Distance between the nodes.
q_{ij} : Transmission queue length.
α : Relative importance of pheromone deposition.
β : Relative importance of congestion.
ρ : Rate at which pheromone evaporates.

Output:

$\mathbf{P_{i,j}}(\tau_{ij}(t))$: Goodness of Congestion
$P_{i,j}(T_{i,j})$: Residual pheromone

BEGIN

Update parameter for Probabilistic path selection
$\tau_{ij}(t) \leftarrow$ Pheromone residual on the link.
$T_{i,j} \leftarrow$ Probability of congestion it discards.
if ($\alpha > \beta$) then
 while($\tau_{ij}(t) > \tau_{ij \in N_b}(t)$)
  Distribute.Traffic(Max.Prob$_{i,j}$($\tau_{ij}(t)$))
 wend
 if(max($\tau_{ij}(t)$)= $\tau_{ij \in N_b}(t)$) then
  Distribute.Traffic(Max.Prob$_{i,j}$($T_{i,j}$))
 end if
else
if ($\alpha < \beta$) then
 while($T_{i,j} > T_{i,j \in N_b}$) then
  Distribute.Traffic(Max.Prob$_{i,j}$($T_{i,j}$))
 wend
 if (max($T_{i,j}$) = $T_{i,j \in N_b}$) then
  Distribute.Traffic(Max.Prob$_{i,j}$($\tau_{ij}(t)$))
 end if
 end if
end if
END

3.5. The Route Discovery

When initialization process starts the single-hop HELLO messages creates the neighbourhood for each node. The HELLO packet contains *source IP address* and *hop count* (Initialized to 0). Whenever a traffic source *s* needs a route to a destination *d*, it broadcasts *route request* packets (RQ) across the network. The RQ packet contains *destination IP address*, *source IP address*, and *broadcast ID*. For the route discovery the solution would be through flooding the RQ packets, we adopt the probabilistic broadcast scheme in combined with the outgoing queue length $Q_{i,j}$ in terms of bits and the distance metric $D_{i,j}$. When a node first receives a packet, with probability p it broadcasts the packet to its neighbours and with probability $1-p$ it discards the packet. The probability value p is calculated as

$$Q_{i,j} = \frac{q_{i,j}}{\sum_{j \in N_b^i} q_{i,j}} \tag{8}$$

$$D_{i,j} = \frac{d_{i,j}}{\sum_{j \in N_b^i} d_{i,j}} \tag{9}$$

$$P_{CONGi,j}^{(t)} = e^{-\sigma(Q_{i,j})} \tag{10}$$

$$P_{Disti,j}^{(t)} = e^{-\sigma(D_{i,j})} \tag{11}$$

$$P_{i,j}^{(t)} = \beta\,(P_{CONGi,j}^{(t)}) + (1-\beta)(P_{Disti,j}^{(t)}), \tag{12}$$

where β lies between 0 and 1.

Table 4: Probability metric of distance and congestion

i, j	$P_{CONGi,j}$	$P_{Disti,j}$
1,2	.86	.76
1,3	.80	.80
1,4	.74	.85
1,6	.69	.71
2,1	.51	.64
2,7	.71	.57
3,1	.56	.60
3,5	.65	.60
4,1	.71	.65
4,5	.51	.56
5,3	.78	.75
5,4	.78	.75
5,6	.67	.81
5,8	.85	.81
6,1	.71	.77
6,5	.79	.81

i,j	$P_{CONGi,j}$	$P_{Disti,j}$
6,11	.51	.58
7,2	.81	.79
7,8	.90	.84
7,9	.74	.74
7,11	.67	.74
8,5	.68	.69
8,7	.68	.69
8,10	.77	.77
9,7	.55	.53
9,11	.65	.69
10,8	.54	.67
10,11	.67	.54
11,6	.69	.65
11,7	.75	.77
11,9	.82	.86
11,10	.82	.86

Table 4: Probability metric on $P_{i,j}$

i, j	$P_{i,j}$ $\beta = .3$	$P_{i,j}$ $\beta = .5$	$P_{i,j}$ $\beta = .7$	i,j	$P_{i,j}$ $\beta = .3$	$P_{i,j}$ $\beta = .5$	$P_{i,j}$ $\beta = .7$
1,2	.79	.81	.83	6,11	.55	.54	.53
1,3	.80	.80	.80	7,2	.79	.80	.80
1,4	.81	.79	.77	7,8	.85	.87	.88
1,6	.70	.70	.69	7,9	.74	.74	.74
2,1	.60	.57	.54	7,11	.71	.70	.69
2,7	.61	.64	.66	8,5	.68	.68	.68
3,1	.58	.58	.57	8,7	.68	.68	.68
3,5	.61	.62	.63	8,10	.77	.77	.77
4,1	.66	.68	.69	9,7	.53	.54	.54
4,5	.54	.53	.52	9,11	.67	.67	.66
5,3	.75	.76	.77	10,8	.63	.60	.57
5,4	.75	.76	.77	10,11	.57	.60	.63
5,6	.76	.74	.71	11,6	.66	.67	.67
5,8	.82	.83	.83	11,7	.76	.76	.75
6,1	.75	.74	.72	11,9	.84	.84	.83
6,5	.80	.80	.79	11,10	.84	.84	.83

3.6. Algorithm P²scm : Node i Processing for selection of path

Algorithm P²scm( )

Input:

d_{ij} : Distance between the nodes.

q_{ij} : Transmission queue length.

Output:

$P_{CONGi,j}^{(t)}$: Goodness of Congestion

$P_{Disti,j}^{(t)}$: Goodness of distance

$P_{i,j}^{(t)}$: Probability of path selection

BEGIN

1. Update parameter for Probabilistic path selection
2. $Q_{i,j}$ ← Relative Pheromone deposition on the link.
3. $D_{i,j}$ ← Relative congestion on the link
4. **if** ($p_{ij}(t) > p_{ij \in N_b}(t)$) then
5. Max.Prob$_{i,j}$ ($p_{ij}(t)$)
6. **end if**
7. **END**

where σ ($\sigma > 0$) is the coefficient. where $\beta \in [0,1]$ is the relative importance of the distance and congestion in the link . When β is equal to .5 the probability of goodness of congestion and distance metric can be calculated. This broadcast scheme helps to discover new routes avoiding congestion areas and adopting less distance probabilistically. If the probability value $P_{CONGi,j}^{(t)}$ is less means more congestion in the path, due to maximum probability is set on $P_{i,j}^{(t)}$ the other parameter like $P_{Disti,j}^{(t)}$ must be high means shorter distance is available in the link. When the link have the same probability, this broadcast scheme by changing the tuneable parameter β as .7 sets a new probability value to find a new path.

During the course of flooding the intermediate node *j* receives a Route Request (RQ) packet. By capturing the *source address* and the *previous hop* node in the *message cache* it first set the reverse path to the source. If a valid route to the destination *d* is available, that is, there is at least one link associated with the pheromone trail greater than the other neighbour node or least distance within the neighbour node, the intermediate node *j* generates a *route reply* (RP). The RP is routed back to the source *s* via the reverse paths. Otherwise, the RQ is rebroadcast.

When the destination node receives RQ it will send a RP to all the neighbours from which it gets a RQ. The intermediate node *j* can maintain multiple loop-free paths by recording all the new paths that have the latest sequence number but a lower hop-count in its routing table, and also send a RP to all the neighbours from which it gets a RQ. During the process of the RP back to the source *s* will assigns an initial pheromone value $\tau_{i,j}^{(t)}$ to the corresponding neighbour node, which indicates a valid route to the destination *d*.

3.7. Description of the Algorithms

Our proposed p²scm algorithm presents the pseudo code of how a node *i* process the out going queue and by using swarm intelligence how it will select the path and based on the routing table, the p²sdt will distribute the data traffic according to the probabilities for each neighbour node j in the routing table[12].

The node *i* first calculate the distance d_{ij} between all its neighbour nodes. By using the smallest distance $mvl_{i,j}$ the node *i* evaluate the relative distance between all its neighbour nodes. P²sdt is also evaluate the $\Delta\tau_{ij}(t)$ which is reciprocal to the distance d_{ij} where the relative pheromone deposition occurs on the link $l_{i,j}$. Now p²sdt evaluate the residual pheromone on the link $l_{i,j}$ from the line 2 of the above algorithm. Secondly p²sdt evaluate the probability of goodness of congestion $(1-\frac{q_{i,j}}{\sum_{j\in N_b^{[i]}} q_{i,j}})$.

In our proposed algorithm from line no 4-10 when α>β it first checks the pheromone ($\tau_{ij}(t)$) deposition in a particular link have the highest value within its neighbour or not. If found yes then the maximum prpbability $P_{i,j}$ will be on that link $l_{i,j}$. Next it checks a pair of paths with highest deposition and same value, if found yes within that, less congested ($T_{i,j}$) path will have the highest probability[13].

Similarly from line no 12 -20 when the α<β, if $T_{i,j}$ have the highest value within its neighbour then the highest probability will be on that link $l_{i,j}$. Next it checks a pair of paths with highest and same value of $T_{i,j}$ or not if found yes within that, more pheromone ($\tau_{ij}(t)$) deposited path will have the highest probability on that link $l_{i,j}$.

4. ANALYSIS

P²sdt at first from node *i* and the two tuneable parameter α>β we observe the higher pheromone on the link l_{ij} will have higher probability of path selection and higher goodness of congestion means small queue data length will have low probability, if the deposition is same on both the link, we observe then p²sdt by using α>β and conditional probability evaluates the less congested path means queue data length is small . In our example node 3 to its neighbour node 1 and 2 from table 1, table 2 and table 3.

Also by making α<β p²sdt consider the less congested path probabilistically means when the queue data length is same on both the link, the higher probability comes with higher pheromone concentration on the link and if the congestion is same within a pair of paths then it evaluates the paths which have more pheromone deposition.

By using two tuneable parameters p^2sdt decides the path probabilistically which have less congested or more pheromone deposition when the nodes have dynamic topology at real time situation.

Similarly P^2scm evaluates the goodness of congestion as well as distance. If $p_{ij}(t)$ > $p_{ij\in N_b}(t)$ means the average value of $p_{ij}(t)$ is maximum within all the links of the neighbouring nodes. Another important feature of P^2scm is that if the probabilistic value of goodness of congestion is low, the probabilistic value of goodness of distance must be high. As we have seen from the table 4.

5. Conclusion

In this paper, we propose a new mathematical model for congestion control. We have developed a single hop congestion aware probabilistic path selection algorithm which employs the swarm intelligence of biological ant. Using distance and congestion metric the algorithm invokes route discovery process with less congestion and shorter distance in a single hop at run time. By probabilistic approach the algorithm can prevent forward data traffic into the congested areas and also finds the shortest distance probabilistically within a pair of nodes. For data transmission, if more congestion occurs in all the links of a node, the algorithms by using the tuneable parameter that can shift the mode of transmission and can easily avoid the congested areas. By using the above stated mathematical models and some hypothetical data we tried to control the congestion at local node level.

References

[1] R.Leung, Jilei Liu, E.Poon, C.A.Chan & Baochun Li, (2001) "MPDSR : a QoS-aware multipath dynamic source routing protocol for wireless ad-hoc networks", In proceedings Local Computer Networks, 26th Annual IEEE Conference, pp 132-141.

[2] S.Lee & Mario Garla, (2001) "Split Multi-path Routing with Maximally Disjoint Paths in Ad Hoc Networks", In proceedings International IEEE Conference on Communication, pp 3201-3205.

[3] Duc. A.Tran & Harish Raghavendra, (2006) " Congestion Adaptive Routing in Mobile Ad Hoc Networks", IEEE transaction on parallel and Distributed systems, 17: pp 1294-1305.

[4] Xiangquan Zheng, Wei Guo, R.Liu & Yongchun Trian, (2004) " A New Dynamic Load-aware Based Load-balanced Routing for Ad Hoc Networks", IEEE, pp 407-411.

[5] Xiaoqin Chen, H.M.Jones & A.D.S.Jatyalath, (2007) " Congestion- Aware Routing Protocol for Mobile Ad Hoc Networks", In proceedings of IEEE conference on vehicular Technology, Doi.10.1109/VETECF.2007.21, pp 21-25.

[6] R.Asokan, A.M.Natarajan & C.Venkatesh, (2008) "Ant Based Dynamic Source Routing Protocol to Support Multiple Quality of Service(QoS) Metrics in Mobile Ad Hoc Networks", International Journal of Computer Science and Security, vol.2 no.3, pp 48-56.

[7] L. Chen & Wendi B. Heinzelman, (2005) "Qos-Aware Routing Based on Bandwidth Estimation for Mobile Ad Hoc Networks", IEEE on Selected Areas in Communication, vol.23, pp 561-572.

[8] L. Zhenyu, M.Z.Kwiatkowska & Costas Constantinou, (2005) "A Biologically Inspired Congestion Control Routing algorithm For MANETs", In Proceedings of the Third IEEE International Conference on Pervasive Computing and Communications , pp 226-231.

[9] A. K. Mondal & P. K. Banerjee, (2008) " The Success of data Transmission in Multipath Routing for MANET", Mobile and Pervasive Computing, CoMPC-2008, pp 48- 51.

[10] R. Beckers, J.L.Deneubourg & S.Goss, (1992) "Trails and u turns in the selection of the shortest path by the ant lasius niger", Journal of Theoretical Biology, pp 397-415.

[11] A. Gunther & Christian hoene, (2005) "Measuring Round trip times to determine the distance between WLAN nodes", In proceedings of Networking 2005, Springer-Verlag, pp 768-779

[12] A.C.S.Jawahar, "Ant Colony Optimization for Mobile Ad-Hoc Networks", ajaychak@eden.rutgers.edu.

[13] S. Soardar, V. Bhattacherjee & D. Giri, (2012) " A Swarm inspired Probabilistic Path Selection with Congestion Control in MANETs ", In Proceedings of second international conference on Computer Science, Engineering and Application, Springer-Verlag, Vol.2, pp 617-626.

11

INTERFERENCE AVOIDANCE TECHNIQUES FOR CLOSED ACCESS SMALL CELLS NETWORKS

Cristian Androne[1] and Tudor Palade[1]

[1]Communications Department, Technical University of Cluj-Napoca
Cluj-Napoca, Romania
[cristian.androne, tudor.palade]@com.utcluj.ro

ABSTRACT

This paper investigates the impact of small cells networks deployment in a typical residential environment. First we will present some aspects with regard to small cells coverage area when deployed in a macrocell sector area. Using the observations with regards to coverage we will further investigate the deployment of a residential femtocell network, implemented in closed access form. The paper proposes two algorithms in order to reduce the interference level at the macrocell user site, considering that the operator owns at least two carrier frequencies which will be used for both the macrocellular and femtocellular layers. The performances of the two proposed algorithms are presented in comparison with the case when no interference mitigation methods are implemented. The obtained results show a significant improvement in the signal to interference and noise ratio, obtained at the user location.

KEYWORDS

Femtocell, Macrocell, Interference Mitigation, SINR, Small Cell, Residential Scenario.

1. INTRODUCTION

In the past few years, due to the saturation of the market caused by the offering of only voice services [1], cellular communication operators have directed their attention towards IP data based services [2]. Because of this, as shown in [1], the average revenues per user have increased for operators that offer enhanced data services. Even so, one important impediment towards this direction is represented by the fact that assuring coverage using the traditional macrocellular networks can be very costly, especially in areas with high density population, which have correspondingly higher demands in terms of service rate [3]. It is estimated that approximately 66% of all voice calls and 90% of all data transfers happen in indoor environments [4], that is why it is crucial to assure a good indoor coverage in this areas, not just for voice services, but also for data which become more and more important.

Recent surveys have shown that approximately 45% of home environments and 30% of business environments suffer because of weak indoor coverage [5]. That is why it is expected that a good indoor coverage and high quality of service will lead to increased revenues for operators and will enhance customer loyalty.

The traditional approach towards what concerns coverage assurance in the indoor environment is represented by the use of outdoor macrocellular base stations (MBSs). This techniques present a series of drawbacks which reveal the need for alternative solutions:

- It is inefficient to assure a good indoor coverage using the traditional approach. For example, in UMTS networks an indoor user will have to use more power in order to compensate the losses caused by the penetrations of the walls. This will lead to an increase in the overall cell noise which could limit the capacity of the network and/or the quality of the services provided.

- A high capacity network will need a high number of base stations, which will lead to increased deployment costs for the operators. Moreover, with the increase of the number of base stations, the planning and optimization of the network will be harder to realize. A high density in base station transmitters will cause interference problems on the downlink, because of the superimposing of the coverage areas.
- In 3G and beyond wireless communication networks, the carrier frequency used will be 2 GHz or higher, therefore the penetrations of the buildings will become an important problem. More than that, the performances of the networks will suffer important downfalls, since the data rates can not be guaranteed for indoor users because of the communication channel conditions. Therefore, the need to have superior modulation and coding schemes is absolutely necessary in HSDPA, WiMAX or LTE deployments .

A detailed description of the heterogeneous network architecture implemented according to the WCDMA standard is given in [6] and [7]. The authors of these papers describe each of the entities which compose the new small cell-macrocell architecture. The main novelty in this situation consists in introducing a new equipment, i.e. the Home NodeB Gateway (3GPP terminology), which has the role of managing the small cell base stations, more exactly: assuring a secure connection between the small cell base station and the core network (CN), providing support for paging and handover procedures, transparent transfer of Layer 3 messages between the User Equipment (UE) and core network A part of the interfaces which were developed in order to communicate with the new entity are defined in [8], [9] and [10]. For other mobile communications standards similar entities were implemented with the same basic role.

In order to compensate the limitations of outdoor macrocells, different approaches have been proposed, in order to enhance the signal level in the indoor environment. A first solution was to introduce antennas inside the buildings. Therefore, the Distributed Antenna System (DAS) was developed, which consists in deploying multiple antennas ,connected to a common base station, in such a way that the resulting coverage would be uniform [4]. Another solution proposed, especially to assure coverage inside tunnels and long corridors, was the deployment of the radiant cable [4]. A more recent approach refers to the implementation of small cells inside buildings in order to enhance both coverage and capacity of the network.

Depending upon the actual deployment scenario, the small cells can take many shapes, as presented by the authors of [11]: residential femtocell, enterprise femtocell and hot-spot or picocell. The residential femtocell, is characteristic for suburban areas, and the femtocell base station (FBS) acts as a cellular network access point which has the role of providing the radio resources to the users of the networks. It is similar in concept to the wireless access point used in local area networks, and therefore it is designed to be implemented by the user in a decentralized manner. It has a low transmit power of maximum 24 dBm [12] and allows a number of up to 5 simultaneous calls and data sessions [4].

A second approach in small cells deployment options refers to the implementation of enterprise femtocells. This type of base stations is characteristic to the office or industrial environment. The femtocell base station has a maximum transmit power of up to 30 dBm [12] and can support a maximum number of 16 simultaneous calls and data sessions [4]. If the residential femtocell has a stand-alone implementation, the enterprise femtocell is normally deployed in a network of femtocells, creating in this way a network that is similar to the macrocellular network.

The last deployment option considers outdoor small cells, often known as picocells. The maximum transmit power for this kind of base station is of 37 dBm [12], and can support a variable number of users depending on the quality of service level which needs to be met. For the rest of the paper we will consider only the case of the residential femtocell, and the term femtocell and small cell will be interchangeably used.

Regardless of their deployment options, small cells offer, according to the authors of [13] and [14], certain advantages which put them in front of the traditional macrocellular networks:

- Because of the low transmit power, small cell base stations can be implemented with a much higher deployment density than macrocells, leading to a superior spectral efficiency.
- Because the small cells are located inside indoor environments, they provide an enhanced signal level, assuring a better and more uniform coverage of the scenario. Due to their low transmit power, and using the basic concept of spectrum reuse, small cells increase the overall network capacity. Also, because they are very close to the mobile terminals, this leads to a longer lifetime of the terminals batteries.
- Due to the fact that some of the indoor users are handled by the small cells, the macrocell base stations can redirect their resources in order to offer better services to the outdoor users.
- Reduced costs for the operator for the implementation and maintenance of the overall network. Because the main characteristics of the small cell concept is it's availability, the mobile user will experience higher quality of service (QoS) levels, at low prices.

One important technical characteristic when it comes to small cells deployments is represented by the way in which the carrier frequencies of the operators are shared among the macrocellular and femtocellular layer, respectively. In this sense, Figure 1 from [15] illustrates the three main options: the dedicated carrier approach, the shared carrier approach and the partially shared carrier approach. Next, the authors of [15], give detailed characterizations of the three implementation options. For our later experiments, we will consider the partially shared carrier approach. Related to the deployment options of small cells (residential femtocell, enterprise femtocell and pico-cell), as well as to the carrier implementation, the access mode to the femtocell defines the way in which the femtocellular layer acts upon the macrocellular layer. Therefore, the authors of [4] talk about two access modes to the small cell: open and respectively closed access mode. The open access mode is characterized by the fact that all users are allowed access to the small cells resources. In this case the femtocell network acts as an extension to the macrocellular layer, enhancing the coverage and the capacity of the overall network.

On the other hand, when considering the closed access mode, the femtocell forbids access to the resources to all users which are not listed in the CSG (Closed Subscriber Group) list [4]. This access list contains the IMSI codes of the users which rightfully belong to the femtocell. This access mode is particularly specific to the residential femtocell, but it is not limited to this deployment option. The small cell base stations operating in closed access mode represent an important interference source upon the other macrocellular users which are not allowed to access the femtocell. This is due to the fact that these users are not connected to the strongest transmitter, but to one that is further away, usually a macrocell. According to recent surveys [4], the residential femtocell owners would like to have their equipment deployed in closed access mode, which can cause significant problems to the macrocellular layer if there aren't any interference mitigation methods implemented. This particular issue will represent the main theme of the paper in the next chapters.

The rest of the paper is organized as follows: in section 2 we will present the main simulation parameters that will be used later in the paper; in section 3 some aspects regarding small cell coverage area and dependency upon different parameters are studied; section 4 describes the residential scenario that will be implemented for the experiments; in section 5 the proposed algorithms for cross-layer interference avoidance are described and their performances analysed within the scenario previously described; finally section 6 draws the main conclusions of the paper and proposes future work for the interference avoidance methods.

2. Small Cells Coverage Aspects

The deployment of small cells is a process that started in 2009 for the third generation mobile communications networks [16], and now-a-days it is in full expansion so that during 2012 femtocells integrated in the LTE networks will become an important factor in commercial mobile networks. One important feature of the small cells base stations is represented by the coverage area that is about to be obtained with these transmitters, which depends on the implementation environment, on radio channel conditions and the availability of the equipment.

In the following chapter we will investigate the impact that small cells networks have upon the network users, from the point of view of their coverage area. The coverage radius of the femtocell base station is an important parameter that affects the capacity of the network on the downlink communication, on one side: the larger the coverage area of the femtocell, the more mobile users it will include managing to free a part of the macrocells resources; and on the uplink communication channel, on the other side, where the interference caused by the macrocell users upon the femtocellular layer can be so significant that it will block the communication at the small cell base station.

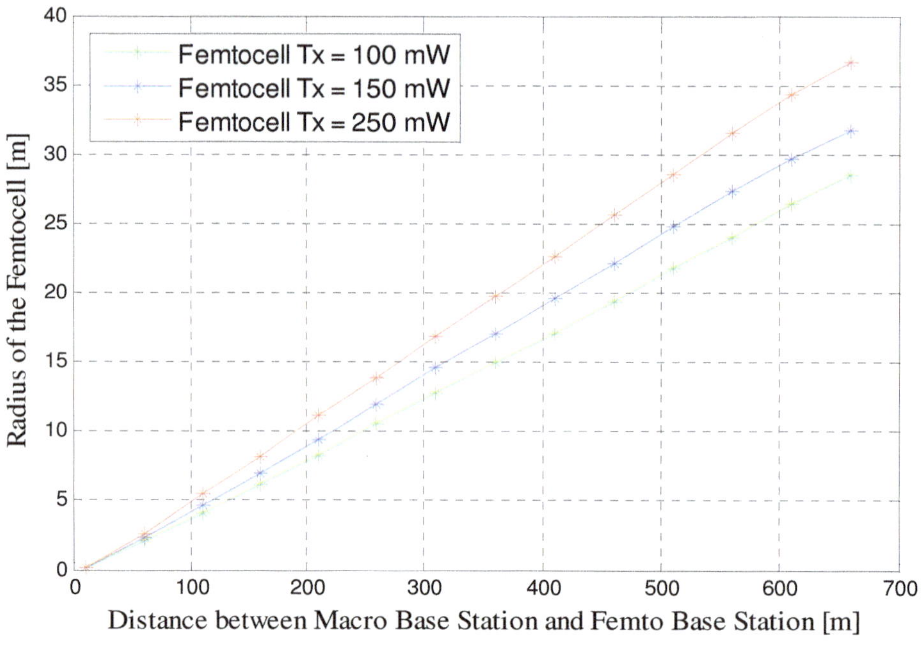

Figure 1. Coverage areas of residential femtocells vs The distance between MBS and FBS.

The larger the coverage area of the femtocell, the larger the minimum distance between the macrocell users and the FBS will be, which will lead to a decrease in the level of interference caused by the macrocellular user upon the femtocellular layer. In the following we will analyze the values of the coverage radii of the small cells base stations obtained for different transmit powers corresponding to the residential femtocells case. In this sense, Figure 1 offers a detailed view of the situation involved.

Because small cells base stations are deployed within the coverage area of a macrocell, most often within a macrocellular sector, the coverage area of a femtocell depends on the level of signal coming from the macrocell base station. So, the higher the ratio between the signal coming from the femtocell and the signal coming from the macrocell, the more users the femtocell will manage to include. Figure 1 illustrates the variations of the coverage areas of the small cells in terms of the distance between the MBS and the FBS, for some typical values of the transmit powers of the residential femtocell.

So, as the FBS is moved away from the MBS, the signal from the latter one decreases, which will lead to an expansion of the small cell coverage area. The effect is more pronounced for a higher transmit power of the FBS.

For the calculation of the coverage areas of the small cells we considered an outdoor environment, without obstacles, for a better understanding of the impact that the distance between the two base stations has. For the link budget calculation only the path losses were taken into account, calculated according to relations given by Table 1.

Path-loss component	Path-loss is modelled as $11.81+38.63\log_{10}(d)$ for macrocell users and $29.94+36.70\log_{10}(d)$ for small cell users where d is the distance from the base station in meters.
Shadow fading component	Shadow fading is modelled as spatially correlated random process with log normal distribution (6 dB standard deviation for the macrocell signal, 4 dB standard deviation for the femtocell signal, spatial correlation $r(x) = e^{x/20}$ for distance x.
Receiver noise power component	The receiver noise power is modelled as $10 \log_{10} (kT\ NF\ W)$ where the effective noise bandwidth is given as $W = 3.84 \times 10^6$ Hz, and $kT = 1.3804 \times 10^{-23} \times 290$ W/Hz. The noise figure at the UE is $NF_{[dB]} = 7$ dB.
Macrocell antenna gain component	The macrocell antenna gain is calculated as $$G(\theta)dB = G_{max} - \min\left[12\left(\frac{\theta}{\beta}\right)^2, G_s\right], \quad -\pi \le \theta \le \pi \quad \text{with}$$ $\beta = 70\pi/180$ angle where gain pattern is 3dB down from peak $G_s = 20$dB sidelobe gain level in dB $G_{max} = 16$dB maximum gain level in dB

Table 1. Simulation parameters.

The propagation model considered in the simulation is characterized by a log-distant path loss model which uses fixed and exponential coefficients defined according to [12]. The variation in amplitude caused by shadowing is often modelled as a log-normal distribution with a standard deviation defined according to the log-distant model. The fading maps have been generated according to the method described in [17] with the values of the standard deviation set according to the recommendation of [12].

Another important parameter that influences the coverage area of a femtocell is represented by the angular offset which is formed between the direction defined by the position of the FBS and the position of the MBS respectively, and the direction of the main lobe of the radiation pattern of the MBS antenna. Figure 2 illustrates the shapes and dimensions of the FBSs coverage areas deployed within a macrocellular sector. From Figure 2, one can observe the effect of the distance between the FBS and the MBS, as well as the effect of the angular offset.

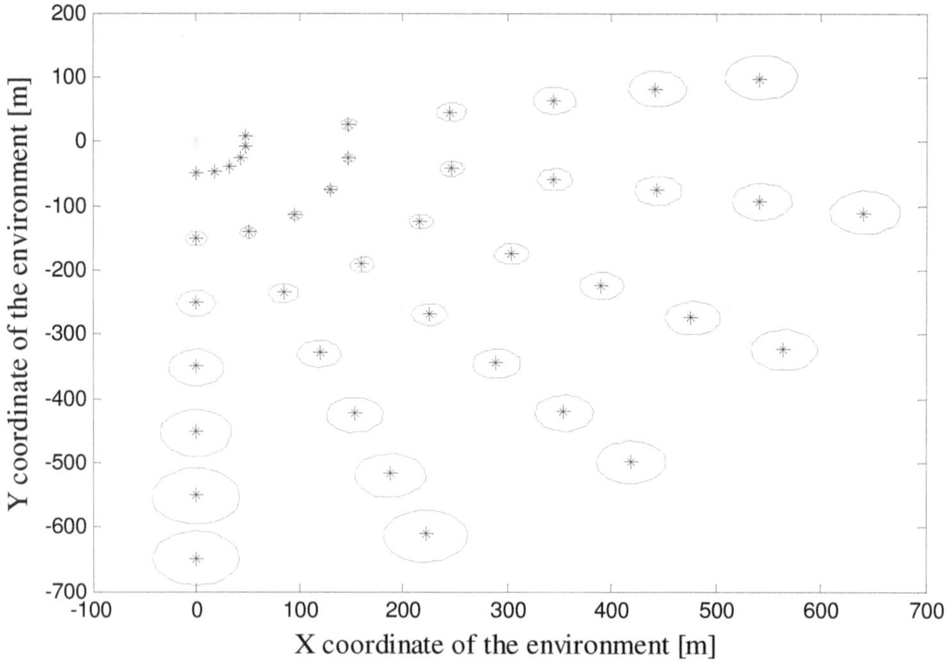

Figure 2. The deployment of small cells inside a macrocellular sector.

The small cell base stations deployed in the scenario have an ordered distribution with respect to MBS position, which is located at the (0,0) coordinates, highlighted by the green star in the figure. The FBSs are deployed starting with a distance of 50 meters relative to the position of the MBS and with a distance of 100 meters between them, until they reach the edge of the macrocell which is at approximately 650 meters considering the same reference point. Their angular distribution within the macrocellular sector is made at 20 degrees angular offsets between the deployment directions.

One may observe an increase in the coverage area of the FBSs as the level of the signal coming from the MBS decreases, due to the distance between the MBS and respectively FBS positions, as well as due to the angular offset from the main lobe of the radiation pattern of the MBS antenna. The transmit power of the small cell base station is considered to be 250 mW. For different values of the transmit power, the results are similar and vary according to the pattern defined by Figure 1.

The red stars in the figure illustrate the actual positions of the small cell base stations. One may observe the fact that the shape of the coverage area is not symmetrical with respect to the position of the FBS, mainly because of the level of the signal coming from the MBS which is stronger towards MBS position and weaker in the opposite direction, which allows the coverage area of the femtocell to extend more in that direction.

The results illustrated by Figure 2 are obtained in an outdoor environment which doesn't take obstacles into account, in order to better show the effect of the angular offset. Also, in this case, for the link budget calculation we take into account only the path losses, defined according to Table 1.

In Figure 3, on the other side, we illustrate the shapes and sizes of the femtocells in which we also take the shadow fading component into account, defined also by Table 1. The deployment arrangement is the same as for the previous case, with distances of 100 meters between the FBSs and angular offsets of 20 degrees between the deployment directions. One may observe a more pronounced irregularity of the coverage areas, due to the shadow fading, defined as a log-normal distribution with a standard deviation of 6 dB for the macrocellular signal and respectively 4 dB for the femtocellular signal [12]. The pattern defined by Figure 2 is still valid here, in the sense that the coverage areas of the small cells increase with distance and with de shifting from the direction of the main lobe of the radiation pattern of the MBS antenna. The transmit power considered here for the small cell base station is of 250 mW.

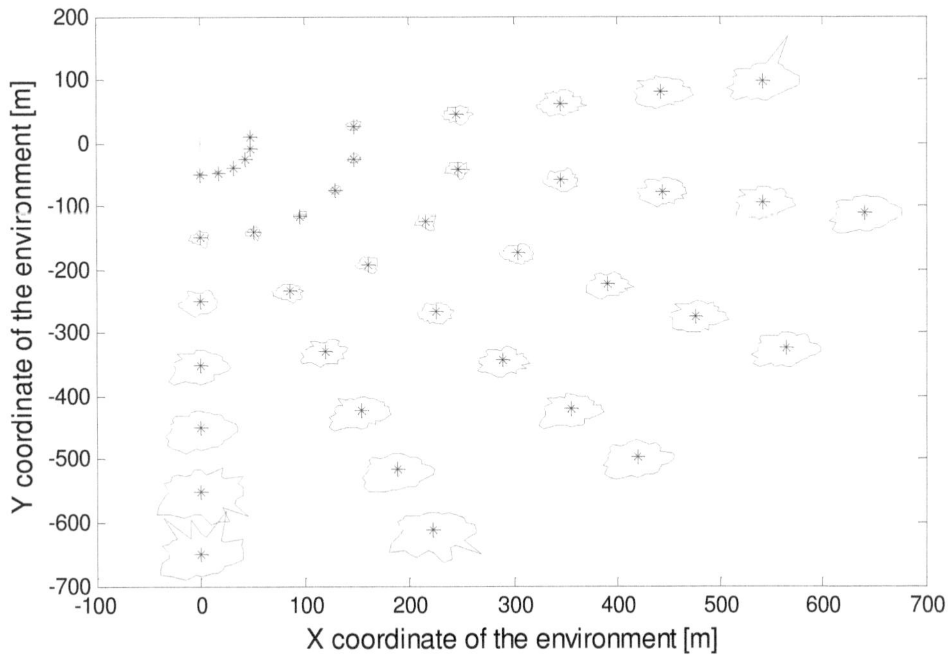

Figure 3. The deployment of small cells inside a macrocelullar sector with shadow fading.

The effect of shadow fading must be taken into account mainly when developing power control algorithms for the FBS, as well as for the mobile terminal, especially when it is handed over from the FBS to the MBS. This irregularity of the femtocells shape can determine weak performances of the power control algorithms if they are not flexible enough to overcome this effect. The higher the influence of the shadow fading upon the received signal, the more irregular the shape of the femtocell will be, which will lead to the necessity of power control and handover algorithms with higher performances. Therefore, Figure 4 illustrates the values of the difference between the maximum and respectively minimum level of the received signal strength from the FBS, on the downlink communication channel, in terms of the value of the shadow fading standard deviation, corresponding to the signal coming from the MBS. In order to fully understand the dependency upon the shadow fading standard deviation, we have illustrated graphs for different values of the shadow fading standard deviation of the signal coming from the FBS as well.

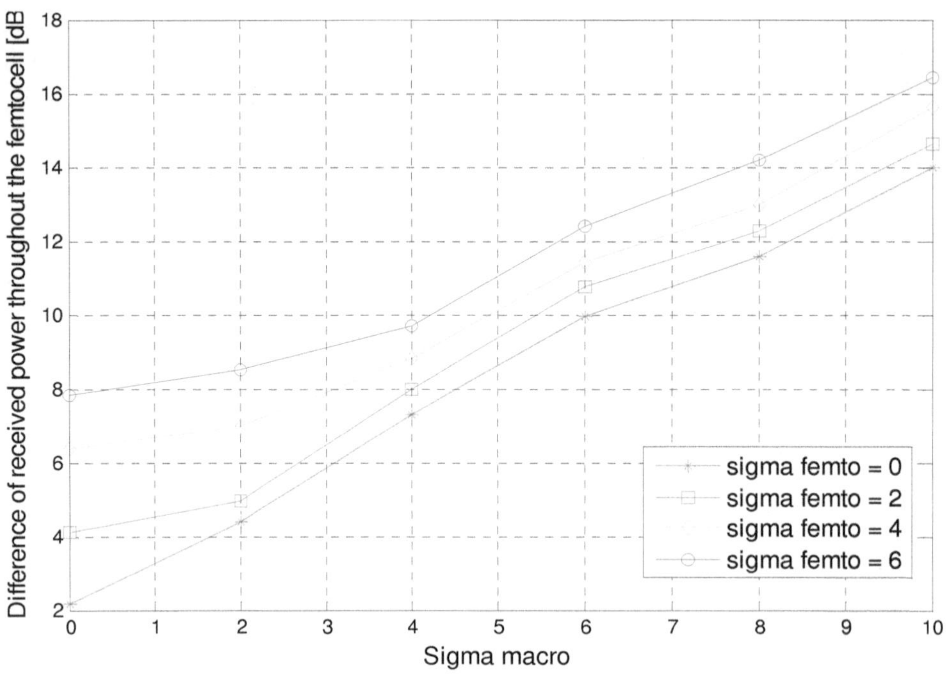

Figure 4. The difference in received signal strength coming from the FBS.

The graphs are obtained for a FBS transmit power of 250 mW and a positioning of it at half the maximum possible distance, i.e. half the distance between the MBS position and the macrocell coverage area edge. According to Figure 4 one may notice that there is a variation of the received signal strength coming from the femtocell even for the case when there is no shadow fading taken into account, an aspect which was validated by Figure 2 as well. The values presented in Figure 4 represent the average values obtained for that certain distance between the MBS and FBS, while varying the angular offset with respect to the direction of the main lobe of radiation of the MBS antenna, so that it sweeps the whole macrocellular sector.

As the levels of the shadow fading standard deviation increase for the signals coming from both macrocell and femtocell, so does the level of irregularity of the femtocell coverage area. For the extreme case when we consider the shadow fading standard deviation for the signal coming from the MBS to be 10 dB, and the one coming from the FBS to be 6 dB, the resulting value of the level difference in received signal strength is 16 dB, with no differences in signal attenuations for different directions.

In Figure 5 we illustrate the difference between the maximum and respectively minimum level of the received signal strength coming from the FBS in terms of the distance between the MBS and FBS. The presented graphs are obtained for the case when we do not consider any shadow fading in the environment, and for the cases when the shadow fading standard deviation of the signal coming from the FBS is 2 dB, while the one from the MBS is 6 dB; the one from the FBS is 4 dB and the one from the MBS is 8 dB and respectively the one from the FBS is 6 dB and the one from the MBS is 10 dB. The average value of each graph validates the aspects presented in Figure 4, proving that the higher the values of the shadow fading standard deviation, the higher the difference between the maximum and minimum levels of the received signal strength.

The evolutions of the graphs are distinct due to the various factors that influence their values. So, for the case when the shadow fading is neglected, at short distances between the MBS and FBS, the level difference is lower, due to the more symmetric shape of the femtocell with respect to the FBS position, following that it will increase with a value of maximum 2 dB, a value that is practically insignificant from the point of view of the power control and handover algorithms. Therefore, for an environment without shadow fading, we can argue that the difference in received signal strength throughout the femtocell coverage area is approximately constant, with respect to the distance between the MBS and FBS.

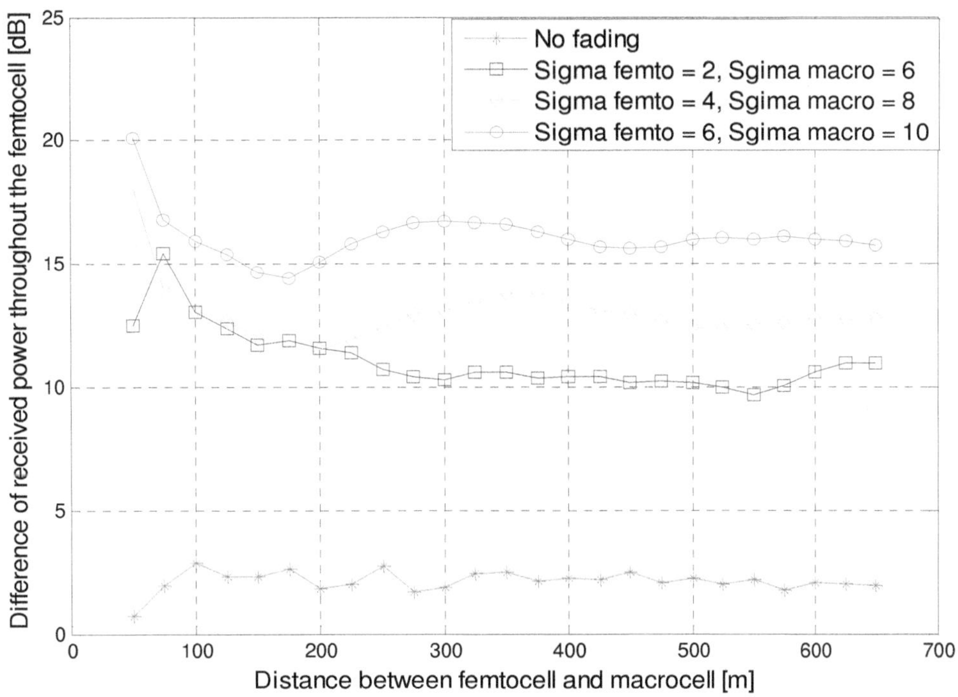

Figure 5. The difference in received signal strength as a function of the distance between the MBS and FBS.

If we consider environments in which shadow fading is present, according to Figure 5 we can argue that the difference in received signal level is higher for shorter distances between the MBS and FBS, due to the pronounced irregularity of the femtocell coverage area. This, corroborated with the fact that the signal level from any transmitter decreases with the square distance, makes the difference between the maximum and minimum received signal strength levels to be high in close vicinity to the MBS.

As the FBS get further away from the MBS, thus resulting in a larger coverage area, the effect of the irregularities decreases, this due also to the fact that the level of the received signal strength has a better constancy at longer distances from the transmitter. It can be observed that after approximately 30% of the maximum distance between the MBS and FBS, the variations tend to remain constant around a specific value.

3. Simulation Scenario

In order to implement the interference mitigation algorithms of which we have spoken earlier in the paper, we need to define the environment scenario in which we will place out the two-layer mobile communications network. As mentioned before, we will consider the deployment of a residential femtocell network in a typical sub-urban environment. For this study we will use the same environment we have used in our previous work, [15] but we will consider a different approach from the user deployment point of view. The general view of the residential environment is presented in Figure 6.

The simulation scenario consists of a street-like environment with houses placed on both sides of a road. The architecture of the houses is typical for the residential areas of the cities in the Transylvania area, Romania. Houses are symmetrically placed on both sides of the street which has a width of 20 meters. For larger scale scenarios we consider an additional backyard and garden for each house.

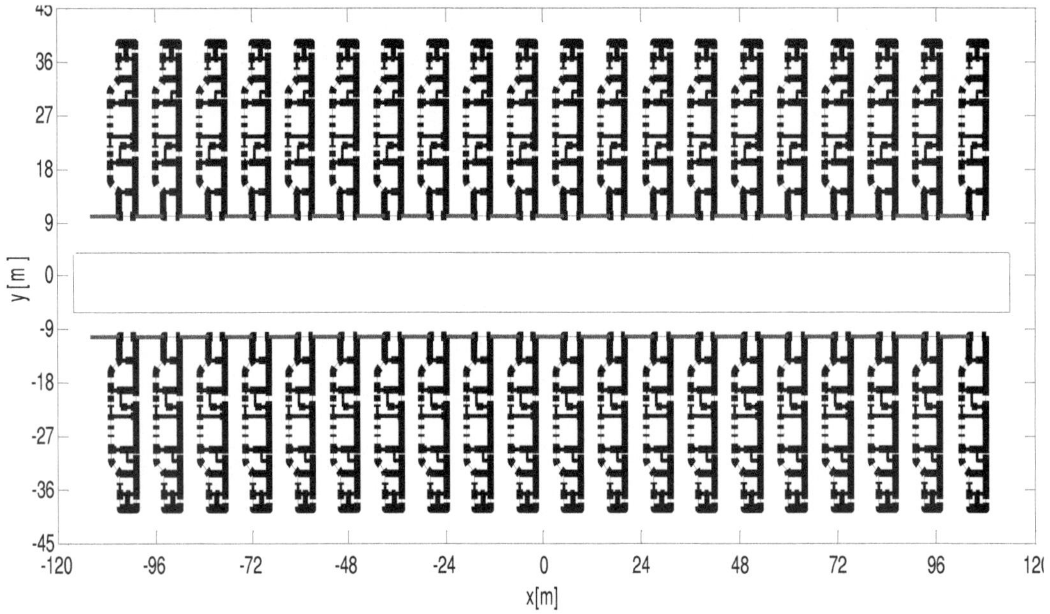

Figure 6. Simulation scenario.

The houses have one floor, with rooms mainly aligned one after the other in a train-like deployment. There are several entrances all of them being positioned on one side of the house. The footprint of the house is about 30 meters in length and 6 meters in width. For the construction of the houses several materials are considered, like concrete, brick, sheet-rock, glass and wood. The houses are equipped with entrance gates which are made of reinforced wood. The attenuations introduced by the construction materials, when considering perpendicular incidence angles of the waves, are presented in Table 2.

In the environment illustrated in Figure 6 we deploy a variable number of FBSs, with the condition that they are placed inside the houses and no more than one FBS is implemented in one house. The houses chosen are random, as well as their position inside the house. In our experiments we realized a sufficient number of simulation runs so that the position dependency of the FBSs influences as little as possible the final results.

The macrocellular layer is composed of 19 MBSs each of them equipped with three sectorial antennas. The MBS inter side distance (ISD) is equal to 1732 meters [12], as for a typical residential environment. The scenario previously presented in Figure 6 is placed at a distance of 750 meters from the central MBS, in a similar manner to the one presented in [15].

House Element (Material used)	Attenuation introduced [dB]
External wall (Concrete)	15
Internal wall (Brick)	10
Internal wall (Sheet-rock)	7
House gate (Reinforced wood)	5
Doors (Wood)	3
Window (Glass)	1

Table 2. Attenuations introduced by the house elements.

The red line in Figure 6 represents the path followed by users that travel through the scenario. As we will later explain in the paper, we will consider several different paths for the users, and their closeness towards the houses varies from 1 meter to a maximum of 10 meters in some experiments.

4. ALGORITHMS DESCRIPTION

As mentioned earlier in the paper, most residential femtocell owners will most likely choose to have their equipment working in closed access mode. In this situation, the FBS will become a powerful interference source that may significantly degrade the performances of the macrocellular layer [15].

Considering the scenario previously described in the paper, a user that is passing through the environment may be seriously affected by the FBSs deployed in the nearby houses. A moving user may receive strong downlink signals from the near-by femtocells, but it can not connect to them, even if the femtocell signal is more powerful than the macrocell signal, because it is not allowed to do that. Therefore, under certain conditions it may even be rejected from the network because of its low channel quality indicators.

In order to avoid such a catastrophic scenario, we have proposed two carrier jumping algorithms which tend to solve the high interference problem by a mechanism which is often known as interference avoidance [4]. The two algorithms can be applied only for operators which have at least two carriers available for usage. The idea which stands behind the two proposals consists in a mechanism that triggers the macrocell to change its carrier once the downlink conditions for its travelling user degrade below a certain threshold.

The mobile terminal periodically reports its channel conditions to the host base station. The decisions regarding the change of carrier are based on these information averaged at every 10 ms. Considering the partially shared carrier deployment option described in [10], which considers that there are at least two carriers available, one which is used by both femtocellular and macrocellular layers and respectively one that is used only by the macrocells, the first algorithm that we propose is characterised by carrier jumping once the channel conditions fall below the considered threshold. Figure 7 illustrates the basic flow diagram of the algorithm.

The algorithm takes decisions at every 10 ms based on the previously three consecutive samples taken. If all of the three bins are below a certain threshold (SINR_thr), the algorithm triggers the MBS to change its carrier for that particular user. The carrier setting is kept for as long as the user is in that area or it is handed over to some other base station. Only if the logical AND of the three decisions is 1, the algorithm will give the carrier jumping command, otherwise the MBS will keep its current carrier.

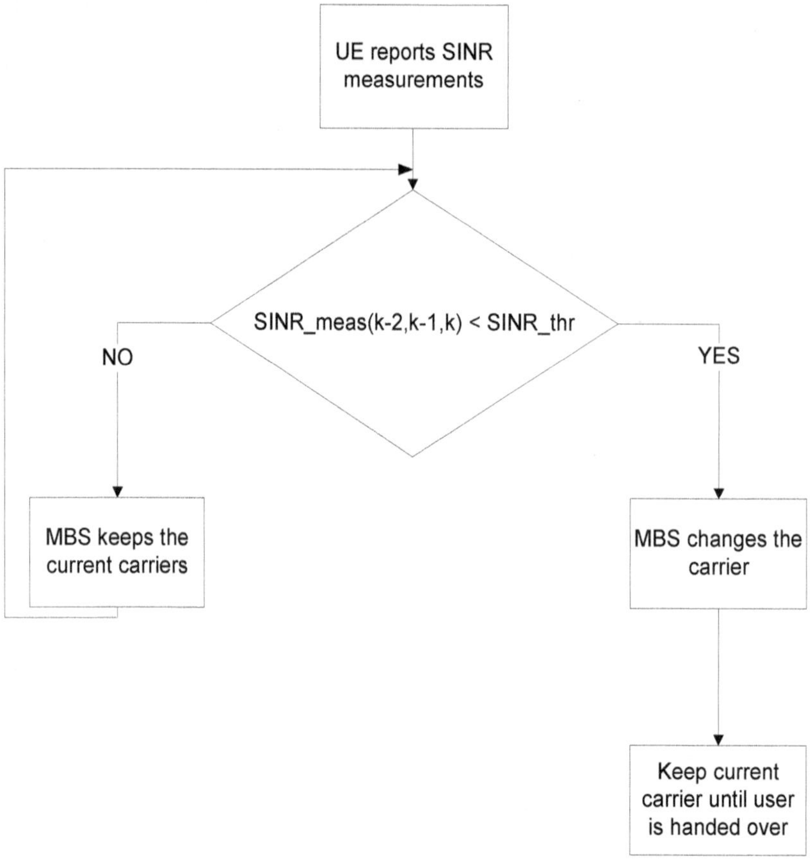

Figure 7. Carrier jumping algorithm1.

The algorithm presents the advantage of assuring a high SINR value for the travelling user. Once the channel conditions fall below the threshold, practically the user will experience a low SINR for 30 ms, and then it will be transferred to a low interference environment, taken that the macrocellular layer was properly planned so that suitable performance indicators are met. However, in high user density environments, if we apply this algorithm and we will transfer all users on another carrier, this will tend to become a dedicated carrier approach with the femtocellular users one carrier and the macrocellular users on another carrier, which from the point of view of spectrum usage is highly inefficient.

Therefore, the second algorithm we propose is based on the same principle, but with the difference that once the macrocell changed its carrier for a particular user, after a predefined period of time it moves back to the original carrier and performs once again the channel conditions check. The carrier is then kept for as long as the channel conditions are not met for three consecutive samples, after that the user being moved again on the second carrier. The full flow diagram of the new algorithm is presented in Figure 8.

The advantage of this algorithm is that the user is not kept permanently on the new carrier, but the macrocell tries to move it back to the original carrier, thus resulting in a higher spectral efficiency. In this case, the second carrier can be seen as a safety band on which users with insufficient channel conditions can be brought in order to avoid a dropped call or data session. Thus most of the macrocells resources can be concentrated on the first carrier, while for the second one the resources can be assigned in such a way that it will support a limited number of users.

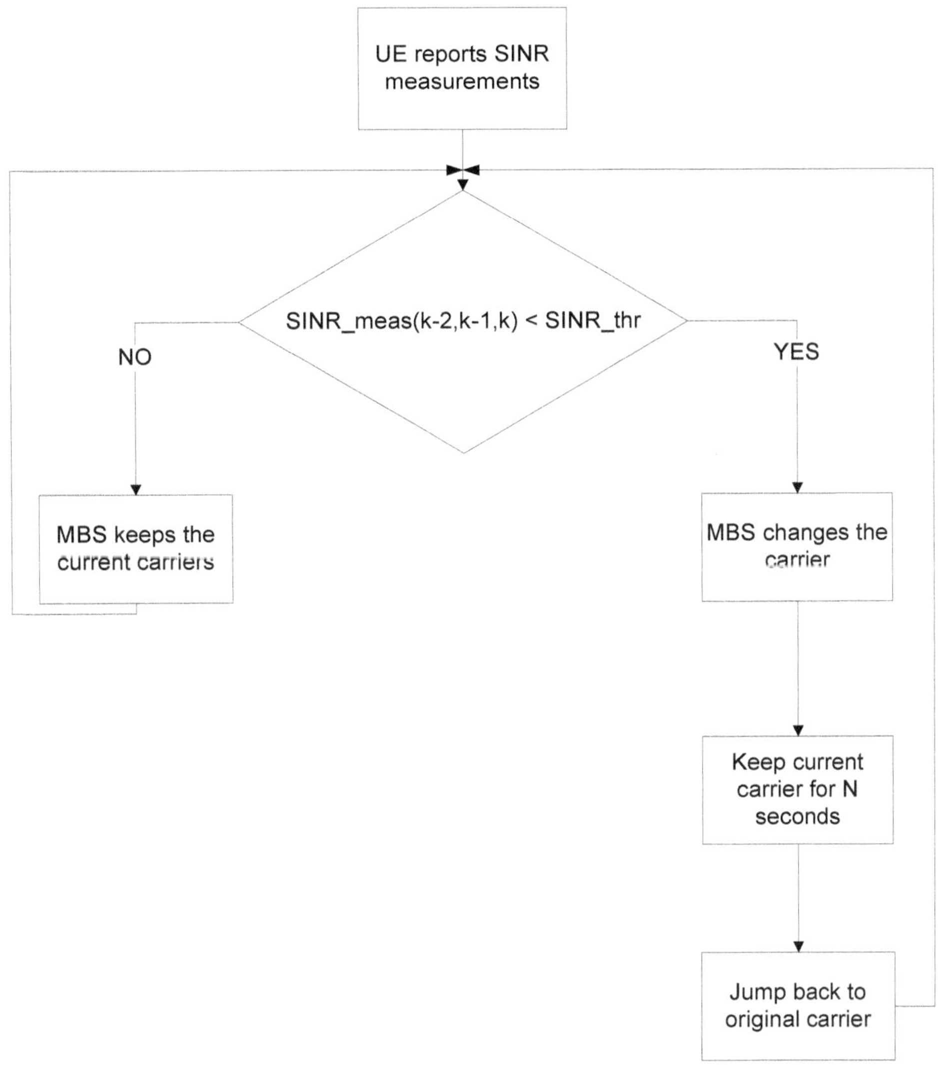

Figure 8. Carrier jumping algorithm 2.

In the following section of the paper we will investigate the performances obtained with the two algorithms compared to each other and with the case when no algorithm is implemented and the moving user is subject to high interference from the femtocellular layer.

5. SIMULATION RESULTS

Considering the two algorithms defined in the previous chapter, in this part of the paper, we will evaluate their performances in terms of parameters like the distance of the travelling user to the houses, the FBS deployment density, the transmit power of the FBSs, the speed of the travelling user and respectively the come-back time for the second algorithm. In order to see the impact of the variance of these parameters, first we will present in Figure 9 the general performances of the four deployment options: a purely macrocelullar network, a network with femtocells in which no interference avoidance techniques have been implemented, a network with femtocells for which Algorithm 1 was implemented and respectively, a network with femtocells for which Algorithm 2 was implemented.

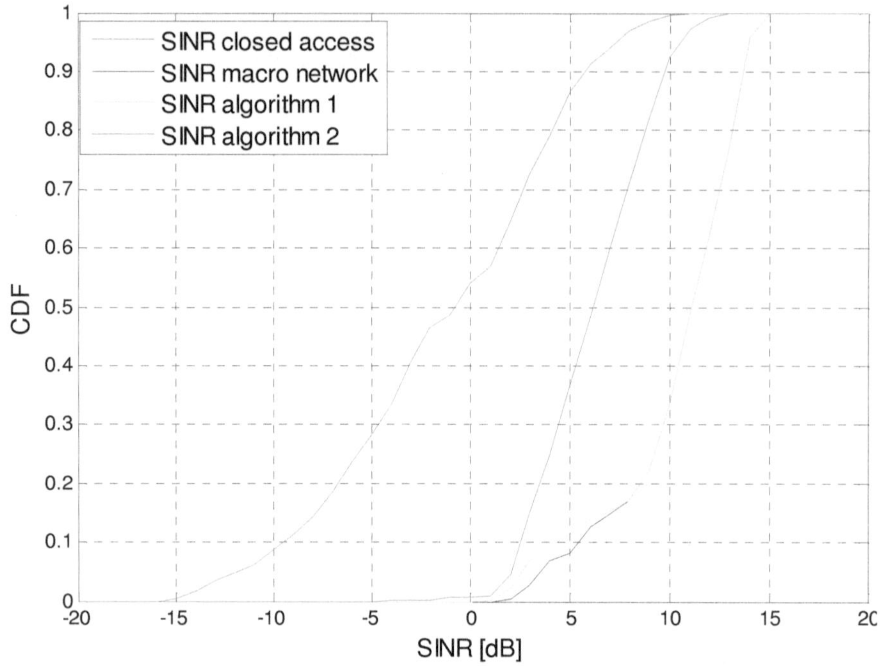

Figure 9. General performances of the deployment options.

One may see that the worst performances are obtained considering the closed access femtocell network with no interference avoidance techniques implemented. As expected, a purely macrocellular network will obtain the best performances in terms of reaching the lowest interference. Algorithm 1 obtains good performances, very close to the purely macrocellular network, because as soon as the network conditions imply a carrier modification, this is realized and the user is kept on that new carrier, limiting the interference.

Algorithm 2 doesn't realize such good performances, but the values obtained assure a proper QoS level for all the users, plus there is an enhanced spectrum efficiency. For this general view we considered a distance of 5 meters from the users travelling path to the houses, the transmit power for the FBSs is considered to be 250 mW and the travelling speed of the users is 1m/s. In the case of Algorithm 2 the come-back time is considered to be 5 seconds, which will be the default values for these parameters in the following simulation scenarios. For the results in Figure 9 the fentocell density considered is of 220 FBSs/km^2 , but for the next experiments we will consider a default FBS density of 440 FBSs/km^2.

Given that we have the average performances of the different deployment options, in the following part of the paper, we will observe the effect of each of the parameters previously enumerated upon the obtained SINR level for the cases when we have no interference avoidance methods, when Algorithm 1 is implemented and, respectively when Algorithm 2 is implemented. Therefore, Figure 10 illustrates the effect of the distance to houses on the SINR level obtained at the locations of a user which moves across the environment along a path similar to the one defined in Figure 6

. We have investigated the obtained SINR level for paths of shapes similar to the one in Figure 6, defined along the environment houses, but for distances to the houses of 1m, 5m and respectively 10m. In this case we consider the default values for all the other parameters previously enumerated. Respecting the general performance chart, the worst case scenario is when there are no interference avoidance methods implemented.

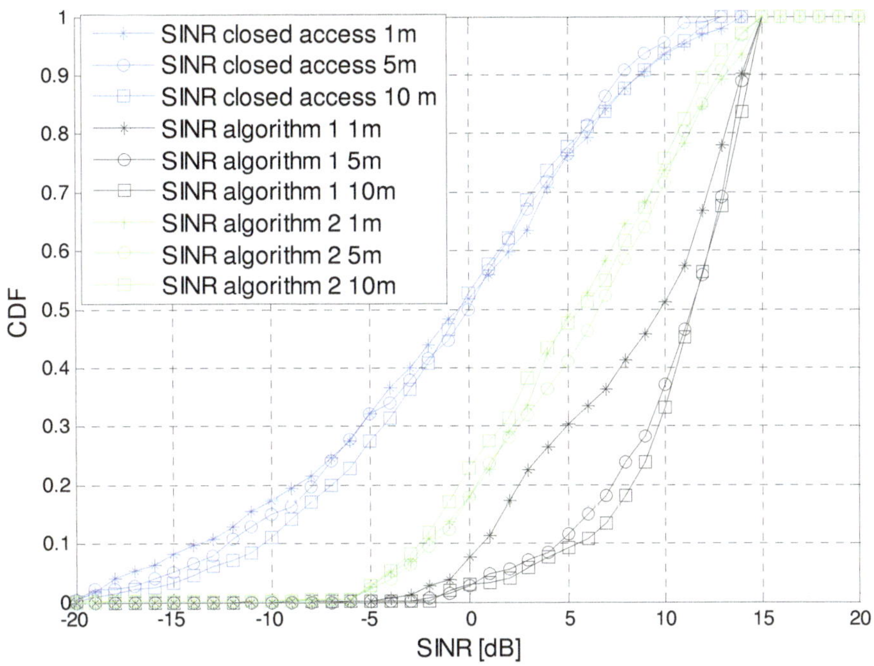

Figure 10. Effect of the distance to houses.

For this case, the closer the path is to the houses, the more powerful the negative impact is. This closeness is visible for the case of the Algorithm 1 implementation, when the distance to the houses affects the performances of the network. As soon as the path is shifted away from the houses, the travelling users obtain higher SINR values.

With the implementation of Algorithm 2, the position dependency of the obtained SINR level is significantly decreased, in this case there is no visible differentiation between the performances obtained for the three distances to the houses, which makes Algorithm 2 a viable solution irrespective of the position of the travelling user. Considering a minimum required SINR level of -6.5 dB [10], in order to assure the desired QoS level for the users, both proposed algorithms meet these requirements and can be successfully implemented in real life scenarios.

Next, we will observe the effect of the FBS deployment density upon the level of the obtained SINR, at the positions of the travelling users. In this sense, Figure 11 illustrates the obtained performances for four different FBSs deployment densities. The lowest density corresponds to the deployment of a FBS in every eight houses, while the highest density corresponds to the deployment of a FBS in every two houses. The rest of the parameters have the default values previously mentioned.

One may notice the high impact of the FBS deployment density upon the obtained level of the SINR, when there are no interference avoidance methods implemented. Basically there is a difference of about 10 dB in the SINR level between the lowest and respectively highest deployment densities. Even with the lowest density, the performances obtained are still below the threshold of -6.5 dB. An important fact that needs to be added here is that we consider that a network performs within the required target if at least 99% of the SINR values obtained at the user position are above the threshold.

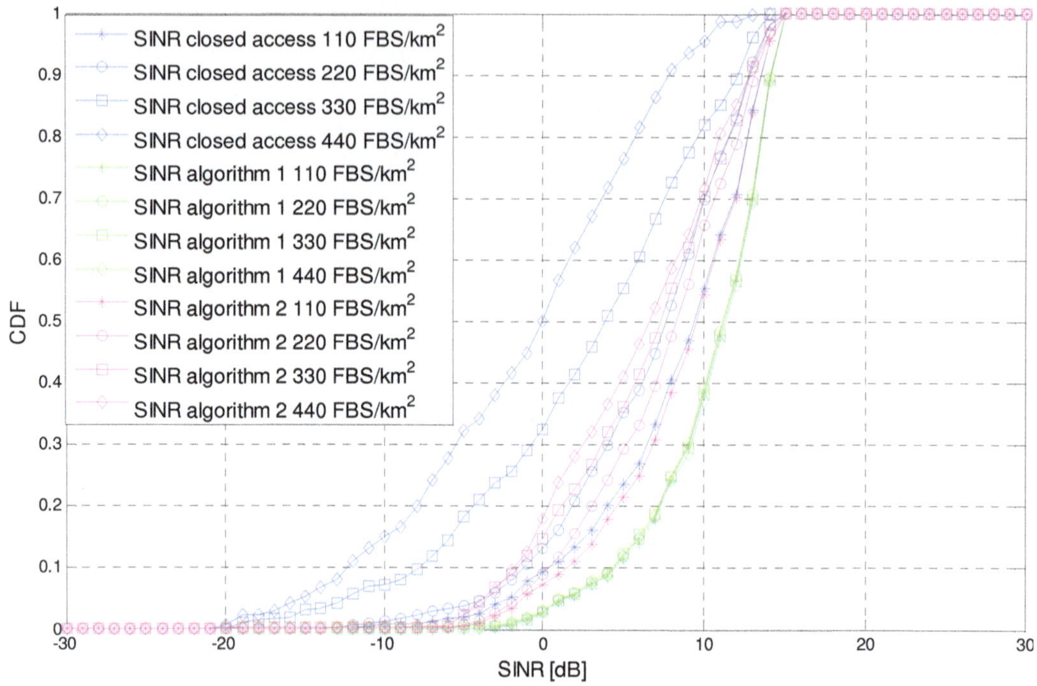

Figure 11. Effect of the FBS deployment density.

The femtocell density has little impact when we consider the implementation of Algorithm 1, simply because it commands the carrier jump, once the channel conditions fulfil the requirements and keeps the user on that carrier for the rest of the session. For the implementation of Algorithm 2, there is an effect of the FBS density, but the differences go down to about 5 dB. Even so, both algorithms meet the QoS requirements.

Another important parameter that influences the level of the obtained SINR level at the user position is the FBS transmit power. In our experiment we have considered the same transmit powers for all the FBS, throughout one simulation case. Therefore, we take into account three transmit power: 100 mW, 150 mW and respectively, 250 mW. These values are within the maximum allowed transmit powers for the residential femtocell, and offer a wide view of the phenomenon involved. Figure 12 presents the effect of the FBS transmit power upon the level of the obtained SINR.

Taking into consideration the trend of the previous results, the worst case scenario is for considering the closed access mode with no interference avoidance methods implemented. Within this set of results, using a FBS transmit power of 250 mW, introduces the highest interferences, and thus the SINR has the lowest values. The rest of the parameters mentioned above have the default values. The differences from one transmit power to another, within this set of results, are about 4 dB from worst to best case, proving once again that in terms of the influence of the FBS upon the macrocellular layer, the dominating factor is the FBS deployment density, rather than their transmit powers. Considering Algorithm 1, we obtain the best results, with little influence of the FBS transmit power, due to the carrier jumping and remaining on the new carrier, reducing thus the impact of the femtocellullar layer. Algorithm 2 manages to obtain suitable results for the SINR level, within the defined threshold of -6.5 dB. Using this latter algorithm the effect of the transmit power of the FBS is mitigated, thus making it suitable for any case scenario.

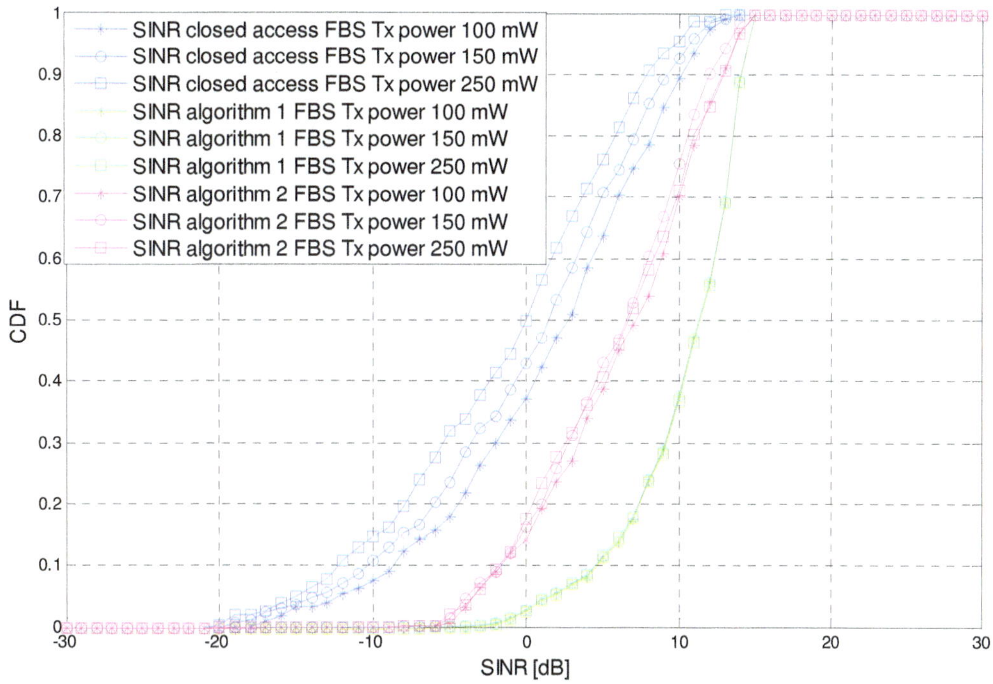

Figure 12. Effect of FBS transmit power.

Because Algorithm 2 manages to fulfil the SINR requirements and offers a better usage of the radio spectrum, in our opinion this is the most suitable option for implementation. Therefore, in the next part of the chapter we will concentrate on this algorithm and investigate how parameters like speed of the user and the come-back time influence its performances. That is why, Figure 13 illustrates the results obtained for the SINR level in terms of the travelling speed of the user. In this case, we consider the default values for all the other parameters.

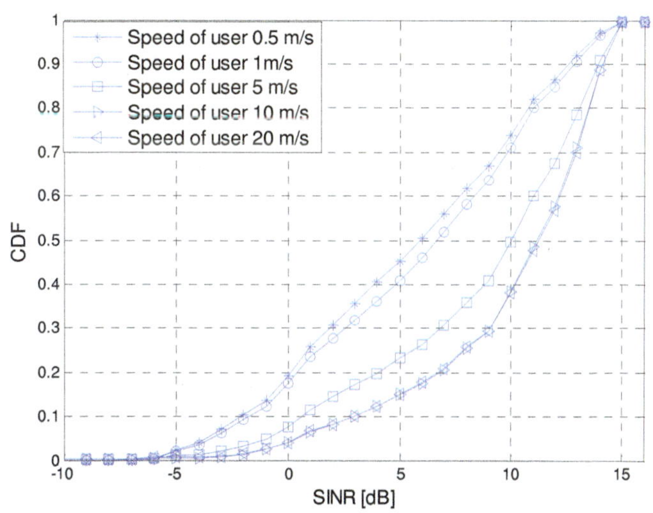

Figure 13. The effect of the user travelling speed.

One important fact that needs to be mentioned here is that for the simulation cases considered and presented in Figure 13, the SINR conditions are fulfilled, considering the same threshold of -6.5 dB. The lowest values are obtained for the case when we consider the travelling speed of 0.5 m/s, simply because the user moves slower and it is subject to interferences for a longer time period. If the travelling speed is increased, in that same amount of time of 5 seconds in which the user is on the second carrier it covers a longer distance, therefore for the same path distance there will be less jumps and less susceptibility to interferences. The best performances are obtained for high speeds, when the user travels its path while being mainly on the second carrier.

Next, we will observe the effect of the come-back time for different values of the travelling speeds of the users. Figure 14 illustrates the effect of the come-back time for four different returning periods: 1 s, 2 s, 5 s and respectively 10 s. One may observe that as the returning period after which there is a jump back to the original carrier, increases so does the levels of the obtained SINR.

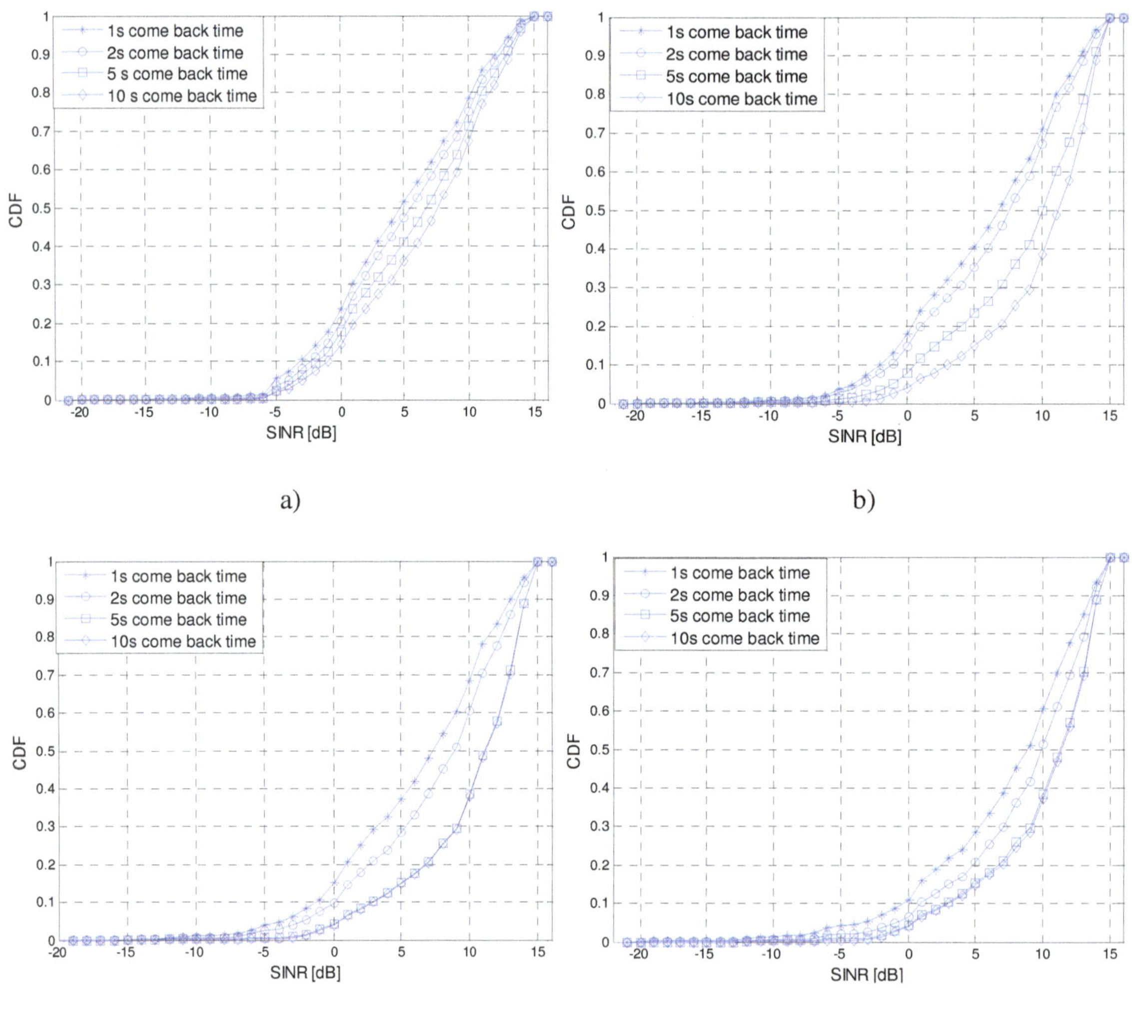

Figure 14. The effect of the come-back time to the original carrier for different travelling speeds: a) 1 m/s; b) 5m/s; c) 10 m/s; d) 20 m/s.

This is due to the fact that if the jump back happens after a shorter period of time, the travelling user will be spending more time on the interference susceptible carrier, thus having lower SINR values. Considering the trend defined by Figure 13, as the speed and come-back time increase, so do the SINR levels. Therefore, the highest performances are obtained for the highest travelling speed of 20 m/s and the highest come-back time of 10 seconds.

6. Conclusions

The paper managed to investigate in the first part some small cells coverage aspects, in terms of the different parameters that concur to the level of the received signal at the user position. Then, considering a residential environment, we proposed two algorithms that aim at mitigating cross-layer interference, from the femtocell base stations to the macrocellular users that are passing by. We analyzed the performances of the proposed algorithms in terms of the obtained SINR level along the path of the travelling user.

The results show a significant improvement of the SINR level when using either one of the proposed algorithms. Because the first algorithm acts based on a single jump per travelling user, once the channel conditions indicate a jump is suitable and remains on the second carrier, from the spectrum efficiency point of view, this not the best option. In this case, the resulting deployment will be similar to a dedicated carrier approach, which is not desirable.

On the other hand, considering the second algorithm, which imposes a return to the original carrier after a predefined period of time, the obtained SINR level is above the defined threshold value, for more than 99% of the time, fulfilling thus the QoS requirements imposed by the operator. It offers a higher spectrum efficiency, in this case the second carrier can be viewed as a safety band on which travelling users can be transferred to in order to avoid the high interference. In this way, the operator can assign fewer resources to this second carrier, redirecting most of them to the first carrier.

The proposed solutions offer important enhancements, but still have the disadvantage that they can be applied only for operators that have at least two carriers available for usage. This could be an important option for operators, considering the future benefits they will obtain.

Acknowledgements

This paper was supported by the project "Doctoral studies in engineering sciences for developing the knowledge based society-SIDOC" contract no. POSDRU/88/1.5/S/60078, project co-funded from European Social Fund through Sectorial Operational Program Human Resources 2007-2013.

References

[1] E. Fox, North American ARPU Growth Outpaces the World: A Look at Wireless Forecast Drivers. Yankee Group, March 2006.

[2] H. Claussen, “Co-Channel Operation of Macro- and Femtocells in a Hierarchical Cell Structure”, International Journal of Wireless Information Networks, Volume 15, Numbers 3-4, pp 137-147.

[3] H. Claussen, L. T. W. Ho, and L. G. Samuel, “Financial Analysis of a Pico-cellular Home Network Deployment”, Procedings of IEEE International Conference on Communications (ICC) 2007, 24–28 June.

[4] Jie Zhang and Guillaume de la Roche, “Femtocells, Technologies and Deployments”, Wiley, 2010.

[5] J. Cullen, ‘Radioframe presentation,’ in Femtocell Europe 2008 , London, UK, June 2008.

[6] TS 43.318, "Generic Access Network (GAN) Stage 2", Rel-5.

[7] TS 44.318, "Generic Access Network (GAN); Mobile GAN Interface Layer 3 Specification", Rel-5.

[8] TS 25.410, "UTRAN Iu Interface: General Aspects and Principles", Rel-5.

[9] TS 25.450, "UTRAN IuPC Interface General Aspects and Principles", Rel-5.

[10] TS 25.419, "UTRAN IuBC Interface: Service Area Broadcast Protocol (SABP)", Rel-5.

[11] Alcatel-Lucent, "The importance of Small Cells in Wireless Networks", Fierce Wireless, April 2011.

[12] 3GPP TR 36.814, "Evolved Universal Terrestrial Radio Access (E-UTRA); Further advancements for E-UTRA physical layer aspects" Version 9, 2010.

[13] C. Androne, T. Palade and E. Puschita, "Open Loop Sensor based System used for Mitigation of Cross-tier Interference in Femtocell Networks ", TELFOR 2010, Belgrade, Serbia.

[14] M. Z. Chowdhury, Y. M. Jang and Z.J. Haas, "Network Evolution and QoS Provisioning for Integrated Fentocell/Macrocell Networks", International Journal of Wireless & Mobile Networks, Vol.2, No.3, August 2010.

[15] C. Androne and T. Palade, "Characterization Of Small Cells Networks Deployment Options And Their Impact Upon Macro-Cellular Networks", Acta Technica Napocensis, Vol.53, No.1, pp. 15-20, March 2012.

[16] Mikko Jarvinen, "Femtocell Deployment in 3rd Generation Networks", Master of Science in Technology thesis, Helsinki University of Technology.

[17]]H. Claussen "Efficient modelling of channel maps with correlated shadow fading in mobile radio systems," Procedings of IEEE Personal, Indoor and Mobile Radio Communication (PIMRC), 2005.

12

AN ENERGY SAVING ALGORITHM TO PROLONG THE LIFETIME OF WIRELESS SENSOR NETWORK

Monika Raghatate[1] and Prof. Dipak W. Wajgi[2]

[1]Department of Computer Engineering, St. Vincent Pallotti College of Engineering and Technology, Nagpur, India
[2]Department of Computer Engineering, St. Vincent Pallotti College of Engineering and Technology, Nagpur, India

ABSTRACT

This paper considers a heterogeneous network of energy constrained sensors deployed over a region. Each Normal sensor node in a network is systematically gathering and transmitting sensed data to the cluster-head, and then cluster head sending data to a base station (via intermediate cluster- heads). This paper focuses on reducing the energy consumption and hence improving lifetime of wireless sensor Networks. Clustering sensor node is an effective topology for the energy constrained networks. So energy saving algorithm has been developed in which clusters are formed considering a subset of high energy nodes as a cluster-head and another subset of powerful nodes is ask to go to sleep. When Cluster heads deplete their energy another subset of nodes becomes active and acts as a cluster head. Proposed approach is implemented in MATLAB, Simulation results shows that it can prolong the network lifetime than LEACH protocol, and achieves better performance than the existing clustering algorithms such as LEACH.

KEYWORDS

Cluster Head, Energy Saving Algorithm, MATLAB, Network Lifetime, Wireless Sensor Network.

1. INTRODUCTION

A wireless sensor networks (WSN) is a network of small devices called as sensor nodes deployed over a geographical area for monitoring physical phenomena like seismic events, vibrations, temperature, humidity and so on [1]. Typically, a sensor node is a small/tiny device which has three basic components: sensing subsystem, processing subsystem and wireless communication subsystem. A sensing subsystem for sensing data from the physical surrounding environment, a processing subsystem for processing data and storage, and a wireless communication subsystem for data transmission. It also has a power source often consists of a battery which has limited energy budget and it is inconvenient or impractical to recharge the battery or nodes may be deployed in an environment where human being cannot reach. For example, in underwater scenario, there is no plugin socket to provide power as per requirement.

Military applications motivated the development of wireless sensor network. However, Now in many application areas including environment, Hospitals, agriculture, habitat monitoring etc. wireless sensor networks are used, due to various limitations arising from their inexpensive nature, weight, limited size, and ad hoc method of deployment; To prolong the lifetime of the network it is necessary to find energy efficient route setup and reliable transmission of data from the sensor nodes to the base station or sink [2]. To improve the performance of the sensor network application, it is necessary to improve the lifetime of network and it can be possible by saving energy of nodes. In some cases it is possible to gather energy from the external environment [3].

However, external power supply sources often show an interrupted behavior so that an energy buffer is needed as well. Energy is a very limited resource and must be used properly. Therefore, energy conservation is a fundamental issue in the design of systems based on wireless sensor networks.

The remainder of the paper is organized as follows. In Section II, review of literature is discussed. Section III describes Mathematical Model. Section IV describes Proposed Approach. In Section V simulation Results are discussed. Finally, Section VI gives concluding remarks and future scope.

2. RELATED WORK

2.1 Leach

Low Energy Adaptive Clustering Hierarchy (LEACH) protocol [4] has proposed by W. R. Heinzelman, A. P. Chandrakasan and H. Balakrishnan. It is one of the most popular hierarchical routing algorithms for sensor networks. The idea is to form clusters of the sensor nodes based on the received signal strength and use local cluster heads as routers to the sink. This reduces energy consumption since the transmissions will only be done by such cluster heads rather than all sensor nodes. Minimum number of cluster heads (CHs) is estimated to be 5% of the total number of nodes. All the data processing such as data fusion and aggregation are local to the cluster. Cluster heads change randomly over time in order to balance the energy depletion of nodes. The working of LEACH is divided into two phases: Set up Phase and steady state phase. Each round starts with a set-up (clustering) phase, where clusters are formed, followed by a steady-state (transmission) phase in which data packets are transferred from normal nodes to cluster heads. After data aggregation, cluster heads will transmit the aggregated data to the Base Station.

2.1.1 Disadvantages

Homogeneous distribution of sensor nodes is assumed.
LEACH is not applicable in large regions.
Distribution of the CH nodes in the network is not uniformed.

2.2 PEGASIS Protocol

S. Lindsey and C. Raghavendra [5] have proposed Power Efficient Gathering in Sensor Information Systems (PEGASIS) protocol. It is an improved version of LEACH. Rather than forming clusters, it forms chains of sensor nodes. Single node is responsible for transferring the aggregated data to the base station. Each node aggregates the collected data with its own data, and then passes the aggregated data to the next ring. PEGASIS is different from LEACH because it employ multi hop transmission and select only one node to transmit the data to the sink or base station. Since it eliminates the overhead caused by dynamic cluster formation, multi hop transmission and data aggregation is employed; PEGASIS achieves better performance than LEACH. However for distant nodes excessive delay is introduced, especially for large networks and single leader can be a bottleneck.

2.3 TEEN Protocol

Threshold sensitive Energy Efficient sensor Network Protocol (TEEN) [6] has proposed by A. Manjeshwar and D. P. Agarwal. In this protocol clusters are formed considering the nearer nodes, with a cluster heads to transmit the collected data to one upper layer. After forming the clusters, cluster heads broadcast two threshold values. First one is hard threshold; it is minimum possible value of an attribute to trigger a sensor node. Hard threshold allows nodes transmit the event, if the event occurs in the range of interest. Therefore a significant reduction of the transmission delay occurs. Unless a change of minimum soft threshold occurs, the nodes don't send a new data packet. Employing soft threshold prevents from the redundant data transmission. Since the protocol is to be responsive to the sudden changes in the sensed attribute, it is suitable for time-critical applications. They have also proposed Adaptive Threshold sensitive Energy Efficient sensor Network Protocol (APTEEN) protocol [7]. The protocol is an extension of TEEN aiming to capture both time-critical events and periodic data collections. The network architecture is same as TEEN. After forming clusters the cluster heads broadcast attributes, the threshold values, and the transmission schedule to all nodes. Cluster heads are also responsible for data aggregation in order to decrease the size data transmitted so energy consumed. According to energy dissipation and network lifetime, TEEN gives better performance than LEACH and APTEEN because of the decreased number of transmissions. The main drawbacks of TEEN and APTEEN are overhead and complexity of forming clusters in multiple levels, implementing threshold-based functions and dealing with attribute based naming of queries.

2.4 SEP Protocol

Stable Election Protocol (SEP) protocol [8] has proposed by G. Smaragdakis, I. Matta and A. Bestavros. This protocol is an extension to the LEACH protocol. It is a heterogeneous aware protocol which is based on weighted election probabilities of each node to become cluster head according to their respective energy. SEP ensures that the cluster head election is randomly selected and distributed based on the fraction of energy of each node considering a uniform use of the nodes energy. In SEP, two types of nodes i.e. two tier in-clustering and two level hierarchies were considered.

2.5 DEEC Protocol

Distributed Energy Efficient Clustering Protocol (DEEC) [9] has proposed by Q. Li, Z. Qingxin and W. Mingwen. This protocol is a cluster based scheme for two level and multi level energy heterogeneous wireless sensor networks. This scheme, selects cluster heads using the probability based on the ratio between residual energy of each node and the average energy of the network. The epochs are different according to their initial and residual energy of being cluster-heads for nodes. The chance of the becoming cluster heads depends on nodes initial and residual energy, the nodes with high initial and residual energy have more chances than the nodes with low energy.

2.6 HEED Protocol

Hybrid Energy Efficient Distributed clustering Protocol (HEED) [10] has proposed by O. Younis and S. Fahmy. The basic scheme of LEACH is extended by HEED using residual energy as primary parameter and network topology features for e.g. node degree, distances to neighbours etc. are used as a secondary parameters to break tie between candidate cluster heads, as a metric for cluster selection to achieve power balancing[13]. The process of clustering is divided into a number of iterations, and in each iterations, nodes which are not covered by any cluster head shows their probability of becoming a cluster head. Since these energy-efficient clustering protocols enable every node to independently and probabilistically decide on its role in the clustered network, they cannot guarantee minimum elected set of cluster heads [12, 13].

2.7 EEICCP Protocol

Shalli Rani et.al has proposed Energy Efficient inter cluster coordination Protocol [11] for the wireless sensor networks, It is a cluster based algorithm in which different levels of clusters are considered on the basis of received signal strength to recognize the distance of the clusters from the BS (base station) and to determine the number of cluster coordinators to make routes for the CHs to transmit the data. The protocol depends upon the fact that some cluster head sends data directly to the base station (Single hop) and the some send by multi hop transmission, but energy consumption in conventional protocols increases due to multi path fading channel which affects the network life time. An attempt has been made to reduce this power loss in to free space model that is d2. The arrangements of the nodes has been done in this way that one cluster always, is very close to the base station *i.e.* in line of sight propagation and that cluster will have head nodes sufficient for all the below clusters which will forward the data of all those clusters. Layers of clusters have been so there is always one cluster coordinator for every lowest cluster. The division of clusters is done from top to bottom.

EEICCP protocol works by starting the election phase in which the cluster heads are elected according to the distance based on RSS. Number of clusters is fixed so as the cluster heads and the cluster coordinators after election of cluster coordinators by the CHs, a cluster id is assigned to each cluster head and the cluster coordinator. This id is transmitted by each cluster to their nodes by the advertisement message. And cluster co-coordinators also pass their own ids. After that the transmission phase begins in which data is transferred to the cluster head and that data is passed to the base station with the help of CCOs. In first round the data is collected by the CH of that cluster which has data to send, then in the other iterations the data is passed to the base station with the help of cluster co-coordinators.

EEICCP outperforms conventional protocols that send data directly to the BS through their respective CHs. Dividing the network into layers of clusters has been proved to be a good approach in reducing the energy to a great extent. Each node has the equal responsibility of receiving data from all other nodes in the cluster and to transmit the aggregating signal to the base station. Simulations show that EEICCP reduces start energy 151 times than both HCR and LEACH, energy of one iteration by 0.2926×10^4 times than HCR and 0.3169×10^4 times than LEACH. Thus total energy reduction is 43% than HCR and 50% than LEACH.

3. Mathematical Model

Many of the research protocols have used the first order radio model as described in [5]. Energy is dissipated while transmitting and receiving the data and energy consumption for the short distance is *d*2 when propagation is in line of sight and *d*4 for the long distance due to multipath fading propagation [1].

Transmission energy and receiving energy are calculated as follows:

$$E_{Tx}(l,d) = \begin{cases} lE_{elect} + lE_{fs}d^2, & d < do \\ lE_{elect} + lE_{mp}d^4, & d > do \end{cases}$$

$$E_{Rx}(L) = E_{elect}L$$

Where d is a difference between transmitter and receiver. And L is a the length of the message in bits,

$$d_0 = \sqrt{E_{fs}/E_{mp}}$$

In proposed approach, two types of sensor nodes, i.e., the High Energy nodes (powerful nodes) and normal nodes are considered. Let us assume that there are 'N' numbers of sensor nodes deployed in an area, where the initial energy of the normal nodes is assumed to be E_0, and m is the percentage of powerful nodes, which own a times more energy than the normal ones. Thus $(m * N)$ powerful nodes are equipped with initial energy of (Eo *(1+α)), and ((1-m) *N) normal nodes are equipped with initial energy of $E0$. The total initial energy of the network is given by:

$$\begin{aligned} Etotal &= N \cdot (1 - m) \cdot Eo + N \cdot m \cdot Eo \cdot (1 + \alpha) \\ &= N \cdot Eo \cdot (1 + \alpha \cdot m) \end{aligned}$$

So, by using heterogeneous network the total energy of the system is increased by a factor of (1+α ·m).

4. PROPOSED APPROACH

According to the proposed approach a heterogeneous network with the sensor nodes having different energy levels and processing power is assumed. Some high Energy nodes (Powerful nodes/Higher nodes) having a times more energy than the normal nodes are deployed in the network. A Subset of Powerful node from the set of higher nodes is selected as a CH based on the fact that they should not be in a communication range of each other. Another subset of higher nodes is asked to go to sleep. The subset of Powerful node which is chosen as a CH broadcast its presence to all normal sensor nodes and starts receiving data from the sensors which are members of their cluster. Normal sensor decides to which cluster they want to belong based on the signal from the broadcast. It is assumed that the stronger the signal, the closer the head is and therefore the head with strongest signal are chosen. All the cluster members send the sensed data to the CH. The CH will send the aggregated data to the Base Station directly or by using some intermediate CH.

When the energy of the CH reaches to zero, another subset of nodes would act as a Cluster Head. This information about the new CH will be sent to all normal nodes and the same process is repeated. The network will be stable or dead when the energy of all cluster heads becomes zero.

4.1 Algorithm

The algorithm is divided into two phases

4.1.1 Selection Phase

Step1: Select a Subset of high energy node as cluster head.
Step2: Cluster Head broadcast a hello message to all normal sensor nodes and starts receiving data from the sensors that have decided to become a cluster member.
Step3: If the Cluster Head energy is depleted completely or reaches to the threshold value i.e. zero, then high energy nodes of another subset will act as Cluster Head & Repeat step 2.
In selection phase a subset of high energy node is selected as a cluster head and other subset of node is asked to go to sleep. The cluster head sends an advertisement message to the normal sensor nodes and the nodes send acknowledge message to the cluster head. Normal sensors decide

to which cluster they wish to belong based on the distance of node from cluster head. Cluster head collects data from all the members of their cluster and send it to base station. When the energy of cluster head becomes zero then other subset of high energy node will acts as a cluster head as same process is repeated.

4.1.2 Data Transfer Phase

Step1: The cluster members send the data to the cluster head.
Step2: Cluster Head will collect the data from all the members of their cluster.
Step3: Cluster Head will transmit the aggregated data to base station through the cluster heads nearer to it.

Transmission begins from the depth first CH which goes to the Cluster head of the next higher level cluster and so on as shown in Figure 1 which is finally sent to the BS by the CH of the last cluster nearest to the BS. In this way the distance travelled is decreased, and energy is saved because more energy is require to send data to the base station ,if it is located far away from it.

4.2 Algorithm for Data Transmission

1. Set k=0, n=1
2. Set j= number of clusters – k
3. Repeat while j>=1
4. If j! =1 then
Cluster [j-1][n]=cluster[j][n]
Else
Base station=cluster[j][n]
End of if structure
5. [Decrement] set j= j-1
Go to step 2
6. End of Step 3 loop
7. Exit

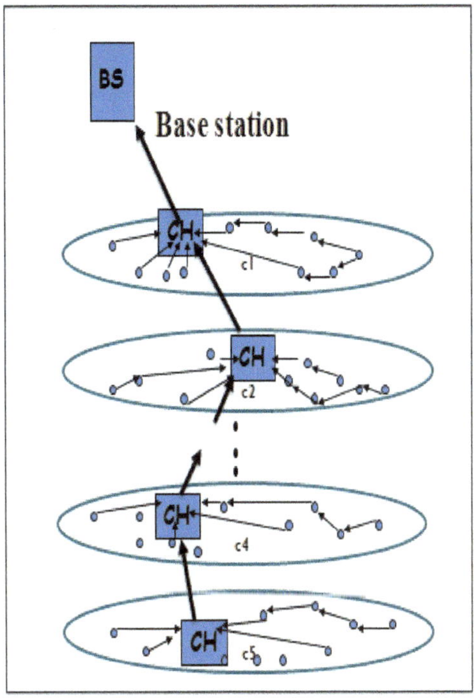

Figure 1. Transfers of Data

5. Simulation Results And Discussion

Based on the analytical work described in Section 4 extensive simulations are carried out in MATLAB. Heterogeneous sensor network of 50 sensor nodes is considered. 44 normal nodes are randomly distributed whereas 6 powerful nodes are deployed deterministically in the 100m*100m area. The base station is located at (50, 130). Table 1. Shows The Simulation parameter.

Table 1. Simulation Parameters

Parameters	Values
Sink	At (50,130)
Cluster Radius	25 m
E_{elect} (Energy consumed in the electronics circuit to transmit or receive the Signal)	50 nJ/bit
E_{fs} (Energy consumed by the amplifier to transmit at a short distance)	10 pJ/bit/m2
E_{mp} (Energy consumed by the amplifier to transmit at a longer distance)	0.0013 pJ/bit/m4
EDA (Data Aggregation Energy)	5 nJ/bit/signal
Message Size	4000 bits
E0, Initial Energy	0.5 J

Figure 2 shows the deployment of nodes in the network, Blue colour nodes are normal nodes which are deployed randomly and the nodes shown in red colour are high energetic nodes which are deployed by using deterministic approach. The network of N nodes is considered from which m percent of nodes are high energetic nodes and remaining nodes are normal nodes, Therefore the high energetic nodes would be m*N and normal nodes would be (1-m)*N. For example if we consider the network of 50 nodes i.e. N= 50 and 12% of high energetic nodes i.e. m=0.12 then normal nodes would be (1-0.12)*50=44 and high energetic nodes would be 0.12*50=6.

Base station is located far away from the nodes, as shown in Figure 2. It is located at (50,130). All cluster heads handover their aggregated data to base station for processing and decision making therefore deployment of base station is very important in wireless sensor network

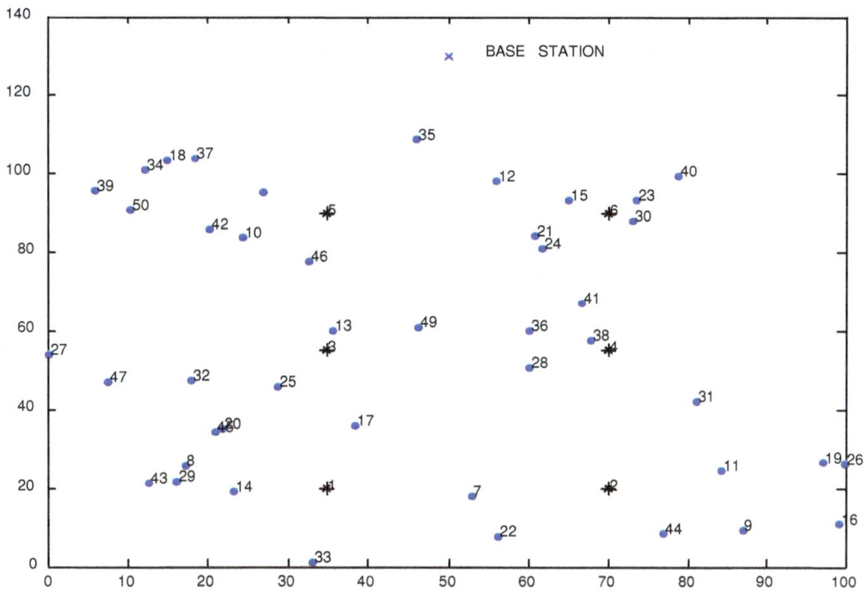

Figure 2. Random Deployment of Normal Nodes

Figure 3 Shows the Cluster Formation considering the subset of high Energy Nodes as a Cluster Head. From the high energy nodes a subset of high energy node is considered as a cluster head e.g. node 1, 4, 5 as shown in figure and other subset of node will go to sleep for some duration.

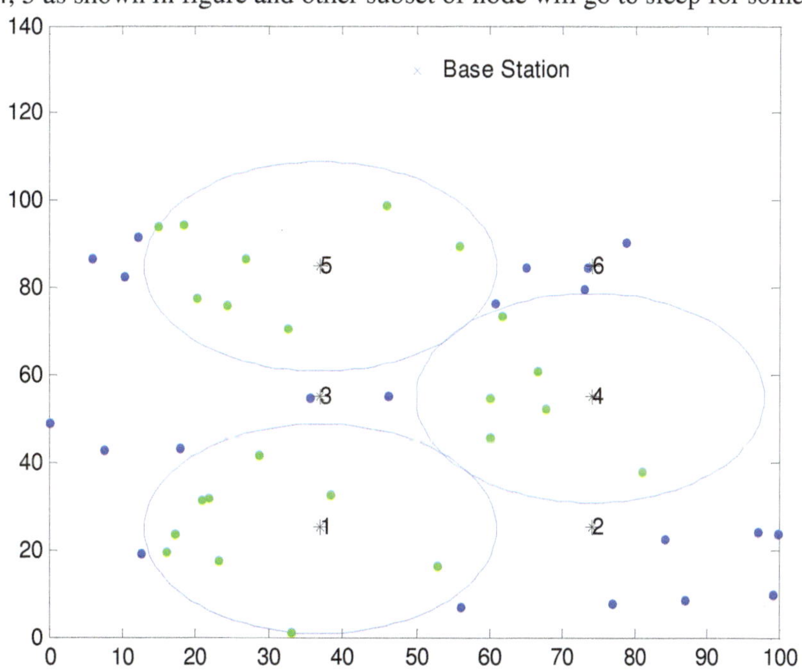

Figure 3. Cluster Formation Considering the First Subset of Nodes as a cluster Head

These cluster head will decide their communication range and form clusters by considering the normal nodes as a cluster members which are in the communication range of that cluster head. Initially round starts from cluster head 1, it will collect data from the members of their cluster and send it to cluster head which is nearer to it i.e. cluster head 4, then cluster head 4 will transfer it to cluster head 5 and finally cluster head 5 will send it to base station.

Similarly second cluster head i.e. cluster head 4 collect data from all cluster member and send it to base station, then in next round cluster head 5 will send gathered data to base station directly. The same process is repeated for number of rounds.

As shown in figure 4, when the energy of cluster heads becomes zero a other subset of higher nodes becomes active and will act as a cluster head and the same process is repeated till the energy of all cluster head becomes zero.

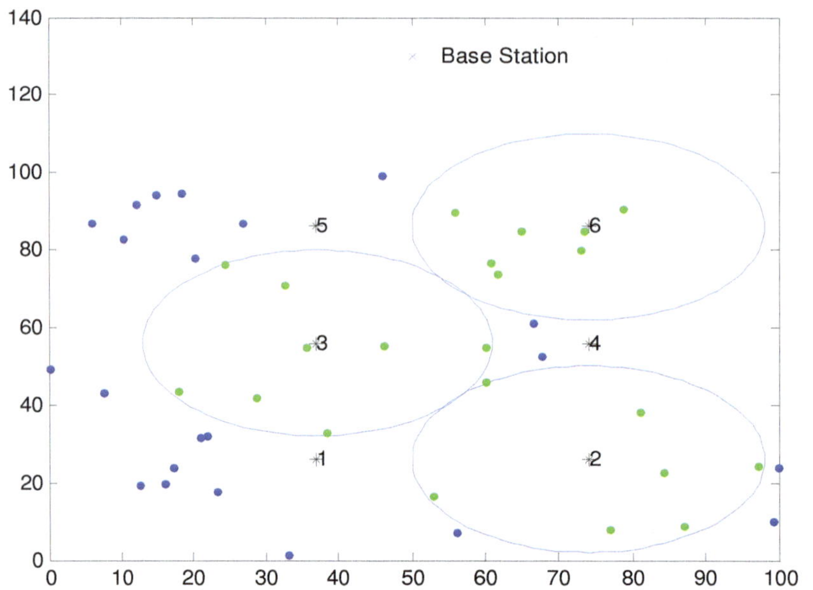

Figure 4. Cluster Formation Considering the Second Subset of Nodes as a cluster Head

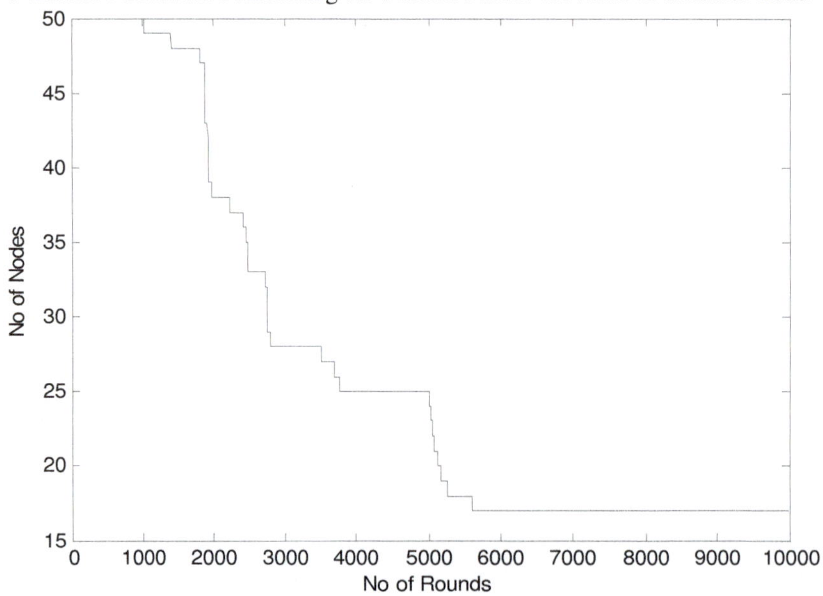

Figure 5. Performance graph of Proposed Approach, showing the number of nodes on y-axis and the rounds on the x-axis

Figure 5. illustrates the detailed view of the behavior of proposed Approach; Graph shows the number of alive nodes per round where number of rounds are shown on x axis and no of nodes on y axis . In this case 10000 rounds and 50 nodes are considered. It is observed that 50 nodes remains in network for 1000 rounds and the death of first node occurs after 1000 rounds.
As the numbers of rounds are increases the number of alive nodes decreases and at round 5806 network become dead because energy of all cluster head becomes zero. Here the cluster heads which are nearer to base station depleted their energy faster than other cluster heads because along with their own data it has to send data of other cluster heads.

Figure 6. illustrates a detailed view of the behaviour of proposed approach and LEACH protocol; it shows the number of alive nodes per round. The number of nodes die in Proposed Approach is much less than LEACH Protocol over the same number of rounds.

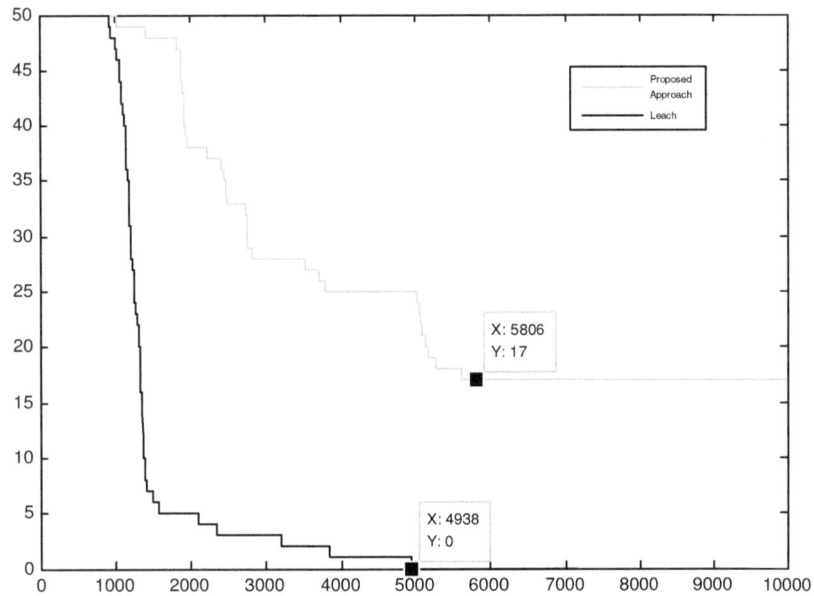

Figure 6. Comparative graphs of Proposed Approach and LEACH Protocol

In proposed Approach death of first node occurs much later than the death of first node in case of LEACH Protocol, and the network works for more number of rounds i.e. 5806 whereas in LEACII it works for only 4938 rounds. From this it can be say that proposed approach prolongs lifetime and shows better performance than LEACH protocol.

6. CONCLUSION

In this paper, a new approach for energy saving and hence prolonging network lifetime in the wireless sensor network has proposed. Algorithms for cluster head selection and data Transmission in wireless sensor network are proposed and implemented. From the simulation results we can say that proposed approach increase network overall energy, saves energy, prolongs lifetime and shows higher performance than LEACH protocol. But the drawback of proposed approach is that when all cluster head dies the network become stable (dead), therefore all sensor nodes are not utililized. Due to this the network does not work till the death of last node.

In future, we planned to design a new approach which will overcome this drawback, and all sensor nodes would be utilized & network will work until the death of last node.

REFERENCES

[1] M. Ye, C. Li, G. Chen, J. Wu., (2005) "EECS: an energy efficient cluster scheme in wireless sensor networks", In IEEE International Workshop on Strategies for Energy Efficiency in Ad Hoc and Sensor Networks (IEEE IWSEEASN-2005), Phoenix, Arizona, April 7–9. .

[2] Sajid Hussain et al., (2005) "Hierarchical cluster-based routing in wireless sensor network", ,jodrey school of computer science acadia university wolfville, pp. 1-33

[3] Zahra Rezaei, Shima Mobininejad, (2012) "Energy Saving in Wireless Sensor Networks", International Journal of Computer Science & Engineering Survey (IJCSES) Vol.3, No.1.

[4] W. Heinzelman, A. Chandrakasan and H. Balakrishnan, (2000) "Energy-Efficient Communication Protocol for Wireless Microsensor NetworkS", Proceedings of the 33rd Hawaii International Conference on System Sciences (HICSS '00).

[5] S. Lindsey, C. Raghavendra, (2002) "PEGASIS: Power- Efficient Gathering in Sensor Information Systems", IEEE Aerospace Conference Proceedings, Vol. 3, 9-16 pp. 1125- 1130

[6] A. Manjeshwar and D. P. Agarwal, (2001) "TEEN: a routing protocol for enhanced efficiency in wireless sensor networks", In 1st International Workshop on Parallel and Distributed Computing Issues in Wireless Networks and Mobile Computing

[7] A. Manjeshwar and D. P. Agarwal, (2002) "APTEEN: A hybrid protocol for efficient routing and comprehensive information retrieval in wireless sensor networks", Parallel and Distributed Processing Symposium, Proceedings International, IPDPS, pp. 195-202

[8] G. Smaragdakis, I. Matta, A. Bestavros, (2004) "SEP: A Stable Election Protocol for clustered heterogeneous wireless sensor networks", In Second Internatiolnal Workshop on Sensor and Actor Network Protocols and Applications (SANPA).

[9] Q. Li, Z. Qingxin, and W. Mingwen, (2006) "Design of a distributed energy efficient clustering algorithm for heterogeneous wireless sensor networks", Computer Communications, vol. 29, pp. 2230-7.

[10] Ossama Younis and Sonia Fahmy, (2004) "Distributed Clustering in Ad-hoc Sensor Networks: A Hybrid, Energy-Efficient Approach", In Proceedings of IEEE INFOCOM, Hong Kong, an extended version appeared in IEEE Transactions on Mobile Computing, 3(4).

[11] Shalli Rani et al, (2013) "EEICCP-Energy efficient protocol for wireless sensor network", Wireless sensor network, 5,127-136, scientific research.

[12] Amar shah, Kirti Patel, Prof. Pinal Patel, "A Survey of the Optimization of clustering techniques in Wireless sensor network", International journal of engineering development and research ISSN: 2321-9939

[13] S. Thiripura Sundari, P. Maheswara Venkatesh, Dr. A. Sivanantharaja, (2013) "A Novel Protocol for Energy Efficient Clustering for Heterogeneous Wireless Sensor Networks", International Journal of Scientific & Engineering Research, Volume 4, Issue 8, ISSN 2229-5518

13

OVERBOOKING IN MOBILE BACKHAUL

Chunming Liu, Bryan Fleming

T-Mobile, Bellevue, USA

{Chunming.Liu, Bryan.Fleming}@t-mobile.com

ABSTRACT

The rapid increase of mobile data subscribers and deployment of broadband radio access networks push mobile operators to upgrade their mobile backhaul from expensive legacy TDM transport to carrier Ethernet for cheaper operational expenditures. The Metro Ethernet Forum has defined a set of Ethernet Virtual Connection services that are adopted to provide scalable Ethernet transport for mobile backhaul. However, these services usually address single cell site backhaul per UNI handoff, neither considering statistical multiplexing gain at a hub site which aggregates backhaul traffics of multiple cell sites, nor supporting Quality-of-Service provisioning in overbooked backhaul pipe. This paper proposed an efficient carrier Ethernet overbooking solution for mobile backhaul. A statistical estimation method has been developed for deriving a safe overbooking factor at a given UNI. Then two efficient transport architectures were proposed to support bandwidth sharing in cellular cluster with overbooked backhaul bandwidth in carrier Ethernet. A novel bandwidth control algorithm has been derived at customer edge device to provide Quality-of-Service for multimedia traffics over overbooked UNI, with SLA policing and protection on high priority services. Experimental data analysis and simulations have showed that our new schemes can benefit mobile operators in resource utilization efficiency, carrier Ethernet cost saving and backhaul performance.

KEYWORDS

Mobile backhaul, overbooking, Carrier Ethernet, UNI handoff, CIR, SLA, Quality of Service

1. INTRODUCTION

In recent years, carrier Ethernet becomes a more popular scheme for the mobile backhaul, because it is packet based transport network and allow flexible Ethernet circuit provisioning with bandwidth scalability and low operational expenditures (OPEX) for mobile operators [1]. The Metro Ethernet Forum (MEF) [9] has defined Ethernet virtual Connection (EVC) services with performance requirements, as a transport association between two or more user-network-interfaces (UNI) [2][9]. There are three types of EVC service including E-Line, E-LAN and E-Tree services, which represent a point-to-point (P2P), multipoint-to-multipoint (MP2MP), and rooted multipoint-to-multipoint EVC service, respectively. Due to simplicity and reliability, P2P EVC services are adopted by most mobile operators to connect their cellular sites to mobile core networks. A carrier Ethernet vendor is responsible to provide such a P2P EVC connectivity following the bandwidth profile and performance requirements defined in a Service Level Agreement (SLA) [2][3] between the carrier vendor and the mobile operator.

Wireless radios become popular to provide mobile backhaul connectivity for cellular sites in cases where Ethernet cable or fiber is not available, or cannot be deployed in a timely or economic manner. Compared to single cell site backhaul that is directly connected with carrier Ethernet cable or fiber, mobile operators normally develop their own wireless radio links, such as microwave [4], to connect multiple cell sites nearby, and deliver aggregated backhaul traffic to a hub site which handoff to carrier Ethernet network at UNI. In order to support different backhaul bandwidth requirements for each cell site in cluster, carrier vendors normally assign each mobile cell site an individual EVC circuit with a distinct VLAN ID and Committed Information Rate (CIR), and then implement parallel multiple EVC handoff over a same UNI at the hub site, as shown in Figure 1.

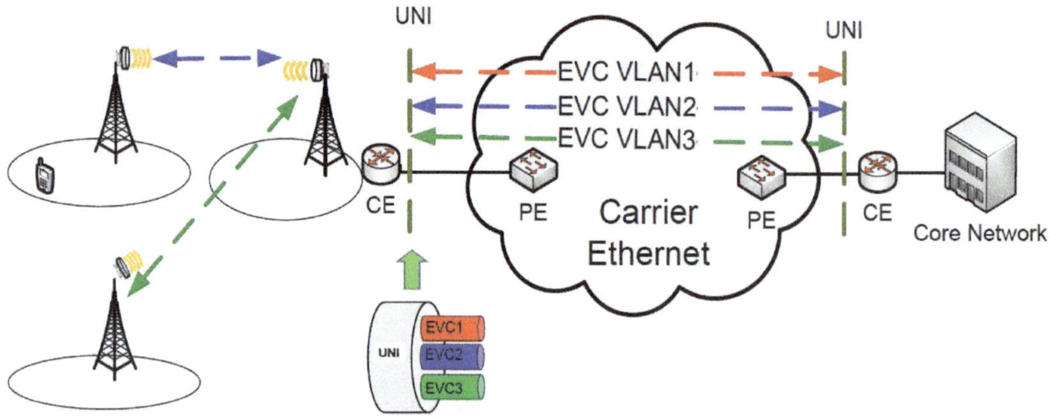

Figure 1. Microwave backhaul with multiple EVC bundling UNI handoffs

However, since all EVC circuits are assigned with fixed CIR values, they cannot take advantages of statistical multiplexing gain at UNI handoff and have no way sharing idle bandwidth among EVCs [1]. Therefore, the overall resource utilization of an UNI is low and mobile operators overpay carrier vendors for unused bandwidth. Authors in [5, 6] proposed delay based methods to decide the minimum transport capacity for which all traffic delay requirements are met. However, multimedia services have different performance requirements and optimal delay performance may result in low bandwidth utilization too [7]. A piecewise linear approach was proposed in [7] to achieve balance between network performance and bandwidth saving. Authors proposed a selective overbooking scheme based on trunk size and usage profile. However, due to different service types and traffic distributions in cell sites, peak bandwidth utilizations at hub site may differ even UNI bandwidth or cluster sizes are same, which means overbooking factors at different hub UNIs may differ from each other, instead of linearly following CIR. A capacity planning scheme for LTE backhaul networks has been proposed in [8] where an overbooking factor calculation method based on traffic forecast, multiplexing gain and peak throughput. However, the overbooking factor is derived from estimation on 50 user peak throughput model, without considering real cellular traffic distributions. Authors neither addressed how to implement carrier Ethernet overbooking scheme, nor investigated bandwidth allocation with QoS for multimedia services when an overbooked UNI becomes congested.

The contributions of this paper are follows: First, we developed a statistical regression method for overbooking factor estimation, based on UNI peak utilization data, cellular traffic distributions, and statistical service outage performance. Then we designed two efficient carrier Ethernet architectures to help bypass CIR binding in EVC and implement mobile backhaul bandwidth sharing among a cellular cluster. Specifically, a mobile operator can adopt either an Ethernet switching or routing scheme to complete backhaul transport in its cellular cluster, while using a single VLAN EVC pipe in carrier Ethernet for bandwidth overbooking. Then to overcome network congestion caused by aggregated burst traffics at UNI, a new bandwidth control algorithm was proposed to provide Quality of Service and fairness for resource allocation. Solid experimental networking data and simulation results showed that the proposed schemes can benefit mobile operators in resource utilization efficiency, backhaul cost saving and quality of service provisioning.

This paper is organized as follows: in section 3, we present the UNI capacity estimation and overbooking derivation methods. Then we describe two efficient Carrier Ethernet architectures for implementing UNI overbooking with carrier Ethernet. In section 4, we present a new bandwidth control scheme at an overbooked UNI to support QoS. Network performance data and simulation results are shown in section 5 with analysis. Finally, we conclude this paper with a brief summary.

2. Acronyms

CBS	Constant Burst Size
CE	Customer Edge device
CIR	Committed Information Rate
CoS	Class of Service
EVC	Ethernet Virtual Connection
MBS	Maximum Burst Size
MEF	Metro Ethernet Forum
PE	Provider Edge device
PIR	Peak Information Rate
SLA	Service Level Agreement
UNI	User-Network-Interface

3. UNI Capacity and Overbooking Derivation

Consider a microwave cluster containing total K cell sites and traffic intensity of a cell site i, X_i, the aggregated traffic intensity at UNI, Y_{uni}, is represented as,

$$Y_{uni} = A_{uni} \cdot \sum_{i=1}^{K} X_i \tag{1}$$

where A_{uni} is the statistical multiplexing gain value at the UNI.

Assume backhaul traffic intensities of cell sites are independent, and there are two major traffic patterns: the Poisson based model and self-similar model [7]. The Poisson-based traffic model has been intensively used to represent cellular voice connections, while the self-similar model is used to represent data services with burst throughputs. We further assume there are M types of voice services, and N types of self-similar services. We adopt an ON-OFF source model to analyze the peak throughput of a voice connection, where the ON and OFF states represent the active and silent conditions of the voice connection, respectively. Both ON and OFF state intervals are assumed to be exponentially distributed, and R_j is a constant packet generation rate of voice class j in the ON state. Due to packet burst characteristics and CIR throttle on backhaul capacity, the throughput ξ_j of self-similar service class j follows truncated Pareto distribution with following probability distribution function [5],

$$f(\xi_j) = \frac{\alpha_j E_j^{\alpha_j} \xi_j^{-\alpha_j - 1}}{1-(E_j/H_j)^{\alpha_j}} \tag{2}$$

where α_j denotes shape parameter, E_j denotes the minimal traffic rate, and H_j denotes the maximum traffic rate of service class j. Then the aggregated throughput ξ_{uni} at the UNI is denoted as,

$$\xi_{uni} = \sum_{i=1}^{K}\sum_{j=1}^{M}\sum_{k=0}^{l_{i,j}} \xi_p(i,j,k) + \sum_{i=1}^{K}\sum_{j=M+1}^{M+N}\sum_{k=0}^{l_{i,j}} \xi_s(i,j,k)$$

$$\leqslant \sum_{i=1}^{K}\sum_{j=1}^{M} l_{i,j} R(j) + \sum_{i=1}^{K}\sum_{j=M+1}^{M+N}\sum_{k=0}^{l_{i,j}} \xi_s(i,j,k) \quad (3)$$

where $\xi_p(i,j,k)$ represents the throughput of connection k of Poisson-based service type j in cell i, $\xi_s(i,j,k)$ represents throughput of connection k of self-similar service type j in cell i, and $l_{i,j}$ represents total connection number of class j in cell i.

From the equation (3), the peak aggregation throughput at UNI is determined by the sum of self-similar traffics following truncated Pareto distribution function, which can be approximated as a Gaussian distribution [5][7], with mean value μ_{peak}, and standard deviation σ_{peak}, which can be derived from peak throughputs at the UNI. When an overbooking ratio O_{uni} is applied at UNI bandwidth which equal to $\sum_{i=1}^{N} CIR(i)$, it is expected to achieve low bandwidth outage probability, i.e., the probability of the case that overbooked bandwidth cannot transport aggregated peak throughput is smaller than or equal to a threshold service outage ratio ε, $0 < \varepsilon < 1$. Then we can get,

$$P(\xi_{uni} \leq O_{uni} \cdot \sum_{i=1}^{N} CIR(i)) \geq \varepsilon \quad (4)$$

When equality holds in (4), the overbooking ratio O_{uni} is minimized, and the overbooked UNI bandwidth can be minimized. Since ξ_{uni} follows normal distribution, then the following relationship is satisfied,

$$O_{uni} = \frac{\sigma_{peak} \cdot Q^{-1}(\varepsilon) + \mu_{peak}}{\sum_{i=1}^{N} CIR(i)} \quad (5)$$

where $Q^{-1}(x)$ is the inverse function of normal distribution $Q(x)$, and $Q(x) = \frac{1}{2\pi}\int_x^{\infty} e^{-\frac{s^2}{2}} ds$.

Applying O_{uni} on existing UNI bandwidth, we get the overbooked UNI bandwidth for the whole cluster as,

$$B_o = O_{uni} \cdot \sum_{i=1}^{N} CIR(i) \quad (6)$$

Because full utilization results in high latency and congestion [8], the real UNI bandwidth after overbooking is set to be larger than overbooked bandwidth derived in equation (6) by dividing a maximum utilization ratio η. Therefore, the feasible UNI bandwidth B_{uni} is set as,

$$B_{uni} = B_o / \eta \quad (7)$$

where $0.5 \leqslant \eta < 1$ and it is normally defined by mobile operators based on backhaul performance requirements [7].

4. EFFICIENT UNI HANDOFF ARCHITECTURE

Since current MEF E-line service does not support overbooking at UNI with parallel EVC circuits, we need to improve the carrier Ethernet transport architecture to implement bandwidth sharing inside the cellular cluster. Considering a CE device at a UNI is either a layer 2 switch or a cell site router (CSR) which supports layer 3 networking functionalities, we proposed two UNI handoff architectures to implement the above goal.

4.1. Q-in-Q Tunneling Architecture

In scenarios that the CE at UNI only supports Layer 2 VLAN switching, a mobile operator can order a single EVC pipe with its contracted carrier vendor for mobile backhaul of a target cell cluster. The EVC pipe is implemented with IEEE 802.1ad stacked VLAN bridging technology (Q-in-Q tunneling) [2], in which the Ethernet carrier configures and delivers P2P mobile backhaul traffic between two PEs based on outer VLAN service tag (S-tag) in carrier Ethernet frames. After the carrier PE handoffs traffic to associated CE in mobile network, the CE implement switching based on inner customer VLAN tag (c-tag) for frame delivery to the destination recipient cell, as shown in Figure 2.

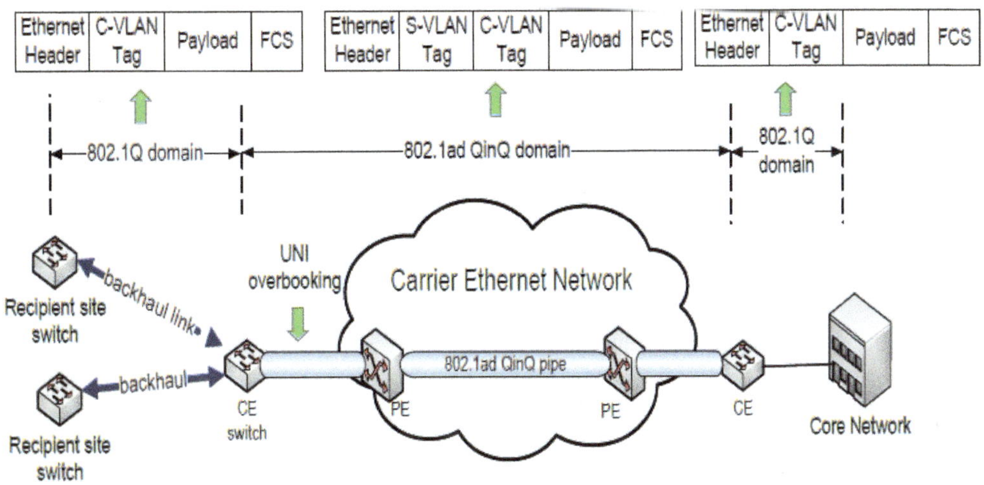

Figure 2. Q-in-Q Tunneling Carrier Backhaul Architecture

The EVC capacity is determined by the overbooked UNI bandwidth defined in equation (7). The mobile operator predefines and encapsulates a VLAN ID associated with one of their destination cell sites with inner customer-tag (C-tag) without necessarily notifying the carrier vendor CIR value of each cell site, and only contract with the vendor with the CIR value of ordered EVC circuit, which tunnel backhaul traffic between MSO and the hub site. The tunneling backhaul scheme goes as follows: any backhaul Ethernet frame generated in MSO or in a cellular cluster is attached with a C-tag VLAN ID which works as an ID marker of the backhaul of a destination cell site. When aggregated backhaul frames arrive at CE devices at UNI, the CE attaches S-tag VLAN ID in each frame as an indicator of the pre-specified carrier Ethernet EVC circuit, and handoffs the Q-in-Q frame to carrier Ethernet network. When the PE device in the carrier network receives the frame, it checks S-tag VLAN ID in the frame and finds out the associated carrier EVC circuit for delivery. Once a CE on the other side of the EVC circuit receives Q-in-Q frame from its associated PE, it detaches the outer S-tag, checks the inner VLAN ID in C-tag for destination site, and delivers the frame to next hop that follows same layer 2 switching scheme based on VLAN ID until it reaches the destination cell site.

4.2. Cross-layer Backhaul Architecture

In scenarios that the CE at UNI is a cell site router (CSR) which supports IP layer functionalities, we adopt a cross-layer solution for cellular cluster backhaul, i.e., a carrier vendor configures a single EVC pipe from PE associated with the MSO to the edge PE associated with a CE at the hub site with overbooked bandwidth. After a UNI handoff, the router at hub site will deliver the payload packet to the destination site by checking IP address in frame payload. Then in the cellular cluster, the backhaul packets are delivered through routing, which implements an end-to-end cross-layer backhaul scheme, as shown in Figure 3.

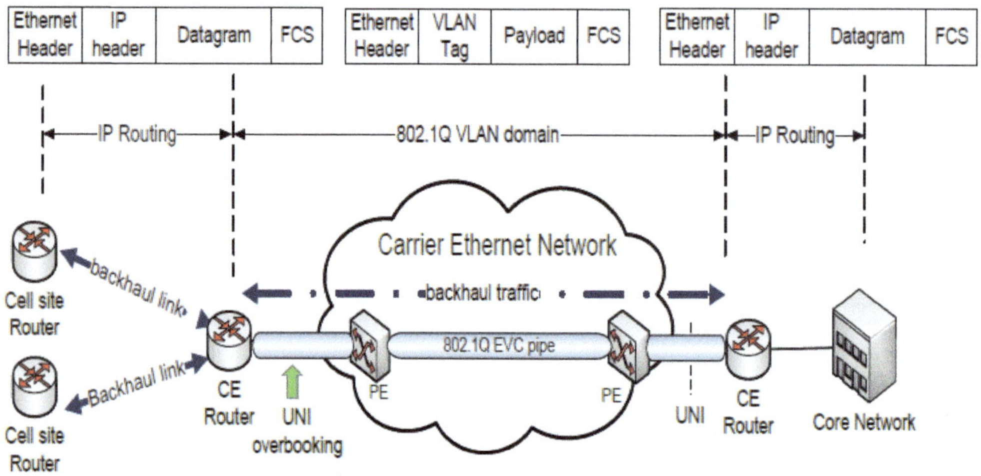

Figure 3. Cross-layer backhaul architecture

The traffic routing in cellular cluster can be implemented by either static routing or dynamic routing. If the cluster size is small, static routing is preferred and the routing table is configured manually in the CSR at hub site by specifying next-hop interface IP address towards the destination cell site. However, when the cluster size increase, the static routing is neither fault tolerant nor efficient for networking configuration and management. Dynamic routing can fix these problems by supporting real-time routing path construction and selection. It is required that the recipient sites have their own routers so that they can communicate with the CSR at hub site with common dynamic routing protocols, such as Open Shortest Path First (OSPF) or Routing Information Protocol (RIP) [1][2], for creation, maintenance, and updating routing paths in their routing table.

4.3. Advantages of Proposed Architectures

Advantages of our proposed architectures: the UNI is implemented as a backhaul aggregation pipe, instead of a group of parallel EVC circuits. And the UNI bandwidth profile is determined by the multiplication product value of the derived overbooking ratio and the sum of CIR values of all cell sites in the cluster. Therefore, the statistical multiplexing gain is used for bandwidth sharing among the cell site cluster, and no more dependency on carrier EVC provisioning. The mobile operator can adjust bandwidth profile more easily based on cluster size and the peak utilization at the UNI. The scalability of mobile backhaul also get improved because it is not necessary for the mobile operator to request additional EVC circuits for newly added cell sites in the cluster.

Furthermore, carrier vendors reduce the complexity in networking configurations and EVC maintenance, through a single EVC circuit provisioning for whole cellular cluster.

5. Bandwidth Control at Overbooked UNI

Due to burst throughput and overbooking at UNI bandwidth, a traffic policing and rate shaping scheme are required for providing stable QoS level even when UNI becomes congested [12,13], as well as maximizing bandwidth utilization [9, 10]. MEF has defined a two-rate, three-color marker (trTCM) algorithm for CE VLAN CoS which can be implemented via two token buckets scheduling [3]. One bucket is used to determine in-profile service frame rate per CoS, following CIR value, while the other bucket is to determine excess service frame rate per CoS, following EIR value. However, in the MEF scheme, each bucket size is a fixed value and egress point does not differentiate service priorities for multimedia traffics. Due to cluster traffic aggregation and UNI bandwidth overbooking, the bandwidth scheduling and rate enforcement need to be re-calculated in real time. Based on service performance requirements, traffics generated in mobile network are classified into three types which are marked with P bits in the Priority Code Point (PCP) field of IEEE 802.1Q tag in Ethernet frames. When aggregated traffics arrive, The CE at UNI checks p bits in frames and puts traffic of each class into an individual forwarding class (FC) queue which implement egress policy with specified CIR, Peak Information Rate (PIR), Constant Burst Size (CBS) and Maximum Burst Size (MBS). Policing rule of each FC queue is set as follows: the specified CIR rate is enforced when egress to the PE device in carrier network, and service traffics above CIR are marked as out-of-profile. The PIR is the maximum rate at which frames are allowed to burst over the CIR and delivered as best efforts. Traffic shaping is implemented through absorbing the traffic burst up to MBS and dropping traffic over PIR after MBS. The CBS in kilobyte is generally determined by the product of maximal tolerable latency of the service specified in the SLA and the assigned CIR rate, as follows:

$$CBS_i = \frac{CIR_i \cdot L_i}{1000 \times 8} \tag{8}$$

where L_i denotes the required latency value of service i, and $i \in \{voice, video, data\}$. Similarly, the MBS in kilobyte is denoted as the product of maximal tolerable latency of the service and the assigned PIR rate,

$$MBS_i = \frac{PIR_i \cdot L_i}{1000 \times 8} \tag{9}$$

The scheduling priorities: the voice service is assigned with highest priority for rate allocation, due to its most strict requirements on throughput, latency, jitter and FLR. The next priority is given to video stream service, due to its high requirements on throughput, latency and FLR. The data service priority is lowest due to best-effort traffic characteristics. However, the data service is sensitive to the minimum throughput to avoid starvation and network management blocking. To implement dynamic rate allocation based on aggregated traffic intensities at UNI, we further divide time into a sequence of slots, in which each slot contains same time length t_{slot} second(s), and the scheduler assign CIR and PIR rates for each service at the beginning of the slot.

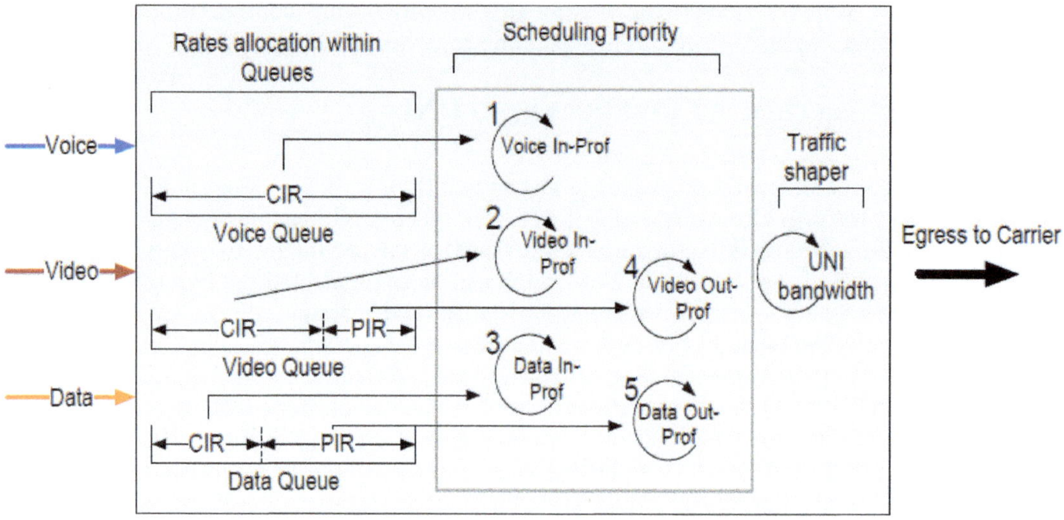

Figure. 4 Egress Traffic Scheduling at Overbooked UNI

Consider $N_{voice}(i)$ voice connections in cluster at time slot i, total CIR of voice stream at CE is assigned with full capacity as follows:

$$CIR_{voice}\ (i) = PIR_{voice}(i) = N_{voice}(i)R_o \tag{10}$$

Since Ro is the packet rate of voice ON state, there is no overbooking adopted for voice traffics and all voice traffics at CE should handoff to the carrier without frame loss. Therefore, there is no excess packet rate over CIR of voice, and PIR_{voice} equal to CIR_{voice}.

Due to bursty video traffics, it is difficult to assign video FC queue with same allocation rate all the time. However, we can assign CIR bandwidth as much as possible to video bursts, if QoS policy of voice and data service can be satisfied. Therefore, the total CIR of video stream at CE at slot i is shown as follows:

$$CIR_{video}(i) = max\ \{B_{uni} - CIR_{voice}(i) - g_{uni}\ B_{uni},\ 0\} \tag{11}$$

where g_{uni} is a guard ratio defined by a mobile operator, normally $0 < g_{uni} < 0.2$, as a percentage of total UNI bandwidth reserved for data service to provide minimum throughput, which help avoiding the data starvation caused by video bursts. The video bursts over CIR will be scheduled for PIR rate which equals to the whole available UNI bandwidth.

For data service, we assign available UNI bandwidth after CIR rate allocations on voice and video service queues, as well as the whole available UNI bandwidth for the data PIR rate allocation:

$$CIR_{data}(i) = B_{uni} - CIR_{voice}(i) - CIR_{video}\ (i) \tag{12}$$

The PIRs allocations for video and data traffic can only provide best-effort access for video and data services after CIR scheduling, and denotes as,

$$PIR_{video}(i) = PIR_{data}(i) = B_{uni} \tag{13}$$

The above rate allocations follow a strict round robin sequence from voice to BE data shown in Figure 4. CIR assignment can guarantee bandwidth assignment while PIR make use of maximum available resource left after CIR assignments. All idle bandwidth left in CIR rate allocation step will become available bandwidth for the PIR allocation in the same slot.

6. Performance Evaluations

In this section, we first evaluate the network performance on overbooked UNI with ns-2 simulation tool [11-13], which has been widely adopted for network simulations and performance evaluations. A dumbbell topology is adopted to simulate the traffic on UNI, as shown in Figure 5. The network scenarios are built from the following elements: (1) Hub CE connects with multiple connections, containing at least one voice, video, and data connections; (2) We assume that congestion only happens when traffic egress from CE to PE; (3) CE supports the trTCM scheduling frame where our proposed dynamic bandwidth control algorithm in section 4 for performance comparison.

Table 1. System Parameters for Simulation

t_{slot}	B_{uni}	η	L_{voice}	L_{video}	L_{data}	FLR_{video}
1s	20Mbps	0.8	8ms	20ms	100ms	10^{-3}

We adopt 3GPP voice model with ON/OFF time interval following exponential distributions: mean active time as 600 millisecond and mean idle time as 400 milliseconds, and active data rate is a constant rate as 24 Kbytes per second. We use a contributed MPEG4 video traffic generator model which is derived from a video trace, with initial seed as 0.3 and rate factor as 60. We adopt FTP application model with TCP transport and file size 100 mega bytes for data service. The frame size of all three traffic model is set as 1024 bytes.

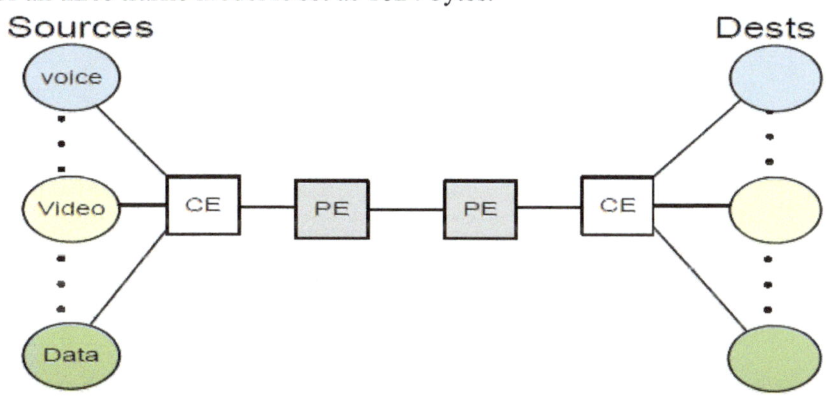

Figure 5. Dumbbell Topology for Network Simulation

Figure 6 shows the result of voice throughput when aggregated video connections increase at the UNI egress point. We can find that total voice throughputs for 5 voice connections and 10 voice connections are about 51 Kbps and 98 Kbps respectively, even when the UNI becomes congested due to large number of video bursts. This is because each voice connection is assigned full bandwidth based on its active state rate at UNI, which guarantee the high priority voice throughput and related Key Performance Index (KPI) performances.

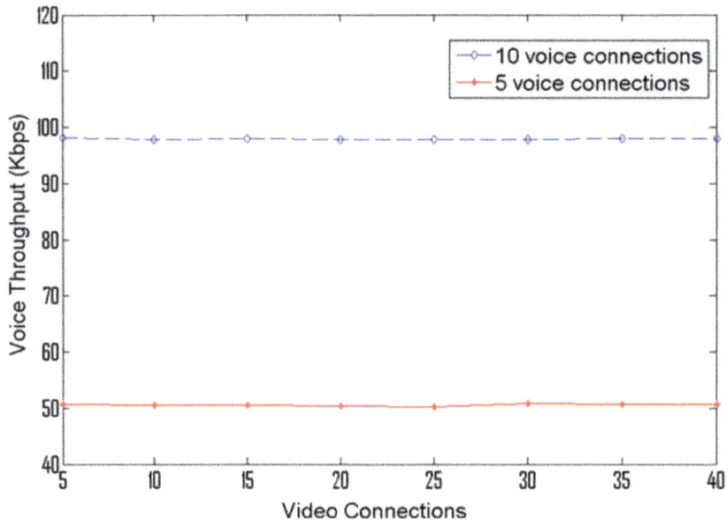

Figure 6. Voice Throughput Performance

Since video service has higher priority for resource allocation over data service, it has been shown in Figure. 7 that data throughput shrinks when video connections increase. However, due to minimum guard CIR bandwidth reserved as 20% of total 20Mbps UNI bandwidth, the data throughput is always above 4 Mbps, which satisfy minimum data throughput requirement of mobile operator for network management and signaling purposes.

Figure 8 shows the frame loss ratio of video traffic when total video connections increase. There are total 5 voice connections and 1 data connection. For static trTCM scheme, we set CIR ratios for voice, video and data as 20%, 40%, and 40% of the UNI bandwidth, respectively. PIR ratio of voice is same as its CIR ratio, and PIR ratios of video and data are both equal to 100% of UNI bandwidth. It is found that video frame loss ratio for static trTCM are almost same as our dynamic bandwidth allocation scheme when video traffic intensity is small. However, the frame loss ratio of static trTCM increases more quickly than our dynamic scheme when heavy video bursts happen. This is because our dynamic scheduling method can make use of most idle bandwidth in each slot for video burst bandwidth allocation after it satisfies requirements of voice bandwidth and minimum data throughput, while in static trTCM, video connections have fixed ratio on CIR scheduling round and cannot compete for idle resource efficiently.

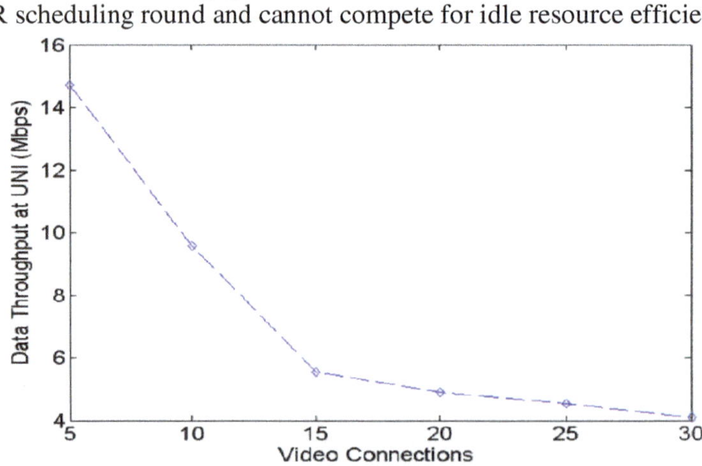

Figure 7. Data Throughput

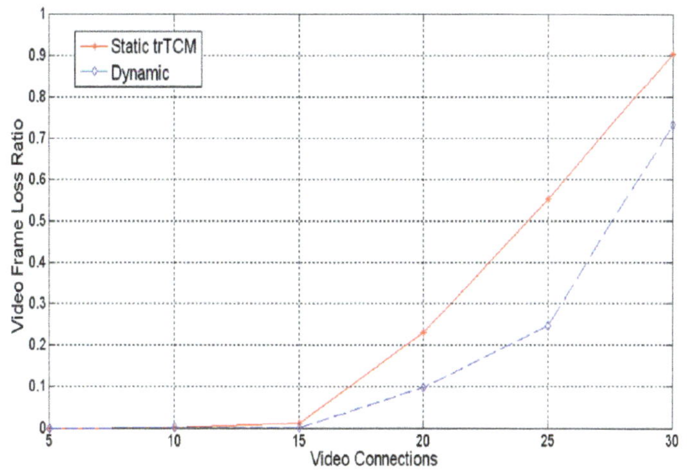

Figure 8. Video FLR

Similarly, Fig. 9 shows the results of video frame latency performance with 5 voice, 10 video and 1 data connections at UNI. We can see that the latency range of our dynamic scheme is smaller than static trTCM method. This is because video connection can get more bandwidth allocation rate in the proposed dynamic scheduling scheme with more CIR and PIR bandwidths. Therefore, the average waiting time in service queue is smaller than the static trTCM method, which determines the video frame latency performance.

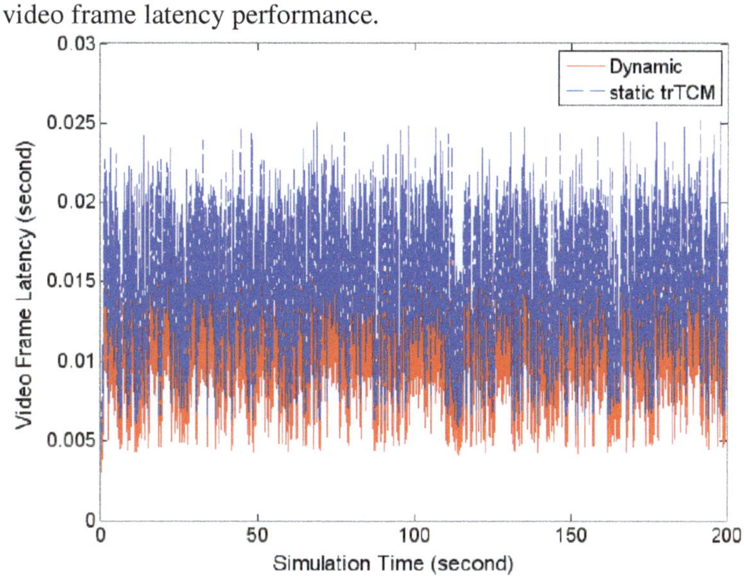

Figure 9. Video Frame Latency

Then we evaluate the overbooking ratio selection with several production microwave clusters in a national cellular network. For proprietary information protection, we neglect locations and names of the selected clusters and only show related performance data, as shown in table 2. For each cluster, the service outage threshold ε is set as 0.001%, and maximum allowable utilization η in equation (7) is set as 0.8. μ_{peak} and σ_{peak} of each UNI are calculated through daily peak UNI throughput data over one year. The peak throughput with 99% confidence interval based on estimation is denoted as T_{peak}. Compared to a linear piecewise relationship between UNI bandwidth and overbooking ratio in [7], we found that peak UNI backhaul throughputs of clusters are different to each other and not strictly follows a monotonically non-increasing relationship with total CIR values of a cell cluster. This demonstrates that traffic intensity and statistical multiplexing gains of clusters are different from each other, even they have same CIR values or cluster size. So it is more appropriate to adopt our statistical estimation method for cluster based overbooking estimation, rather than using the linear piecewise based scheme which derived overbooking ratio only based on UNI CIR value.

Table 2. Real Network Utilization Performance & Overbooking Ratio

Cluster	C1	C2	C3	C4	C5
K	2	3	3	4	4
$\sum_{i=1}^{K} CIR(i)$	100	150	150	200	200
T_{peak}	40.7	33.1	76.3	21.9	80.0
O_{uni}	40.11%	23.76%	52.62%	10.85%	44.43%
B_{uni}	60	50	100	30	110

We further investigate cost saving through overbooking. Figure.10 shows a pricing example from the Carrier Ethernet Market website [14] which is monotonically increasing with UNI bandwidth. It is worthy to notice that some carrier vendors provide discrete UNI bandwidth options for UNI handoff. Once we apply overbooking, the UNI bandwidth can be reduced without deteriorating service quality. The new UNI bandwidth may in same or different pricing range, compared to original UNI bandwidth. For example, when UNI bandwidth drops from 200 Mbps to 100 Mbps, the new pricing can be saved by 40.1%, as shown in Figure 11. In this case, the mobile operator can save transport cost from UNI overbooking. But when UNI bandwidth changes from 50 Mbps to 35 Mbps, the new pricing is same as the old one. Then for the cost perspective, the mobile operator would rather keep original UNI bandwidth from carrier for the cellular cluster than do UNI overbooking.

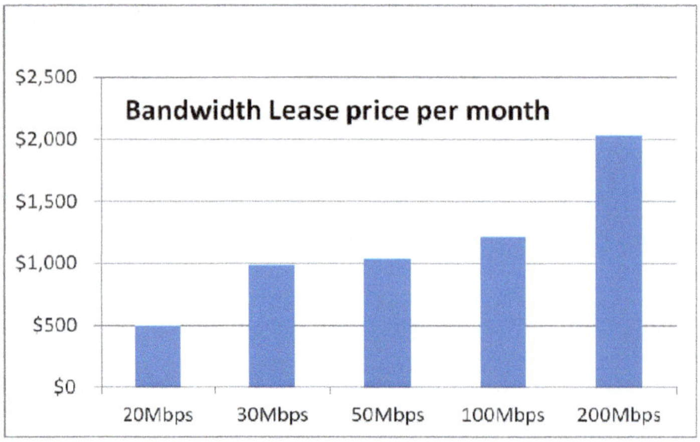

Figure 10. UNI Pricing

Full UNI (Mbps)	UNI band overbooking (Mbps)	Cost Saving (%)
200	100	40.1%
100	50	14.8%
50	35	0%

Figure 11. Cost Saving Map

7. CONCLUSION

This paper proposes an efficient overbooking framework for mobile backhaul. We first develop a statistical estimation method to derive a safe overbooking factor for a given UNI. Then two novel transport architectures are proposed to help cellular cluster backhauls with overbooked UNI bandwidth in carrier Ethernet. A novel bandwidth control algorithm has been derived at customer edge device to provide Quality-of-Service for multimedia traffics over overbooked UNI, with SLA policing and protection on high priority services. Experimental networking data analysis and simulations show that our new schemes can benefit mobile operators in resource utilization efficiency, carrier Ethernet cost saving and backhaul performance.

REFERENCES

[1] P. Briggs, R. Chundury, and J. Olsson, "Carrier ethernet for mobile backhaul," Communications Magazine, IEEE, vol. 48, no. 10, pp. 94–100, 2010.

[2] K. Fouli and M. Maier, "The road to carrier-grade ethernet," Communications Magazine, IEEE, vol. 47, no. 3, pp. S30–S38, 2009.

[3] F. Brockners, Metro Ethernet Services and standardization, Metro Ethernet Forum Std., 2005.

[4] S. Little, "Is microwave backhaul up to the 4g task?" Microwave Magazine, IEEE, vol. 10, no. 5, pp. 67–74, 2009.

[5] A. W. M. Stasiak, M. Glabowski and P. Zwierzykowski, Modelling and dimensioning of mobile wireless networks: from GSM to LTE. John Wiley & Sons.

[6] X. Li, et al., "Dimensioning of the lte s1 interface," in Proc. IEEE Wireless and Mobile Networking Conference, 2010 Third Joint IFIP, 2010, pp. 1–6.

[7] F. Huang, C. Deccio, R. Ball, M. Clement, and Q. Snell, "A piecewise linear approach to overbooking," in Proc. IEEE High Performance Switching and Routing Conference (HPSR'04), 2004.

[8] A. Checko, L. Ellegaard, and M. Berger, "Capacity planning for carrier ethernet lte backhaul networks," in Proc. IEEE Wireless Communications and Networking Conference (WCNC'12), 2012.

[9] Metro Ethernet Forum, "Metro Ethernet Services – A technical Overview", http:/ / metroethernetforum.org/metroethernet-services.pdf

[10] C. Liu, C. Zhou, N. Pissinou and K. Makki, "Adaptive Quality of Service Provisioning in IEEE 802.11 Wireless Local Access Network," International Journal of Sensors, Wireless Communications and Control, Vol. 2, pp.117-129 2012.

[11] ISI, "The network simulator – ns2," http://www.isi.edu/nsnam/ns/

[12] C. Liu, C. Zhou, "An improved interworking architecture for UMTS-WLAN tight coupling," IEEE Wireless Communications and Networking Conference, Vol. 3, pp. 1690-1695, 2005.

[13] C. Liu, C. Zhou, "Providing quality of service in IEEE 802.11 WLAN," IEEE International Conference on Advanced Information Networking and Applications, vol. 1, pp. 817-824, 2006.
[14] Carrier Ethernet Market, http://www.carrier-ethernet-market.com/en/

14

ENERGY EFFICIENT MAC PROTOCOLS FOR WIRELESS SENSOR NETWORK: A SURVEY

Eleazar Chukwuka[1] and Kamran Arshad[2]

School of Engineering, University of Greenwich Chatham Maritime ME4 4TB United Kingdom

[1]ce632@gre.ac.uk [2]k.arshad@gre.ac.uk

ABSTRACT

Wireless Sensor Network (WSN) is an attractive choice for a variety of applications as no wired infrastructure is needed. Other wireless networks are not as energy constrained as WSNs, because they may be plugged into the mains supply or equipped with batteries that are rechargeable and replaceable. Among others, one of the main sources of energy depletion in WSN is communications controlled by the Medium Access Control (MAC) protocols. An extensive survey of energy efficient MAC protocols is presented in this article. We categorise WSN MAC protocols in the following categories: controlled access (CA), random access (RA), slotted protocols (SP) and hybrid protocols (HP). We further discuss how energy efficient MAC protocols have developed from fixed sleep/wake cycles through adaptive to dynamic cycles, thus becoming more responsive to traffic load variations. Finally we present open research questions on MAC layer design for WSNs in terms of energy efficiency.

KEYWORD

Wireless sensor networks, MAC protocols, Energy conservation

1. INTRODUCTION

Wireless sensor network is considered to be one of the most influential technologies of the current century [1]. WSN materialised due to the progresses made in micro-electromechanical systems (MEMS) [2] [3] [4] which combines advanced communications and signal processing capabilities [5]. Consequently this led to the production of power constrained low cost tiny sensor nodes [1] [6]. These tiny sensor nodes have capabilities to sense, process and communicate with a remote user through a gateway called the sink. WSN supports ubiquitous computing which is the third generation computer evolution [7]. However, its capabilities, though great, are limited as a result of energy constrains.

Composition of a WSN is made up of the environment that is to be monitored, sensor nodes that are spatially and randomly deployed and the sink node, which is the main interface between the nodes and the user, as shown in Figure1

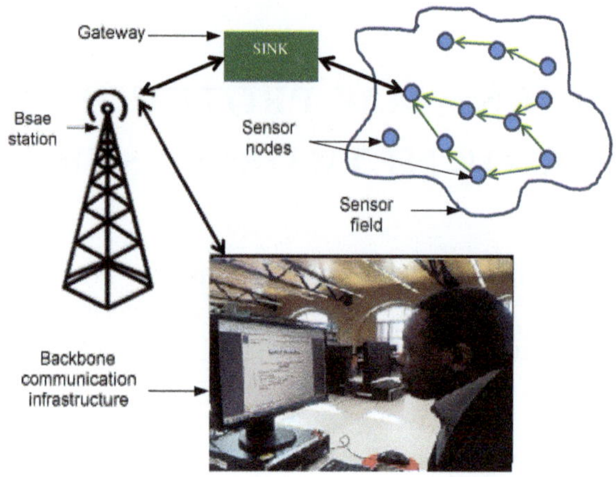

Figure 1: Wireless sensor network architecture

WSN can be deployed for many applications ranging from agricultural [8][9], environmental [10], healthcare delivery [11], military [12], security [13], surveillance [14] [15], home automation [16] and so forth. Smart home concept is also based on WSNs while sensor nodes can be implanted in the human body. These require nodes with good longevity which depends on the source of power supply (usually batteries) among others [17]. These batteries are non-rechargeable due to the nature of deployment and the sensing environment. Hence, it might possibly not be expedient or pragmatic to change the batteries [18]. Furthermore, difficult and adverse terrains like long bridges and high rise civil structures will pose risk to human monitoring (e.g. see Figure 2).

Figure 2: Nodes monitoring a bridge [19]

Enormous research has gone into designing and reviewing energy efficient MAC protocols, reflecting the importance attached to the development of WSNs. Contention-based and schedule-based MAC protocols appear to be the most popular categorisation used in the reviews [20] [21] [22]. In [21] first category is the contention access in which nodes compete to access the channel randomly and conflicts are resolved by distributed conflict resolution technique [23] [24] [25]. The second is contention free access which offers controlled access to the channel. Nodes only gain access to the medium based on a defined schedule [18] [26]. However the above categorisation is not application specific. A classification according to application was used in [27] which categorised protocols according to the specific problem they aim to solve. But [28] improved on that by focusing on mission critical assignments. Vital performance objectives were used to categorise the protocols into two; data transport delay and reliability, while energy efficiency was put in the back burner. The authors of [29] considered both energy efficiency and data delivery. They separated MAC protocols into four groups; asynchronous, synchronous, frame slotted and multichannel protocols. Synchronous protocols contend with reducing delay and improving throughput. Asynchronous protocols battle with establishing proficient communication between nodes with dissimilar sleep and active plans. Frame slotted protocols assign time slots to nodes to enhance throughput. While this addressed the issue of collision, the challenge is channel utilization in low traffic network. Multichannel mechanism enhances capacity of the network but cross channel communication together with distributed channel assignment is an issue.

These reviews are either application specific or according to the challenges addressed. In this article we show how MAC protocols have been developed to respond to network demand fluctuations. Evolution of MAC protocols is outlined to show the shift from just energy efficiency to efficient utilisation of all resources (energy and bandwidth) together with efficient data delivery. Most of the protocols use duty cycles and contention window to preserve energy. In the early designs, both duty cycle and contention windows were fixed. This survey paper details how, adaptive and subsequently, dynamic duty cycles are being employed in response to network demand fluctuations, enhanced data delivery and channel utilisation thereby conserving energy. Method of medium access was used to group the protocols into four categories, which are controlled access, random access, slotted and hybrid protocols. This categorisation highlights not just the challenges addressed by each group but also indicates the inroads made and how dynamically WSN has been more responsive to network demands.

The rest of this article is organised as follows; a summary of sources of energy inefficiency in WSNs is presented in section 2. MAC protocol classifications and why WSNs differs from traditional wireless networks are detailed in section 3. A detailed survey of existing State-Of-The-Art (SOTA) literature is presented in section 4. Challenges to WSNs in terms of energy efficient MAC are summarised in section 5 while section 6 concludes this article.

2. BACKGROUND

WSN is a network of sensor nodes that are randomly and spatially deployed to monitor a phenomenon. Nodes need a source of power supply for them to function; this is usually provided by a battery. The battery might not be replaceable or rechargeable owing to the hazardous terrain in which WSN are normally deployed. Constrained by the power supply, nodes have to optimise their energy consumption but first sources of energy inefficiencies have to be identified. In this section, these sources are identified with radio communication being identified as the most power hungry process. MAC layer controls communication among nodes therefore, we classify MAC protocols used in WSN and show why MAC protocols used in traditional wireless networks cannot be used in WSN.

2.1. Sources of energy inefficiency in WSN

Sources of energy inefficiency in WSNs came as a result of different operations of sensor nodes. Events have to be detected therefore sensing [30] [31] [32] is one of the main sources of energy depletion. Another is routing which determines how the sensed data is relayed back to the sink [33] [34]. Processing of sensed data is also an energy draining process in WSNs. Transmitting all raw data consumes considerable energy as well as transmission bandwidth and hence not ideal. To preserve energy, processing of information in the nodes was introduced in [32] [35].

2.1.1. Radio Communications

Communication is a major source of power drain in WSNs [27] that involves transmission and reception of data packets. Keeping the radio *on* consumes power while keeping the radio on the sleep mode saves energy but increases latency and offers low spectrum utilisation [36] [37]. WSNs operate in the ISM (Industrial Scientific and Medical) band which is being shared by other wireless devices hence prone to interference and energy wastage.

This survey will focus on energy efficient MAC protocols because communication is the biggest source of energy inefficiency in sensor networks [27] and any reduction in its energy budget will significantly enhance the network lifetime of WSNs. The major sources of energy depletion during communication are transmission with highest drain, followed by receiving and idle modes [38]. The sleep mode has the least energy demand as can be seen from Figure 3 with data in

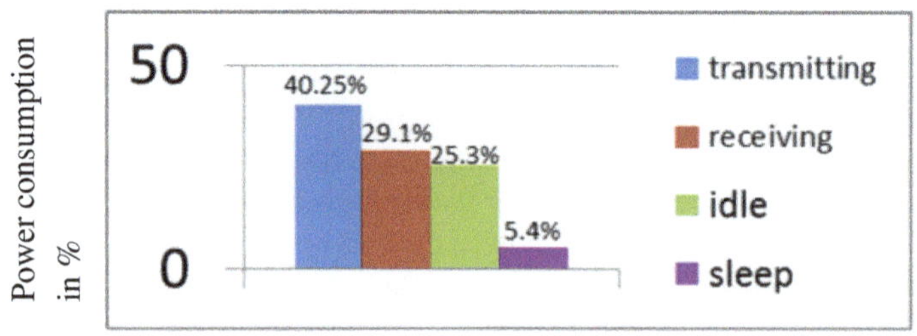

Figure 3: Average power consumption of different node states (modes)

Therefore any protocol that is able to keep transmitting, receiving and idle states to the barest minimum will greatly enhance the network lifetime of WSN. Collision, idle listening, overhearing and control overheads are all causes of energy inefficiencies in MAC protocols [39] which should be reduced.

2.2. MAC Protocols

There are different MAC protocols used in the conventional wireless network. MAC protocols used in wired environment cannot be used in wireless environment because collision occurring at the receiver is to be avoided. In wired network the sender detects collision but since signal strength is virtually the same throughout the wired medium, this does not pose any significant problem. However in wireless network the signal strength depreciates in inverse proportion to the square of the distance according Friis free space equation. Another issue is "hidden" terminals, which is as a result of a node "Green" being within transmission distance of nodes "Blue" and "Red"Figure 4, which are not within the radio distance of each other.

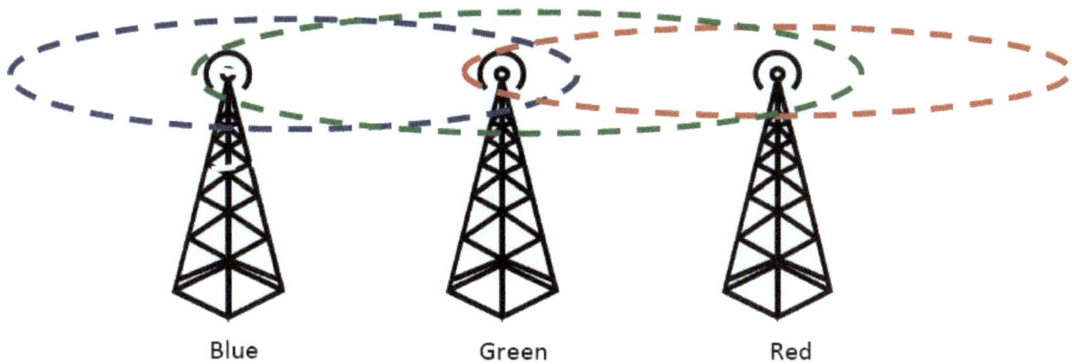

Figure 4: Hidden terminal

Differences between WSN and conventional wireless networks are decentralised control and very long run time in WSNs [27]. Hence ALOHA will not be suitable in WSNs because of collisions [23] [40] and the attendant power drain. CDMA will involve multifarious computing while FDMA require complicated hardware [21], SDMA is usually used in conjunction with other algorithms [41] so does not stand alone. Obviously most of these protocols were not designed with energy constraints as their main concern [42], this is because conventional wireless networks are not as energy constrained as WSNs.

In this article, we group different MAC protocols into four main categories; Controlled Access (CA), Random Access (RA), Slotted Protocols (SP) and Hybrid Protocols (HP). Other categories like low power listening, FDMA, CDMA, beacon enabled etc exist, but this paper will focus on the above categorisations. Next section will present a review of the previous work done.

3. SURVEY OF EXISTING STATE-OF-THE-ART (SOTA)

In this section, we highlight current SOTA related with energy efficient MAC protocols for WSN. A number of MAC protocols have been developed for WSNs to address the challenges mentioned in section 2. Earlier protocols were energy efficiency biased but recent techniques in addition, also address others issues like latency, throughput and spectrum utilisation. However no single protocol has been able to address all the sources of power inefficiencies in the radio communication. As an outcome of the survey, we noticed that most of the progresses were made at the expense of throughput, bandwidth utilisation or increased latency. A protocol that can extend the network lifetime of WSN with a less or no trade off will be of great benefit to the society in view of the numerous applications of WSN. Through this categorisation, we show the evolution of MAC protocol from fixed cycles to dynamic cycles as one of the contributions of this article. We first review CA protocols followed by RA then SP and finally HP.

3.1. Controlled Access Protocols.

In CA protocols, nodes are allocated time slots using TDMA or FDMA in combination with TDMA. In each time slot a node has access to the shared medium and can transmit without collision. As can be seen in Figure 5, in TDMA, nodes are allocated different times such that at time t_2, node N_3 has access to the medium. Receiver nodes are synchronised with their sender nodes to wake up at the same time [43]. This protocol enhances energy efficiency by avoiding collision and overhearing. However a lot of overhead is incurred in synchronisation, which together with clock drift is an issue with this protocol [44] [45].

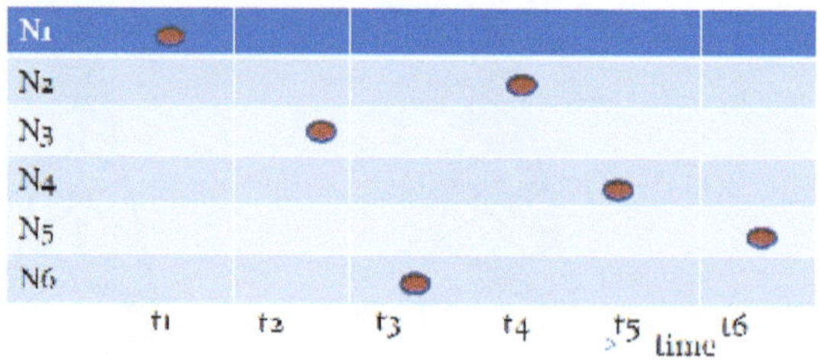

Figure 5: Controlled access protocol

3.1.1. Review of Controlled Access Protocols

CA protocols are energy efficient because they avoid collisions which are common with RA protocols (section 3.2). But it is not as scalable and adaptive to time slot assignments and fluctuations associated with WSNs as the later. Also, it is prone to idle listening under low traffic which wastes energy. A protocol, Latency MAC (LMAC) [46], was introduced to reduce idle listening through increasing data arrival prediction accuracy, by increasing sleep cycle which keeps nodes in low power sleep state for longer periods. But this leads to low throughput, high latency and low channel utilisation. For light traffic networks, LMAC offers good energy savings with long sleep cycles. If data arrival rate is predictable, and transmissions occur at a fixed rate, authors of [47] offer an improvement by using time schedule to turn on the radios. For an event occurring every 10s, the radios are scheduled to come on every 10s, stay on for the duration of the communication and then go back to sleep. This approach avoids idle listening by ensuring that nodes remain in power saving sleep mode, when no transmission is required, which saves energy. However, energy is wasted if there are no transmissions in any interval. This principle is not so efficient since it leads to unnecessary delay in latency or loss of packets due to randomness of data arrival rate. By incorporating adaptive TDMA protocol, [48] [49] enhances the energy efficiency of [47], by making nodes with no packets to transmit to go back to sleep without waiting for its entire slot duration to elapse. Instead of fixed, an adaptive sleep/wake schedule was employed [49]. As the node traffic changes, the schedule adaptively changes according to the traffic. Nodes with light traffic sleep more while nodes with heavy traffic have low duty cycle. If a node has more packets than can be transmitted in one duty cycle, the duty cycle will be extended to accommodate the entire packets, and reversed when packets are less. Thus improving on latency, throughput and channel utilisation, but incurring overhead. By varying the duty cycle according the network need, energy that would have been wasted with fixed cycles which do not consider whether there is traffic or not is saved.

Another protocol that is dynamic and uses TDMA-based protocol was proposed in [50]. Nodes that have nothing to send or receive during their active cycle go back to sleep immediately to save energy [48]. This is an improvement on protocols like low-energy adaptive clustering hierarchy (LEACH) that remains active during the listen frame even though there are no data to transmit. For nodes with more packets than can be sent within the scheduled duty cycle, cluster heads, dynamically assign different time slots to these nodes in accordance with their needs [49]. Thus protocols are being designed to be more application sensitive and responsive without compromising energy efficient. Energy depletion of the cluster head because of heavy traffic is mitigated by round-robin-based algorithm used for efficient rotation of cluster headship. Energy is saved for nodes with low data traffic while enhancing channel utilisation and for heavy traffic nodes; latency is reduced while there is increase in throughput.

3.1.2. Summary of Controlled Access Protocol.

Generally CA protocols conserve nodes energy by avoiding collision since all the nodes are allocated timeslots during which they can transmit. For events that occur at a regular intervals, [47] saves energy by ensuring that node's wake up coincides with event occurrence. But it is not scalable and idle listening addressed in [50] still occurs. Adaptive and dynamic protocols were presented [48] [49] [50]. These adaptively turn ON or off nodes according to the network traffic load demands. This technique saves energy that could have been wasted in idle listening for fixed duty cycles and enhances channel utilisation.

3.2. Random Access (RA) Protocol

RA protocols are less compounded than the CA protocols and also they can be completely distributed thus endangering more scalability [51]. CSMA/CA is used by the nodes to access the medium with no master-slave relationships but all nodes compete to gain access to the channel. Less processing and smaller memory are required in RA because of no need to schedule all the nodes thereby reducing control overhead which is the main source of energy drain in CA. Invariably, the rate of collision is higher and actually the main concern in RA protocols. CSMA/CA, though has good scalability, consumes more power and offers low bandwidth utilisation during heavy traffic. Also RA uses preamble sampling or low power listening which occupies the channel for longer time than data packets while hidden stations' preambles keep colliding.

3.2.1 Review Random Access Protocols

The probability of collision remains constant with fixed contention windows. This means that all the sensors will compete during each successive contention window after a collision. Collision entails loss of packets and retransmission which waste energy. A technique to enhance the contention window was proposed in [25]. The probability of collision was reduced by halving contending probability which is the likelihood that a sensor will wake-up and contend to access the medium. Hence for any collision, the number of sensors that will be contending in the subsequent contention period will be reduced by half, thus generating a probability sequence $\frac{1}{2},\frac{1}{4},\frac{1}{8}$... and simple probability computation reduces overhead.

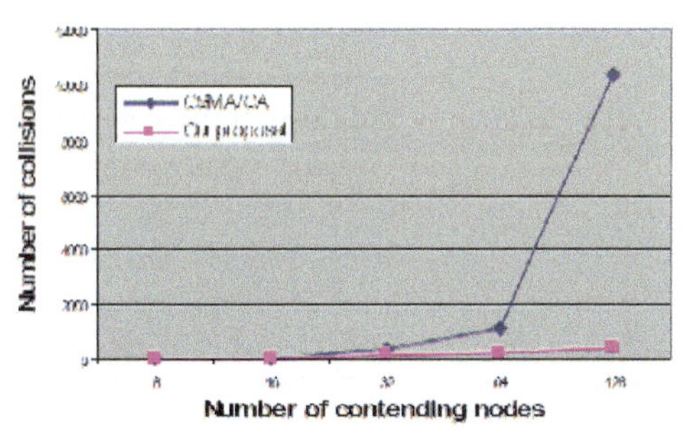

Figure 6: CSMA\CA vs protocol presented in [25]

Figure 6 shows that collisions increase as the number of nodes increase. The protocol proposed in [25] is more energy efficient as it reduces the number of collisions considerably. Less collision preserves energy since there will be less retransmissions. Nonetheless, collisions still occur and many nodes waste energy being awake and contending for channel access. Also in large networks, contending windows will be quite large, wasting energy, increasing latency and lowering bandwidth utilisation. NanoMAC[24] is a non-persistence CSMA/CA energy saving scheme. Contending nodes do not need to continuously listen to the medium, but sleep randomly in the contention window only sensing after backoff. Energy wasted by idle listening is conserved sine nodes listen randomly to the medium. However energy is wasted in carrier sensing and in collision of control packets. More energy is saved by similar protocol, High Efficient Sensor MAC [52] which minimises idle listening by allowing longer sleep periods. Nevertheless idle listening is not completely eliminated since nodes stay awake in case there are packets to transmit. All the nodes hear any on-going transmission and these constitute sources of energy inefficiency. A concern not addressed by [24] and [52] is collision still occurring after the back-off time. Algorithm proposed to reduce collision after back-off time was proposed in [53]. In this algorithm, a node randomly selects back-off period, and notifies others which then select their corresponding back-off time avoiding overlapping. Collision is thus eliminated thereby saving energy but overhead increases which consume more energy.

With reduced overhead and collision, [54] improved on the energy conservation of ML-MAC. The paper proposed a protocol called Adaptive Energy Efficient MAC protocol for Wireless Sensor Networks (AEEMAC). It reduces overhearing by causing nodes with no packets to send to go to sleep upon receipt of Clear-To-Send (CTS) destined for other nodes. There are three optimisation stages in the protocol. The first optimisation tries to reduce idle listening of standard S-MAC protocol by incorporating the duration of the communication in the control packets. When anode overhears the CTS, it knows the duration of the communication and goes to sleep until the end of the communication. The second optimisation considers the actual traffic load of the network and if there are no nodes with packets to send during the active cycle, the nodes will go back to sleep immediately. The last optimisation inserts RTS in ACK packets, reducing overhead and collision while engendering good channel utilisation. But if there are no packet bursts at the beginning of cycle and no multi-hop communications, the protocol will not be effective in saving energy.

3.2.2. Summary of Random Access Protocols

RA protocol is more scalable and do not incur much overhead costs, nonetheless, it is prone to collisions. Fixed probability of collision due to rigid contention window was addressed by [25]. It reduced the probability of contention by half whenever collision occurs while [24] uses probability, *p-1,* to refrain from contending for the medium. In both instances, energy was conserved by reducing collision though overhead was incurred. To enhance energy efficiency [52] reduced idle listening but suffer from repeated collisions after backoff. Collisions occurring even after backoff timer in [24] and [52] were addressed by [53]. None interfering backoff timer was employed to improve on the above protocols. Energy was saved as collisions were eliminated but control overhead increased. Dynamic combinations of control packets were used by [54] to reduce overhead thus increasing energy conservation [53]. Further energy preservation was achieved by making nodes that are not involved in transmission to go into energy saving sleep mode thus reducing idle listening.

3.3. Slotted Protocol (SP)

SP are used to avoid partial collision which is as a result of packets colliding with a part of another packet. Partial collision has the same effect as full collision as all the packets are lost. However, with SP, frames are divided into slots with duration longer than that required for a packet transmission. Stations are allowed to transmit only at the start of each slot hence collisions can only occur at the beginning of slot. The vulnerable time is equal to one frame duration [41].

3.3.1. Review of Slotted Protocol

A Self-Reorganising Slot Allocation protocol was proposed in [55] in which a TDMA MAC frame is kept by each cluster independent of other clusters. Inter cluster collision is avoided by carrier sensing. Whenever the medium is sensed busy, a *carrier sense-collision* is declared which informs the cluster head that there is an overlapping slot. This and hidden terminal issue were resolved by cluster heads reorganising slot allocations after each TDMA frame whenever any of these occurs. This approach increases overhead and there is also the probability that two cluster heads will embark on reorganisation and end up with overlapping slots all the time since slots are assigned independently. Frame scaling includes empty slots and increases frame duration while time slots are allocated to nodes irrespective of whether they have data to send or not. These reduce spectrum efficiency and increase latency. Most TDMA protocols assume that event detection is deterministic but in reality this is not so hence the need for an adaptive protocol that will take into consideration the non-deterministic nature of events. Dynamic Slot Assignment protocol was proposed in [56] and [57], to minimise the effect of nodes occupying the channel when they have no data to transmit. Cluster heads allocate time slots dynamically only to nodes with packets to transmit to cover transmission of all the packets. This saves idle energy and improves bandwidth employment. Nevertheless, if network traffic is heavy, the number of slot requests may exceed the available slots leading to loss of packets.

3.3.2. Summary of Slotted Protocols

In [55] cluster heads assign slots using TDMA to avoid collision and save energy but inter-cluster collisions might occur as a result of coinciding slots. Slot allocation in this protocol is fixed and universal leading to slot allocation to nodes without data to send. A dynamic protocol was introduced in [56] and [57] which dynamically assign slots to only nodes with data to send thereby improving on slot allocation of [55]. This eliminates the energy wastage, and improves throughput and bandwidth efficiency.

3.4. Hybrid Protocols (HP)

In low traffic, TDMA of CA protocol offers low channel utilisation, while in heavy traffic, CSMA of RA protocol is beset with collisions. Hybrid protocols were developed to combine the advantages of the CSMA, TDMA and other energy efficient MAC protocols to maximise energy efficiency, improve latency and spectrum utilisation.

3.4.1. Review Hybrid Protocols

Since no MAC protocol addressed all the sources of energy inefficiencies, hybrids of some of them were introduced to maximise the benefits of various protocols. Zebra-MAC (Z-MAC) [58] was developed to mitigate the short comings of the CSMA and TDMA based protocols while harnessing their advantages. Nodes perform carrier sensing prior to accessing the medium but priority is always given to nodes that own the slot. Each node is assigned a time slot but if it does not have any data to send other nodes will contend for the channel after a predefined set time. Only the slot owner and its one-hop neighbours can contend for the medium in high contention level (HCL). But all nodes can contend in the low contention level (LCL). Explicit congestion notification (ECN) messages are broadcast by a node, upon sensing heavy traffic on the network, to its two-hop neighbourhood to avoid hidden terminal problem. The protocol dynamically uses CSMA and TDMA in light and heavy traffics respectively. Since CSMA is more energy efficient in low traffic by avoiding idle listening, the protocol saves energy. The use of TDMA in heavy traffic reduces collision hence energy is preserved, thus engendering high channel utilisation. Idle listening, waiting for set time to elapse and clear channel assessment (CCA) all contribute to energy depletion and low throughput. Synchronisations of nodes within two hops and switching between TDMA and CSMA have overhead cost also.

A centralised hybrid scheme that uses both the principle of modified slotted contention-based and contention free protocols to preserve energy, was proposed in [59]. It improved on the energy consumption of BMA [60] by reducing control overhead. The cluster heads broadcast a schedule for all nodes with data to send while nodes without data go to sleep [56] and [57]. Synchronisation and defining of the superframe structure were done with beacons to reduce overhead and save energy. No node is required to know its ordering number or to synchronise with its one-hop or two-hop neighbours like in Z-MAC [58]. Compliance period and reservation were intruded to reduce overhearing and collision thereby conserving energy, while dynamic slot allocation improves channel efficiency. However, the protocol dealt only with intra-cluster collisions, a protocol that dealt with both intra [59] and inter-cluster collision was proffered in [61]. CA based TDMA, which reduces energy waste due to collision was used for intra-cluster medium access. RA based CSMA was used for inter-cluster spectrum access among cluster heads thus reducing energy inefficiency of control overhead. However in order to be responsive to network demands, adaptive sleep/active cycle was employed. Nodes on active mode if they have no packets to send or receive go back to sleep immediately, whereas those with more packets have their active cycle increased. Multi-hop communication, which improves energy efficiency, was used in [61] for transmission from cluster heads to the sink. But cluster heads are permanently on active mode and this will lead to energy wastage. Since TDMA is used in intra-cluster communications, during low traffic, there will be low channel utilisation. On the other hand, during heavy inter cluster communications, collisions will increase.

An emergency response (ER-MAC) hybrid protocol that works on a similar principle as Z-MAC, but saves more energy by avoiding contention by a node that owns a slot, was developed in [62]. Also it improves on [61] by eliminating permanently on cluster heads thus saving more energy. It is a multi-hop tree protocol that can be applied in events like patient monitoring, wild fire and intruder detections. In these applications, there may be no activity for a long time and suddenly, there might be an event sensed by different nodes that might require immediate reporting, thus bursts of packets. Packets are queued and prioritised in the protocol, and high priority packets are transmitted before low priority packets.

Though slot owner with high priority packet are exempted from contention, maintaining an update of the time remaining before a packet deadline expires implies lots of overhead control cost. The use of fixed frame in this protocol means that, it is not adaptive to the dynamics of network load traffic variations. Queue-length aware MAC (Queue-MAC) [63] is a multi-hop beacon enabled hybrid MAC protocol that addressed the issue of fixed cycle of [62]. It incorporates a dynamic duty cycled TDMA while the CSMA duty cycle remains fixed. This allows frames to be dynamically adjusted to make room for the transmission of more packets within a frame. Similarly, CSMA and TDMA are used interchangeably according the volume of traffic. Accordingly, making the protocol suitable for applications with fluctuating traffic and saving energy that would have been wasted for idle listening and collisions. Nonetheless, beacon, ACK packets and updating of the queue length indicator table will lead to increase in overhead energy cist.

3.4.2. Summary of Hybrid Protocols

Hybrid techniques were developed to optimise the gains of different protocols which are combined into one single protocol. Z-MAC [58] uses a combination of CSMA and TDMA to enhance energy efficiency, greater spectrum utilisation and throughput. But it involves

synchronisation within two-hop neighbourhood, CCA and idle listening which are all sources of energy wastages. An improvement on the overhead cost of [58] was made in [59] by using beacons for synchronisation and specifying superframe structure thus reducing overhead. Nevertheless, exchange control packets causes collisions and low spectrum usage, while long direct communication with the sink wastes energy. Inter and intra cluster link was the focus of [61], using TDMA for intra cluster link, and CSMA for inter-cluster communication which also improved on [59]. But having cluster heads permanently on the active mode wastes energy. Contention by slot owner with priority packets in Z-MAC was eliminated [62]. Nonetheless fixed frame of the scheme means that nodes with more packets than can be sent in one superframe cannot finish the transmission in one cycle. This was addressed in Queue-MAC [63] by the introduction of dynamic superframe with variable TDMA slot. The flexibility allows nodes to send all their packets in one superframe since the superframe can be adjusted according to the load demands of nodes. Nonetheless, overhead is incurred by the use of beacons and ACK.
Inroad has been made in developing energy efficient MAC protocols but challenges still abound as we have highlighted in this survey. Collision in RA when traffic is heavy has not been comprehensively addressed. Though dynamic cycle is improving on the energy wasted in idle listening in CA protocols under light traffic it requires nodes to access the medium for a short time in case it has packets to transmit.

4. CHALLENGES

Sources of energy inefficiencies in WSN have been highlighted in section 2 and a detailed review of WSN MAC protocols is presented in section 3. The challenges throw up from the survey are outlined in this section. Source of constant power supply has remained a major challenge in the development of WSNs. Suitable energy harvesting algorithm has not been developed and a steady or constant source of power supply for WSNs will significantly enhance the network. There have been improvements on channel utilisation and energy preservation by adaptive and dynamic protocols but the overhead is still high. Cross channel communication is an issue with channelization protocols as nodes have to switch between the channels. Switching between protocols is also a challenge in the implementation of hybrid protocols. Contention based protocols though provides good scalable network with efficient bandwidth utilisation, it wastes energy under heavy traffic and leads to network degradation. On the other hand control packets in CA waste energy. Complete eradication of these can transform WSN and its application. Death of nodes means that a network wide scalability should be introduced.

Routing consumes almost same amount of energy as communication. Routing topologies often make use of relay nodes like parent nodes or cluster heads. This causes uneven energy depletion on nodes at the root level or cluster heads.Another challenge is continuous sensing of same event. Some events do not require continuous sensing since the outcome does not change constantly. Data aggregation is being employed to reduce redundancy sampling but this may require nodes to perform data aggregation before forwarding packets. Possibility of node failure is a challenge as this will cause loss of all previous data from the source node en route to the sink. Also many nodes may sense the same event and take different routing parts to forward the data, which means that even data aggregation may not stop the transmission. Unneeded sampling requires further investigation as the sensing wastes both routing and communication energy. One other source issue is the energy used for maintaining the routing table as nodes need to know the next node nearest to the sink, reducing this will enhance network lifetime.

5. CONCLUSIONS AND FUTURE WORK

This paper reviewed MAC protocol for WSNs, sources and causes of energy inefficiencies and their consequences on the network. WSNs demand tailored MAC protocols because of energy constrains which is not as much an issue with many other wireless networks. Network lifetime extension through energy proficient MAC protocols is reviewed by grouping MAC protocols into four categories according to the methodology of medium access deployed. From the available literature, MAC protocols have evolved from fixed duty cycle or contention windows to adaptive, dynamic and flexible protocols. This has greatly enhanced energy efficiency especially when different protocols are merged in hybrid algorithms.

Most protocols involving clusters use direct communication link between the cluster head and the sink. Research into using a more energy efficient multi-hop communication protocol between the cluster heads and the sink will help conserve the energy, and accordingly extending the lifetime of the WSN. Most of the research on energy efficient controlled access protocol has come at a cost of control packets overhead. By investigating this research area further, an enhanced energy efficiency protocol may be developed that can revolutionise WSN's power consumption. Another challenge with CA protocols is that they are not scalable and this can be investigated further. Though CDMA involves complex computing, but also it can eliminate collision, developing a technique for wide application of this in WSN will boost energy efficiency, and hence, network lifetime of WSNs. This will be more so since processing consumes less energy than communication. Like CDMA, FDMA might improve on the energy budget of WSNs with appropriate protocol. TDMA appeared to be the leading base protocol for energy efficiency in WSNs

6. REFERENCES

[1] C. Zhu, C. Zheng, L. Shu and G. Han, "A survey on coverage and connectivity issues in wireless sensor networks," Journal of Network and Computer Applications, vol. 35, pp. 619-632, 2011.

[2] K. Majumder, S. Ray and S. K. Sarkar, "A Novel Energy Efficient Chain Based Hierarchical Routing Protocol for Wireless Sensor Networks," 2010.

[3] Wiley Series on Parallel and Distributed Computing, Algorithms and Protocols for Wireless Sensor Networks, A. Boukerche, Ed., New Jersey: John Wiley & Sons, Inc, 2009, pp. 437-519.

[4] K. Pahlavan and P. Krishnamurthy, Network Fundamentals: Wide, Local and Personal Area Communications, 1st ed., Chichester: John Wiley & Sons Ltd, 2009, pp. 559-591.

[5] W. L. Tan, W. C. Lau and O. Yue, "Performance analysis of an adaptive, energy-efficient MAC protocol for wireless sensor networks," Journal of Parallel and Distributed Computing, vol. 72, pp. 504-514, 2012.

[6] M. A. S. Jr, P. S. Barreto, C. B. Margi and T. C. Carvalho, "A survey on key management mechanisms for distributed Wireless Sensor Networks," Computer Networks , vol. 54, p. 2591–2612, 2010.

[7] T. Srisooksai, K. Keamarungsi, P. Lamsrichan and K. Araki, "Practical data compression in wireless sensor networks: A survey," Journal of Network and Computer Applications, vol. 35, pp. 37-59, 2011.

[8] W. Chebbi, M. Benjemaa, A. Kamoun, M. Jabloun and A. Sahli, "Development of a WSN Integrated Weather Station node for an Irrigation Alert Program under Tunisian Conditions," in 8th International Multi-Conference on Systems, Signals & Devices, 2011.

[9] M. López, J. Gómez, J. Sabater and A. Herms, "IEEE 802.15.4 based Wireless monitoring of pH and temperature in a fish farm," 2010.

[10] S. Verma, N. Chug and D. V. Gadre, "Wireless Sensor Network for Crop Field Monitoring," in ITC 2010, 2010.

[11] A. Hadjidj, A. Bouabdallah and Y. Challal, "Rehabilitation Supervision Using Wireless Sensor Networks," in WoWMoM 2011 International Symposium, 2011.

[12] M. P. Durisic, Z. Tafa, G. Dimic and V. Milutinovic, "A Survey of Military Applications of Wireless Sensor Networks," in MECO 2012, Bar, Montenegro, 2012.

[13] L. Buttyan, D. Gessner, A. Hessler and P. Langendoerfer, "APPLICATION OF WIRELESS SENSOR NETWORKS IN CRITICAL INFRASTRUCTURE PROTECTION: CHALLENGES AND DESIGN OPTIONS [Security and Privacy in Emerging Wireless Networks]," vol. 17, no. 5, pp. 44-49, Oct 2010.

[14] T. Seco, J. Berm´udez, J. Paniagua and J. Castellanos, "Dynamic and HeterogeneousWireless Sensor Networks for Virtual Instrumentation Services: Application to Perishable Goods Surveillance," in MASS 2011 , 2011.

[15] S. Trigui, A. Koubaa, M. Ben Jamaa, I. Chaari and K. Al-Shalfan, "Coordination in a Multi-Robot Surveillance Application using Wireless Sensor Networks," in MELECOM, 2012.

[16] A. Kamilaris, V. Trifa and A. Pitsillides, "HomeWeb: An Application Framework for Web-based Smart Homes," in ICT 2012, 2012.

[17] Y.-L. Tsou and S. Berber, "Design, Development and Testing of a Wireless Sensor Network for Medical Applications," in IWCMW, 2011.

[18] I. T. Leemt, M. Wu and C. Kim, "A MAC scheme for avoiding inter-cluster collisions in wireless sensor networks," in Advanced Communication Technology (ICACT), 2010 The 12th International Conference , 2010.

[19] "Technology," [Online]. Available: http://www.decentlab.com/index.php?id=2. [Accessed 18 Aug 2012].

[20] V. Ngo and A. Anpalagan, "A detailed review of energy-efficient medium access control protocols for mobile sensor networks," Computers and Electrical Engineering , vol. 36, p. 383–396, 2009.

[21] S. A. Gopalan, D.-H. Kim, J.-W. Nah and J.-T. Park, "A SURVEY ON POWER-EFFICIENT MAC PROTOCOLS FOR WIRELESS BODY AREA NETWORKS," in Proceedings of IC-BNMT, 2010.

[22] O. Bazan and M. Jaseemuddin, "A Survey On MAC Protocols for Wireless Adhoc Networks with Beamforming Antennas," IEEE COMMUNICATIONS SURVEYS & TUTORIALS, vol. 14, no. 2, pp. 216-239, 2012.

[23] A. F. Molisch, Wireless Communications, 2nd ed., Chichester: John Wiley, 2011, pp. 365-385, 731-750.

[24] J. Vidhya, G. Kalpana and P. Dananjayan, "Energy Minimization using NanoMAC Protocols in Multihop Wireless Sensor Network," in INTERNATIONAL CONFERENCE ON "CONTROL, AUTOMATION, COMMUNICATION AND ENERGY CONSERVATION -2009, , 2009.

[25] H.-C. Le, H. Guyennet and N. Zerhouni, "A New Contention Access Method for Collision Avoidance in Wireless Sensor Networks," in Proceedings of the Sixth International Conference on Networking, 2007.

[26] H. Feng, L. Ma and S. Leng, "A Low Overhead Wireless Sensor Networks MAC protocol," in Computer Engineering and Technology (ICCET), 2010 2nd International Conference , 2010.

[27] A. Bachir, M. Dohle, T. Watteyne and K. K. Leung, "MAC Essentials for Wireless Sensor Networks," IEEE COMMUNICATIONS SURVEYS & TUTORIALS, vol. 12, no. 2, pp. 222-248, 2010.

[28] P. Suriyachai, U. Roedig and A. Scott, "A Survey of MAC Protocols for Mission-Critical Applications in Wireless Sensor Networks," IEEE COMMUNICATIONS SURVEYS & TUTORIALS, vol. 14, no. 2, pp. 240-246, 2012.

[29] P. Huang, L. Xiao, S. Soltani, M. W. Mutka and N. Xi, "The Evolution of MAC Protocols in Wireless Sensor Networks: A Survey," in IEEE COMMUNICATIONS SURVEYS & TUTORIALS, ACCEPTED FOR PUBLICATION, 2012.

[30] A. Ghaddar, T. Razafindralambo, I. Simplot-Ryl, S. Tawbi and A. Hijazi, "Algorithm for data similarity measurements to reduce data redundancy in wireless sensor networks," 2010.

[31] W. C. I. Wassell, "Energy-efficient signal acquisition in wireless sensor networks: a compressive sensing framework," 2012.

[32] G. Anastasia, M. Contib, M. D. Francescoa and A. Passarella, "Energy conservation in wireless sensor networks: A survey," Ad Hoc Networks, vol. 7, no. 3, pp. 537-568, May 2008.

[33] Z. Aliouat and M. Aliouat, "Effective Energy Management in Routing Protocol for Wireless Sensor Networks," in NTMS, 2012.

[34] A. Kleerekoper and N. Filer, "Trading Latency for Load Balancing in Many-to-One Wireless Networks," in WTS, 2012.

[35] F. Aiello, F.L.Bellifemine, G.Fortino, S.Galzarano and R.Gravina, "An agent-based signal processing in-node environment for real-time human activity monitoring based on wireless body sensor networks," Engineering Applications of Artificial Intelligence, vol. 24, p. 1147–1161, 2011.

[36] F. Wang and J. Liu, "On Reliable Broadcast in Low Duty-Cycle Wireless Sensor Networks," vol. 11, pp. 767 - 779, May 2012.

[37] H. Yoo, M. Shim and D. Kim, "Dynamic Duty-Cycle Scheduling Schemes forEnergy-Harvesting Wireless Sensor Networks," IEEE COMMUNICATIONS LETTERS, vol. 16, no. 2, pp. 202-204, Feb 2012.

[38] N. K. Ray and A. K. Turuk, "A Review on Energy Efficient MAC Protocols for Wireless LANs," in Fourth International Conference on Industrial and Information Systems, ICIIS 2009, , Sri Lanka, 2009.

[39] N. S. Mazloum and O. Edfors, "DCW-MAC: An energy efficient medium access scheme using duty-cycled low-power wake-up receivers," 2011.

[40] T. S. Rappaport, Wireless Communications Principles and Practice, 2nd ed., Upper Saddle River: Prentice Hall, 2002, pp. 46-138, 395-417, 447-488.

[41] E. Zeinal-Hamadani, Multiplexing & Medium Access Control, 2011.

[42] C. Cirstea, "Energy Efficient Routing Protocols for Wireless Sensor Networks: A Survey," in 17th International Symposium for Design and Technology in Electronic Packaging (SIITME, Timisoara, 2011.

[43] O. Yang and W. B. Heinzelman, "Modeling and Performance Analysis for Duty-Cycled MAC Protocols with Applications to S-MAC and X-MAC," IEEE TRANSACTIONS ON MOBILE COMPUTING, vol. 11, no. 6, pp. 905-921, 2012.

[44] S. A. Gopalan and J.-T. Park, Energy-Efficient MAC Protocols for Wireless Body Area Networks: Survey, IEEE, 2010.

[45] F. Hamady, M. Sabra, Z. Sabra, A. Kayssi, A. Chehab and M. Mansour, "Enhancement of the S-MAC Protocol for Wireless Sensor Networks," 2010.

[46] A.-H. Lee, M.-H. Jing and C.-Y. Kao, "LMAC: An Energy-Latency Trade-off MAC Protocol for Wireless Sensor Networks," in International Symposium on Computer Science and its Applications, 2008.

[47] G. Amato, A. Caruso and S. Chessa, "Application-driven, energy-efficient communication in wireless sensor networks," in Computer Communications, 2009.

[48] C. Ceken, "An energy efficient and delay sensitive centralized MAC protocol for wireless sensor networks," Computer Standards & Interfaces, vol. 30, pp. 20-31, June 2007.

[49] B. Zhang, X. Wang, S. Li and L. Dong, "An Adaptive energy-efficient Medium Access Control Protocol For Wireless Sensor Networks," in Fifth International Conference on Mobile Ad-hoc and Sensor Networks, 2009.

[50] T.-H. H. P.-Y. Yen, "Adaptive time division multiple access-based medium access control protocol for energy conserving and data transmission in wireless sensor networks," IET Communications, vol. 5, no. 18, pp. 2662-2672, 2011.

[51] D. Messina, M. Ortolani and G. L. Re, "Adaptive Collision Avoidance through Implicit Acknowledgments in WSNs," in 2008 IEEE Asia-Pacific Services Computing Conference, 2008.

[52] S. Rashwand, J. Misic, V. Misic, S. Biswas and M. M. Haque, "A Novel Asynchronous, Energy Efficient, Low Transmission Delay MAC Protocol forWireless Sensor networks," in 2009 29th IEEE International Conference on Distributed Computing Systems Workshops, 2009.

[53] A. Samanta, D. Bakshiy, A. Mukherjeez and M. Nasipuri, "ENERGY EFFICIENT WIRELESS SENSOR MAC PROTOCOL FOR COLLISION AVOIDANCE," in 2009 First International Conference on Networks & Communications, 2009.

[54] A. Roy and N. Sarma, "AEEMAC: Adaptive Energy Efficient MAC Protocol for Wireless Sensor Networks," in India Conference (INDICON), 2011.

[55] T. Wu and S. Biswas, "A Self-Reorganizing Slot Allocation Protocol for Multi-cluster Sensor Networks," 2005.

[56] S. U. Hashmi, J. H. Sarker, H. T. Mouftah and N. D. Georganas, "An Efficient MAC Protocol with Correlated Connection Arrival and Variable Slot Assignment in Wireless Sensor Networks," 2010.

[57] S. U. Hashmi, J. H. Sarker, H. T. Mouftah and N. D. Georganas, "An Efficient TDMA Scheme with Dynamic Slot Assignment in Clustered Wireless Sensor Networks," 2010.

[58] I. Rhee, A. Warrier, M. Aia, J. Min and M. L. Sichitiu, "Z-MAC: A Hybrid MAC for Wireless Sensor Networks," IEEE/ACM TRANSACTIONS ON NETWORKING, vol. 16, no. 3, pp. 511-524, Jun 2008.

[59] H.-W. Cho, M.-H. Cho, J.-M. Chung and W.-C. Jeong, "A Centralized Hybrid MAC Protocol for Wireless Sensor Networks," 2007.

[60] J. Li and G. Y. Lazaroul, "A Bit-Map-Assisted Energy-Eff icient MAC Scheme for Wireless Sensor Networks," in IPSN'O4, Berkeley, 2004.

[61] M. Tan, L. Tang, H. Chang and H. Tian, A Hybrid MAC Protocol for Wireless Sensor Network, IEEE, 2009.

[62] L. Sitanayah, C. J. Sreenan and K. N. Brown, "ER-MAC: A Hybrid MAC Protocol for Emergency Response Wireless Sensor Networks," in 2010 Fourth International Conference on Sensor Technologies and Applications, 2010.

[63] S. Zhuo, Y.-Q. Song, Z. Wang and Z. Wang, "Queue-MAC: A queue-length aware hybrid CSMA/TDMA MAC protocol for providing dynamic adaptation to traffic and duty-cycle variation in wireless sensor networks," pp. 105-114, 2012.

[64] E. M. Shakshuki, T. R. Sheltami, H. Malik and C. Yang, "Investigation and Implementation of Border Nodes in S-MAC," in International Conference on Network-Based Information Systems, 2009.

[65] D. Saha and M. A. Matin, "A New Uni-Scheduling Algorithm in S-MAC based Wireless Sensor Networks," 2011.

[66] M. N. Sudha, M. Valarmathi and A. S. Babu, "Energy efficient data transmission in automatic irrigation system using wireless sensor networks," Computers and Electronics in Agriculture, no. 78, pp. 215-221, 2011.

[67] M. K. Jhaa, A. K. Pandeyb, D. Pala and A. Mohanc, "An energy-efficient multi-layer MAC (ML MAC) protocol for wireless sensor networks," International Journal of Electronics and Communications, vol. 65, pp. 209-216, 1 March 2010.

[68] S. Ray, I. Demirkol and W. Heinzelman, "ADV-MAC: Analysis and optimization of energy efficiency through data advertisements for wireless sensor networks," Ad Hoc Networks, vol. 9, p. 876–892, Oct 13 2010.

15

A REVIEW OF SECURITY ATTACKS AND INTRUSION DETECTION SCHEMES IN WIRELESS SENSOR NETWORK

Yassine MALEH[1] and Abdellah Ezzati[2]

Emerging Technologies Laboratory (VETE), Faculty of Sciences and Technology
Hassan 1st University, Settat, MOROCCO

ABSTRACT

Wireless sensor networks are currently the greatest innovation in the field of telecommunications. WSNs have a wide range of potential applications, including security and surveillance, control, actuation and maintenance of complex systems and fine-grain monitoring of indoor and outdoor environments. However security is one of the major aspects of Wireless sensor networks due to the resource limitations of sensor nodes. Those networks are facing several threats that affect their functioning and their life. In this paper we present security attacks in wireless sensor networks, and we focus on comparison and analysis of recent Intrusion Detection schemes in WSNs.

KEYWORDS

Wireless sensor Networks, Security, attack, Denial of Service (DoS), Intrusion Detection Systems (IDS), IDS Architectures, Cluster-based IDS, Anomaly-based IDS, Signature based IDS&Hybrid IDS

1. INTRODUCTION

Recent advances in wireless and micro electronic communications have enabled the development of a new type of wireless network called wireless sensor network (WSN).Wireless sensor networks are associated with vulnerable characteristics such as open-air transmission and self-organizing without a fixed infrastructure [1]. Consequently security of wireless sensor networks (WSN) is the most challenge for this type of network [2]. Intrusion Detection Systems (IDSs) can play an important role in detecting and preventing security attacks. This paper presents a review of the security attacks in wireless sensor network and analyzed some of the existing IDS models and architectures. Finally a comparative study and a discussion of IDS models will be presented.

2. RELATED WORK

Wireless sensor networks are not immune to the risks of destruction and decommissioning. Some of these risks are identical to those in Ad-Hoc networks, and others are specific to the sensors. Several articles [6][7][8][9][10] have presented security attacks and issues in WSNs. Intrusion detection system (IDS) defined as the second line of defense after cryptography, allows the detection and prevention of internal and external attacks.

In [18, it is presented a Rule-based IDS called also Signature-based. Most of the techniques in these schemes follow three main phases: data acquisition phase, rule application phase and intrusion detection phase.In [19], it is proposed two approaches to improve the security of clusters for sensor networks using IDS. The first approach usesamodel-based on authentication, and the second scheme is called Energy-Saving.IN [21] a hybrid intrusion detection system (HIDS) model has been anticipated for wireless sensor networks.This paper does not promote a solution. Rather, it is a comparative study of existing model of intrusion detection in wireless sensor networks. Our aim is to provide a better understanding of the current research issues in this field.

3. Security Goals in WSN

We can classify the security goals into two goals:mainand secondary. The main goals include security objectives that should be available in any system (confidentiality, availability, integrityandauthentication). The other category includes secondary goals (self-organization, secure localization, Time synchronization andResilience to attacks) [3] [4].

- Confidentiality (Forbid access to unwanted third parties)
- Authentication (Identity verification and validation)
- Availability (Service has to be always available)
- Integrity (Data is exchanged without malicious alteration)
- Self Organization(Every sensor node needs to be independent and flexible enough to be self-organizing and self-healing)
- Secure localization (Sensor network often needs location information accurately and automatically)
- Time synchronization (Sensor radio may be turned off periodically in order to conserve power)
- Resilience to attacks (The covenant of a single node must not violate the security of the whole network).Figure1belowsummarizessecurity goals for wirelesssensor network.

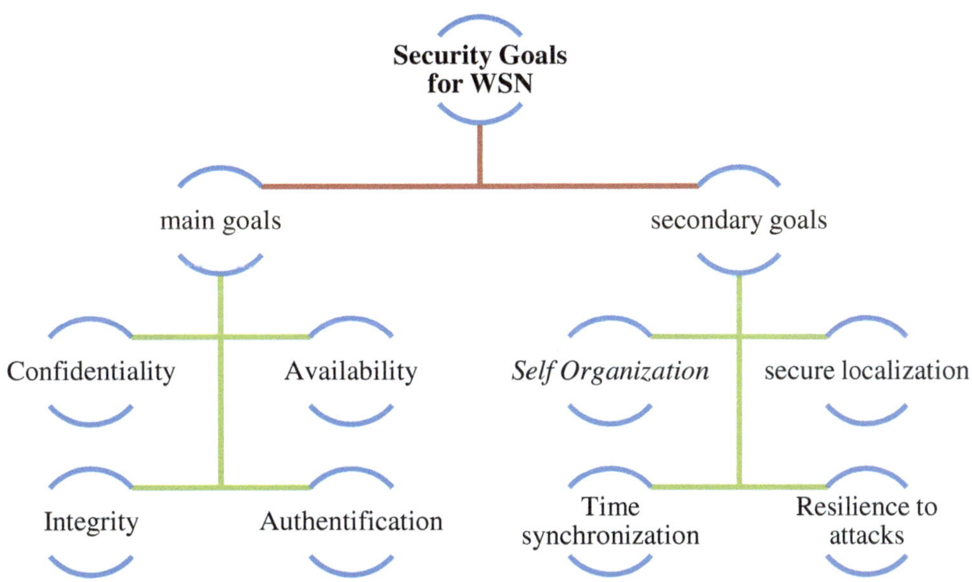

Figure1.Security Goals for WSN

4. SECURITY ATTACKS IN WSN

The different characteristics of wireless sensor networks (energy limited, low-power computing, use of radio waves, etc...) expose them to many securitythreats.We can classify the attacks into two main categories [5]: Active and Passive. In passive attacks, attackers are typically camouflaged, i.e. hidden, and tap the communication lines to collect data. In active attacks, malicious acts are carried out not only against data confidentiality but also data integrity.Several papers havepresentedthesecurityattacks in WSN [6][7][8][9][10].

- **Spoofed, altered or replayed routing information**

May be used for loop construction, attracting or repelling traffic, extend or shorten source route.

- **Selectiveforwarding**

In this attack, the attacker prevents the transmission of some packets. They will be removed later by the malicious node.

- **Worm holeattack:**

Thewormholeattackrequiresinsertion of atleasttwomaliciousnodes.Thesetwonodesareinterconnectedbyapowerful connection for examplea wired link. The malicious node receives packets in one section of the network and sends them to another section of the network.

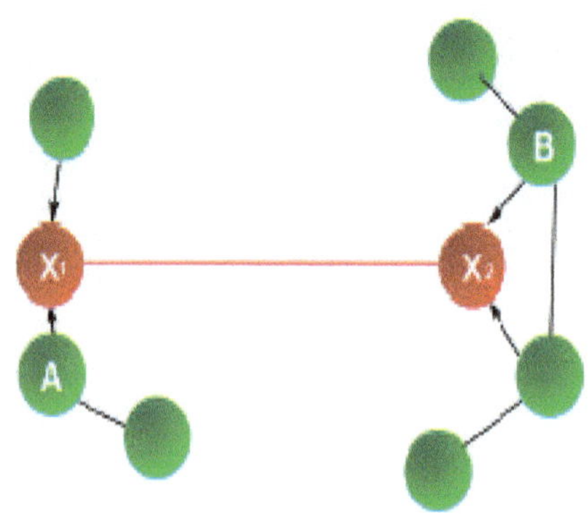

Figure 2. Worm hole attack

- **Sybil attack:**

A malicious node presents multiple identities to the other nodes in the network. This poses a significant threat to routing protocols and will cause the saturation of the routing tables of the nodes with incorrect information.

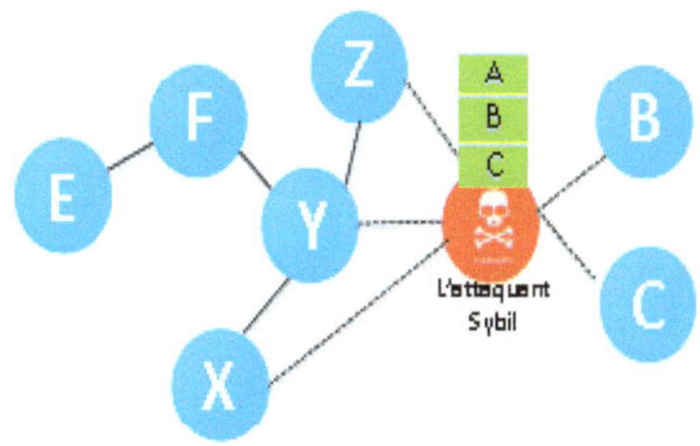

Figure 3. Sybil attack

- **Black hole attack:**

The attackinvolves inserting amaliciousnode in the network. This node, by various means, will modify the routing tablesto force themaximumneighboringnodespassingtheinformationthroughhim.Thenlike a black holeinspace,allthe information that willgoinitwill never beretransmitted.

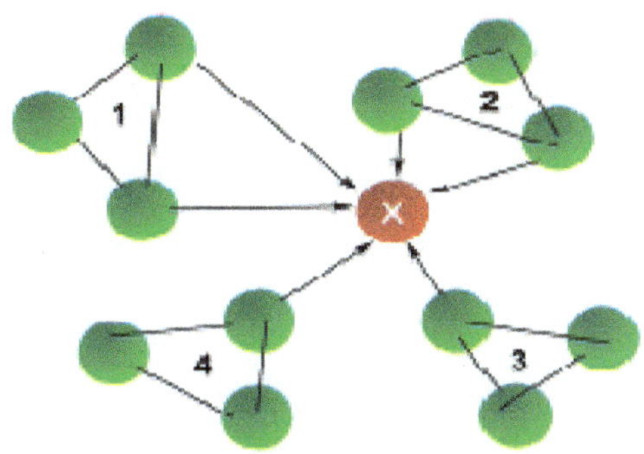

Figure 4. Black hole attack

- **Hello Flooding:**

Discoveryprotocolson WSNs useHELLOmessagestypesto discover itsneighboring nodes. In an attacktypeHELLOFlooding, an attacker will usethismechanismto saturate thenetworkandconsumeenergy.

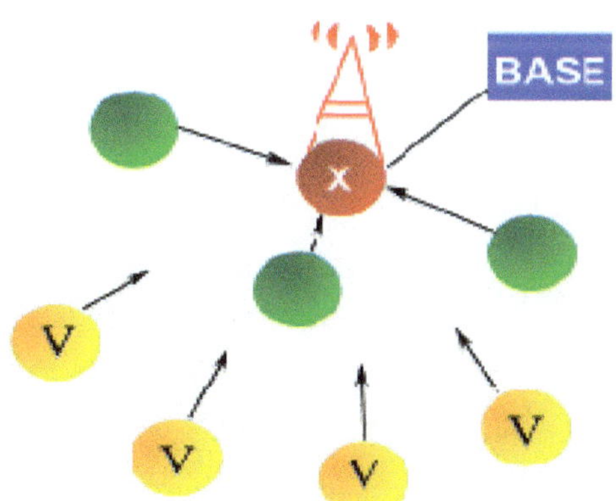

Figure 5. Hello flooding attack

- **Acknowledgementspoofing**

In this attack, the attacker tries to convince the sender that the weak link is strong or that a dead node is alive. Therefore, all packets passing through this link or this node will be lost.

- **Denial-of-Service Attacks**

A denial-of-service (DoS) targets the availability and capacity reduction of network services. Physical constraints of the sensor networks and the nature of their deployment environment, make them vulnerable to DoS attacks more than any other type of network. In this section we will review important DoS scenarios for each layer of the WSN. In [11] Wang *et al.* (2006) have classified the DoS attacks that could target each layer of the WSN.

Layer	Attacks	Defense
Physical	Jamming	Spread-spectrum, priority messages, lower duty cycle, region mapping, mode change
Link	Collision	Error-correction code
	Exhaustion	Rate limitation
	Unfairness	Small frames
Network	Spoofed routing information, andselective forwarding	Egressfiltering, authentication, monitoring
	Sinkhole	Redundancychecking
	Sybil	Authentication, monitoring, redundancy
	Wormhole	Authentication, probing
	Hello Flood	Authentication
Transport	Session Hijacking.	aggregationdata
	SYN flooding	Package authentication
Application	Data Corruption. Repudiation	Authentification

Table 1. Various DOS attacks on WSNs and their countermeasures

5. COUNTERMEASURES

To counter the attacks threatened networks wireless sensors, several research teams are trying to find appropriate solutions. These solutions must take into account the specificities of wireless sensor networks. We need to find simple solutions to secure the network while consuming the least possible energy and adapt these solutions to a low power computing. In the range of these solutions include mechanisms such as data partitioning, the use of appropriate cryptographic methods, intruder detection by location or even the confidence index. Wood and Stankovic [12] studied DoS attacks and possible defense. In [13][14]a suite of optimized security protocols for wireless sensor network is presented. SPIN (Security Protocol for Information via Negotiation) has two security mechanisms: SNEP and TESLA. SNEP provides data confidentiality and data authentication. TESLA provides source authentication in multicast scenarios by using MAC chaining. It is based on loose time synchronization between the sender and the receivers. INSENS (Intrusion Tolerant routing for wireless sensor networks) this protocol allows the base station to draw an accurate map of the network that will establish the routing tables for each node [15]. Du,et al. [16] propose LEAP+ (Localized Encryption and Authentication Protocol), a key management protocol for sensor networks.

6. INTRUSION DETECTIONSYSTEMSINWSN

Afterthe concept ofintrusiondetection(ID), which was established in 1980,two major variants of intrusion detection systems (IDS) have emerged, Host intrusion detection systems (HIDS) and network intrusion detection systems (NIDS) [17]. Intrusion detection is an approach that is complementary with respect to mainstream of security mechanisms such as cryptography and access control [18]. Intrusion detectioncan be defined as Intrusion detectioncan be defined astheautomatic detection andalarmgenerationtoreportthatan intrusion hasoccurredoris in progress. Inthis section we describethe architecture ofIDSinWSNs. IDS cannot takepreventive action, since they arepassivein nature, they can only detect intrusion and generate an alarm. The following figure presents the four main components of IDS [19].

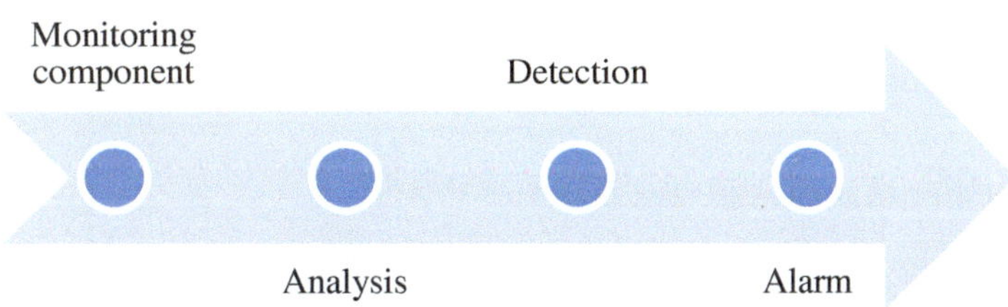

Figure 6.IDS components

There are two distinct technologies of IDS:

- Network Intrusion Detection System (NIDS). These systems are designed to intercept and analyze packets circulating in the network. All communication in the wireless network are conducted on the air and a node can hear the traffic passing from a neighboring node (promiscuous mode) [36]. Therefore, the nodes can mutually check the network traffic. This technology applies this concept, IDS listens for traffic and individually examine each packet.

- Host intrusion detection systems (HIDS). Analysis only data on the node where the IDS is installed. Any decision is based on information collected at this node. These IDSs use two types of sources to provide information about the activity: the log files (file that records all activity on a system in standby), and audit trails (Incoming / outgoing packets node , etc).

6.1 The challenging of designing IDS for WSN

The IDS solutions developed for wired networks cannot be applied directly to sensor networks, view the difference between these two types of networks, this is why it is necessary to introduce an intrusion detection system that meets the special features of sensor networks[20]. The design of this kind of system for wireless sensor network must satisfy the following properties:

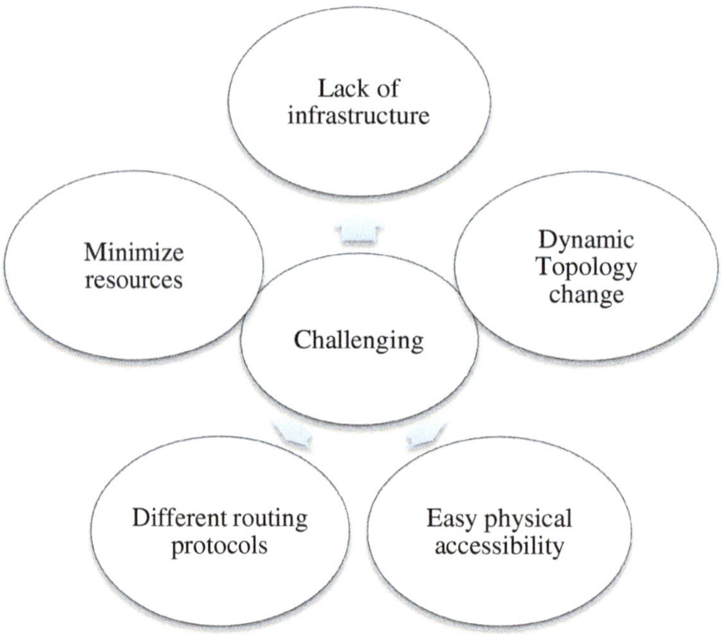

Figure 7. Challenging of designing IDS for WSN

6.2 The requirements of designing IDS for WSN

In wireless sensor networks, the IDS must satisfy the following properties [21]:

- Localize auditing: IDS forwirelesssensor networks mustworkwithlocaldataandpartialaudits, because in WSN there arenocentralizedpoints(apart from thestationbase) thatcancollectglobaldataauditing.
- Minimize resources: IDS must use a minimum number of resources for networks.Communicationbetweennodesforintrusiondetectionshouldnotsaturatetheavailable bandwidth.
- Trust no node: Unlikewired networks, nodes sensors canbecompromisedeasily, IDS must nottrustanynode.
- Bedistributed: means that thecollectionand analysis ofdata should beinseveral locations. Moreoverthedistributedapproachalsoappliestotheexecution of the algorithmofdetectionandalert correlation.
- Besecure: IDSmust be able towithstand attacks.

Figure 8 below summarizes requirements of designing IDS for WSN.

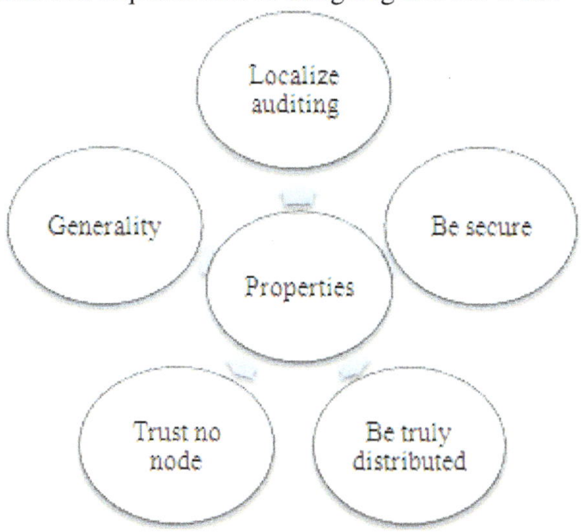

Figure 8.Requirements of designing IDS for WSN

6.3 Architectures for IDS in Wireless Sensor Network

The nature of wireless sensor networks makes them very vulnerable to attack.The Mobile nodes are randomly distributed, there are no physical obstacles for the adversary, therefore, they can be easily captured, and attacks can come from all directions and target any node. To tackle these additional challenges, several possible IDS architectures exist including standalone IDS, distributed and cooperative IDS and hierarchical IDS [22].

6.3.1 Standalone IDS

In this category, each node operatesas independent IDSandisresponsibleforthe detection ofattacksagainsthim.Therefore, theIDSdo not cooperateanddo not shareinformation with each other.This architecturerequires that eachnodeiscapable of executing and running IDS.

6.3.2 Distributed and Cooperative IDS

In this architecture (Zhang et al., 2003), each node has an IDS agent and makes local detection decisions by itself, all the nodes cooperate to create a global detection process. The distributed and cooperative IDS architecture is more suitable for a flat network configuration than a cluster-based multilayered one.

6.3.3 Hierarchical IDS

In this category the network is divided into clusters with cluster-heads. In each cluster, a leader plays the role of cluster-head. This node is responsible for routing in the group and must accept messages from members of the cluster indicating something malicious. Similarly, the cluster-head must detect attacks against other cluster-heads in the network. At the same time all cluster-heads can cooperate with central base station to form global IDS.

6.4 Some open research in IDS

Cross-Layer IDS: Using a cross layer IDS, we could not only pass information between layers but also coordinate mechanisms to prevent threats at all layers.
Dynamic IDS: The IDS that would protect mobile nodes, as in VANET networks.
Internet of Things IDS: There should be mechanisms that could manage all the objects of our everyday life that have an IP address and be connected to the Internet.

7. INTRUSION DETECTION MODELS FOR WSN

Due toarchitecturaldifferencebetweenwiredandwireless networks, their IDSs cannot be used interchangeably. There are specific techniques for WSN [23]. In this section, we analyze and discus some proposed IDSs for WSN.

7.1 Rule-based IDS

Rule-based IDS called also Signature-based IDS, articulates on a database of stored prior rules of security attacks [24]. Most of the techniques in these schemes follow three main phases: data acquisition phase, rule application phase and intrusion detection phase (Silva et al., 2005) [25].The algorithmincludesthreestepsfor detecting intrusions. In the first step monitor nodes monitors the data. In the second step detection ruleswillberankedin order of severity, to the collected information to flag failure. The third step is the intrusion detection phase, where the number of failure flagged is compared to the expected number of the occasional failures in the network.

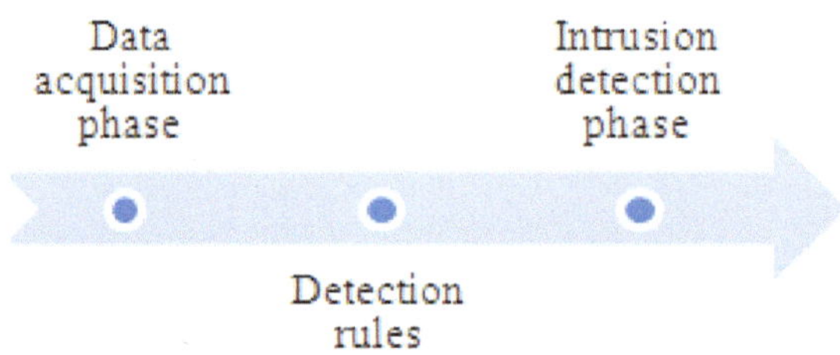

Figure 9. Steps for detecting intrusion in rule based IDS

7.2 Cluster-Based IDS

Su, et al. [26] hasproposed two approaches to improve the security of clusters for sensor networks using IDS. The first approach usesamodel-based on authentication, which can resisttoexternalattacks. Its basic technique is to add a message authentication code (MAC) for each message. Whenever a node wants to send a message, it adds to it a timestamp and a MAC is generated by a key-pair or individually depending on the key role of the sender (cluster-head, member -node, or base station). So that the receiver can verify the sender, the security mechanism is used LEAP. The second scheme is called Energy-Saving. This approach focuses on the detection of misbehavior both in Member nodes (MN) and cluster-head nodes (CH). When misbehavior is detected, the CH broadcasts a warning message encrypted with the cluster key to restrain this specific node.

7.3 Hybrid IDS

In the Hybrid Approach, both techniques (Cluster-Based and Rule-Based) arecombinedto form Hybrid detection technique.Hybrid detection exploitsthe advantages of bothapproachesprovides simplicity, high safety, low consumption of energy [27] [28].The Hybrid Intrusion Detection System achieves the goals of high detection rate and low false positive rate.

8. ANALYSES AND DISCUSSION

Comparing analysis, for the advantages and drawbacks of different models:
Rule-Based : The rule based model is simple, clear levels, anddesigned for a large-sized WSNs. Signature-based IDS need more resource than anomaly-based IDS, and regular updatingofthe database with new attack signatures.

Cluster-based: The cluster-based model requires Low Energy Consumption, provideshigh level of security. Because of Centralized routing data delivery is guaranteed. In cluster-based IDS Message retransmission frequency is high, and the centralized routing may not always use best available path for routing.

Hybrid model: Hybrid model are designed for large and sustainable WSN. Thismodelusestwomechanisms, anomaly-based and signature-based, so it requireshighconsumption of energy.

Table 2 gives the comparison and characteristics of different IDSs.

IDS Model	Network architecture	Detection technique	Handled attacks	Energy consumption	Advantages	Drawbacks
Anomaly based IDS		Anomaly Based	Masquerade, routing attacks, Sinkhole and blackhole	Low	Capable of detecting new attacks	Misses wellknown attack
Rule-based IDS	Distributed	Signature based	Black hole, selective forwarding, Sink hole, DOS	Low	Detects all those attacks having signatures	Cannotdetect new attacks
Cluster-based IDS	Hierarchical	Anomaly Based		Low	Low Energy Consumption Data delivery is guaranteed	Message retransmission frequency is high, Increased Traffic
Hybrid IDS	Hierarchical	Anomaly based	Selective forwarding, sinkhole, Hello flood and wormhole attacks	Medium	Can detect both existing and new attacks	Requires more computation and resources

Intrusion detection in Routing Attacks	Distributed	Anomaly based	DoS, Sinkhole and wormhole attacks	High	Consider resource Parameters (energy and reliability)	High resource Requirement, Increased Traffic

9. CONCLUSION

This article shows how well a security sensor networks is a challenge for researchers and developers of information technology.Our goal was to present the existing security attacks in WSN, focusing on intrusion detection systems (IDS), and examine existing approaches of intrusion detection in WSN.Our goal was to present the existing security mechanisms for WSN, specifically focusing on intrusion detection systems (IDS), and consider existing approaches to provide a fairly comprehensive and effective model. We are now working on our own model that incorporates all the advantages of the approaches proposed for a global model of intrusion detection in WSN.

REFERENCES

[1] Gang Zhao, "Wireless Sensor Networks for Industrial Process Monitoring and Control: A Survey", Network Protocols and Algorithms, ISSN 1943-358, Vol. 3, No. 1, 2011.

[2] G. Padmavath, D. Shanmugapriya, "A Survey of Attacks, Security Mechanisms and Challenges in Wireless Sensor Networks", (IJCSIS) International Journal of Computer Science and Information Security, Vol. 4, No. 1 & 2, 2009.

[3] QusayIdreesSarhana, "Security Attacks and Countermeasures for Wireless Sensor Networks: Survey", International Journal of Current Engineering and Technology ISSN 2277 – 4106, June 2013.

[4] Z. BENENSON, M. CHOLEWINSKI, C. FREILING, "Vulnerabilities and Attacks in Wireless Sensor Networks", Laboratory for Dependable Distributed Systems, University of Mannheim, 68131 Mannheim, Germany, 2010

[5] E.Çayırcı and C.Rong ,"Security in Wireless Ad Hoc and Sensor Networks", ISBN: 978-0-470-02748-6, 2009.

[6] P. Mohanty, S. Panigrahi, N. Sarma and S. Satapathy, "Security issues in wireless sensor network data gathering protocols: a survey", Department of Computer Science and Engineering Tezpur University, Tezpur, India 2010.

[7] Q. IdreesSarhan, "Security Attacks and Countermeasures for Wireless Sensor Networks: Survey", International Journal of Current Engineering and Technology, 2013.

[8] A.Singla, R. Sachdeva, "Review on Security Issues and Attacks in Wireless Sensor Networks", International Journal of Advanced Research in Computer Science and Software Engineering, Volume 3, Issue 4, April 2013.

[9] V. Soni1, P.Modi, V.Chaudhri, "Detecting Sinkhole Attack in Wireless Sensor Network", International Journal of Application or Innovation in Engineering & Management, Volume 2, Issue 2, February 2013.

[10] K. Sharma, M.Ghose, "Wireless Sensor Networks: An Overview on its Security Threats", IJCA Special Issue on "Mobile Ad-hoc Networks" MANETs, 2010.

[11] K. Sun, P. Peng, P. Ning, and C. Wang, "Secure Distributed Cluster Formation in Wireless SensorNetworks",in Proceedings of the 22nd AnnualComputer Security Applications Conference (AC-SAC'06), Pages: 131-140, December 2006.

[12] A. Wood and J. Stankovic, "Denial of service insensornetworks",IEEE Computer, pages 5462,October 2002.

[13] Ullah, Fasee, "Analysis of security protocols for Wireless Sensor Networks", Dept. of Comput. Sci., City Univ. of Sci. & Inf. Technol., Peshawar, Pakistan, Computer Research and Development (ICCRD), 2011.

[14] A. Perrig, R. Szewczyk, J.D Tygar, V. Wen abd D. Culler, "SPINS:Security Protocols f or Sensor Networks", Departement of electrical engineering and Computer Scinces, University of California, Berkley, USA 2002.

[15] J. Deng, R. Han, S. Mishra, "INSENS: Intrusion-Tolerant Routing in Wireless Sensor Networks", University of Colorado, Department of Computer Science, 2002.

[16] S. Zhu, S. Setia, and S. Jajodia, "LEAP+Efficient security mechanisms for large-scaledistributed sensor networks",ACMTransactionson Sensor Networks (TOSN), Volume 2, Issue 4, November 2006.

[17] S. Saha, Md. Safiqul Islam, Md. SakhawatHossen, "A New OHD Based Intrusion Detection System for Wireless Sensor Network", IK2206: Internet security and privacy, 2010.

[18] F. Amini, "Simulation and Evaluation of Security and Intrusion detection in IEEE 802.15.4 Network",university of Manitob 2008.

[19] Nabil Ali Alrajeh, S. Khan, and Bilal Shams, "Intrusion Detection Systems in Wireless Sensor Networks: A Review", International Journal of Distributed Sensor Networks, Volume 2013.

[20] Hassen Mohammed AbduallahAlsafi, 2 Saeed Salem Basamh, "A Review of Intrusion Detection System Schemes in Wireless Sensor Network", Journal of Emerging Trends in Computing and Information Sciences, 2013.

[21] Murad A. Rassam, M.A. Maarof and AnazidaZainal, "A Survey of IntrusionDetection Schemes in Wireless Sensor Networks", American Journal of Applied Sciences 9 (10): 1636-1652, 2012.

[22] Andreas A. Strikos, "A full approach for Intrusion Detection in Wireless Sensor Networks", School of Information and Communication Technology , Stockholm, Sweden ,March 1, 2007.

[23] R. Roman, J. Zhou, J. Lopez, "Applying Intrusion Detection Systems to Wireless Sensor Networks", Proceeding of the 3rd IEEE Consumer Communications and Networking Conference, 2006.

[24] Roosta, Tanya, Sameer Pai, Phoebus Chen, Shankar Sastry, and Stephen Wicker. "Inherent security of routing protocols in ad-hoc and sensor networks." In Global TelecommunicationsConference, 2007. GLOBECOM'07. IEEE, pp. 1273-1278. IEEE, 2007.

[25] A. da Silva, M. Martins, B. Rocha, A. Loureiro, L. Ruiz, and H. Wong, "Decentralized intrusion detection in wireless sensor networks", international workshop on Quality of service & security in wireless and mobile networks, 2005.

[26] C.-C. Su, K.-M.Chang, Y.-H.Kuo, and M.- F. Horng, "The new intrusion prevention and detection approaches for clustering-based sensor networks", in 2005 IEEE Wireless Communications and Networking Conference, WCNC 2005: Broadband Wirelss for the Masses - Ready for Take-off, 2005.

[27] A.Abduvaliyev, A.KPathan, J. Zhou, R. Roman and W. Wong, "On the Vital Areas of Intrusion Detection Systems in Wireless Sensor Networks", Communications Surveys & Tutorials, IEEE Volume15 , Issue 3, 2013.

[28] Mr. Ansar I SheikhMr.PankajKewadkar, "Approach towards Intrusion Detection System for Wireless Sensor Network", International Journal of Advanced Research in Computer and Communication Engineering Vol. 2, 2013.

16

Fault Detection and Recovery in Wireless Sensor Network Using Clustering

Abolfazl Akbari[1] , Arash Dana[2] , Ahmad Khademzadeh[3] and Neda Beikmahdavi[4]

[1]Dept. of Computer Engineering, Islamic Azad University Ayatollah Amoli Branch, Amol, Iran
[2]Islamic Azad University, Central Tehran Branch, Tehran, Iran
[3]Iran Telecom Research Center, Tehran, Iran
[4]Islamic Azad University Ayatollah Amoli Branch, Amol, Iran

Abstract— *Some WSN by a lot of immobile node and with the limited energy and without further charge of energy. Whereas extension of many sensor nodes and their operation. Hence it is normal.unactive nodes miss their communication in network, hence split the network. For avoidance split of network, we proposed a fault recovery corrupted node and Self Healing is necessary. In this Thesis, we design techniques to maintain the cluster structure in the event of failures caused by energy-drained nodes. Initially, node with the maximum residual energy in a cluster becomes cluster heed and node with the second maximum residual energy becomes secondary cluster heed. Later on, selection of cluster heed and secondary cluster heed will be based on available residual energy. We use Matlab software as simulation platform quantities. like, energy consumption at cluster and number of clusters is computed in evaluation of proposed algorithm. Eventually we evaluated and compare this proposed method against previous method and we demonstrate our model is better optimization than other method such as Venkataraman, in energy consumption rate.*

Key words: Sensor Networks, clustering, fault detection, fault recovery.

1. INTRODUCTION

Recent advances in MEMS (Micro-electro-mechanical systems) and wireless network technology have made the development of small, inexpensive, low power distributed devices, which are capable of local processing and wireless communication, a reality. Such devices are called sensor nodes. Sensors provide an easy solution to those applications that are based in the inhospitable and low maintenance areas where conventional approaches prove to be impossible and very costly. Sensors are generally equipped with limited data processing and communication capabilities and are usually deployed in an ad-hoc manner to in an area of interest to monitor events and gather data about the environment. Examples include environmental monitoring- which involves monitoring air soil and water, condition based maintenance, habitat monitoring, seismic detection, military surveillance, inventory tracking, smart spaces etc. Sensor nodes are typically disposable and expected to last until their energy drains. Therefore, it is vital to manage energy wisely in order to extend the life of the sensors for the duration of a particular task. [1-6]. Failures in sensor networks due to energy depletion are continuous and may increase. This often results in scenarios where a certain part of the network become energy constrained and stop operating after sometime. Sensor nodes failure may cause connectivity loss and in some cases network partitioning. In clustered networks, it creates holes in the network topology and disconnects the clusters, thereby causing data loss and connectivity loss [10]. Good numbers of fault tolerance solutions are available but they are limited at different levels. Existing approaches are based on hardware faults and consider hardware components malfunctioning only. Some assume that system software's are already fault tolerant as in [7, 8]. Some are solely focused on fault detection and do not provide any recovery mechanism [9]. Sensor network faults cannot be approached similarly as in traditional wired or wireless networks due to the following reasons [11]:

1. Traditional wired network protocol are not concerned with the energy consumptions as they are constantly powered and wireless ad hoc network are also rechargeable regularly.
2. Traditional network protocols aim to achieve point to point reliability, where as wireless sensor networks are more concerned with reliable event detection.
3. Faults occur more frequently in wireless sensor networks than traditional networks, where client machine, servers and routers are assumed to operate normally.

Therefore, it is important to identify failed nodes to guarantee network connectivity and avoid network partitioning. The New Algorithm recovery scheme is compared to Venkataraman algorithm proposed in [10]. The Venkataraman algorithm is the latest approach towards fault detection and recovery in wireless sensor networks and proven to be more efficient than some existing related work. It solely focused on nodes notifying its neighboring nodes to initiate the recovery mechanism. It can be observed from the simulation results that failure detection and recovery in our proposed algorithm is more energy efficient and quicker than that of Venkataraman algorithm. In [10], it has been found that Venkataraman algorithm is more energy efficient in comparison with Gupta and Younis [15].

Therefore, we conclude that our proposed algorithm is also more efficient than Gupta and Crash fault detection algorithm in term of fault detection and recovery.

This paper is organized as follows: Section 2 provides a brief review of related work in the literature. . In section 3, we provided a detail description of our clustering algorithm.

In section 4, we provided a detail description of our proposed solution. The performance evaluation of our proposed algorithm can be found in Section 5, Finally, section 6 concludes the paper.

2. RELATED WORK

In this section we will give an overview about existing fault detection and recovery approaches in wireless sensor networks. A survey on fault tolerance in wireless sensor networks can be found in [12]. A detailed description on fault detection and recovery is available at [11]. WinMS [13] provides a centralized fault management approach. It uses the central manager with global view of the network to continually analyses network states and executes corrective and preventive management actions according to management policies predefined by human managers. The central manager detects and localized fault by analyzing anomalies in sensor network models. The central manager analyses the collected topology map and the energy map information to detect faults and link qualities. WinMS is a centralized approach and approach is suitable for certain application. However, it is composed of various limitations. It is not scalable and cannot be used for large networks. Also, due to centralize mechanism all the traffic is directed to and from the central point. This creates communication overhead and quick energy depletions. Neighboring co-ordination is another approach to detect faulty nodes. . For Example, the algorithm proposed for faulty sensor identification in [16] is based on neighboring co-ordination. In this scheme, the reading of a sensor is compared with its neighboring' median reading, if the resulting difference is large or large but negative then the sensor is very likely to be faulty. Chihfan et. al [17] developed a Self monitoring fault detection model on the bases of accuracy. This scheme does not support network dynamics and required to be pre configured. In [14], a fault tolerance management architecture has been proposed called MANNA (Management architecture for wireless sensor networks). This approach is used for fault diagnosis using management architecture, termed as MANNA. This scheme creates a manager located externally to the wireless sensor network and has a global vision of the network and can perform complex operations that would not be possible inside the network. However, this scheme performs centralized diagnosis and requires an external manager. Also, the communication between nodes and the manager is too expensive for WSNs. In Crash fault

detection scheme [18], an initiator starts fault detection mechanism by gathering information of its neighbors to access the neighborhood and this process continue until all the faulty nodes are identified. Gathering neighboring nodes information consumes significant energy and time consuming. It does not perform recovery in terms of failure. Gupta algorithm [15] proposed a method to recover from a gateway fault. It incorporates two types of nodes: gateway nodes which are less energy constrained nodes (cluster headers) and sensor nodes which are energy constrained. The less energy constrained gateway nodes maintain the state of sensors as well as multi-hop route for collecting sensors. The disadvantage is that since the gateway nodes are less energy constraint and static than the rest of the network nodes and they are also fixed for the life of the network. Therefore sensor nodes close to the gateway node die quickly while creating holes near gateway nodes and decrease network connectivity.
Also, when a gateway node die, the cluster is dissolved and all its nodes are reallocated to other healthy gateways. This consume more time as all the cluster members are involved in the recovery process. Venkataraman algorithm [10], proposed a failure detection and recovery mechanism due to energy exhaustion. It focused on node notifying its neighboring nodes before it completely shut down due to energy exhaustion. They proposed four types of failure mechanism depending on the type of node in the cluster. The nodes in the cluster are classified into four types, boundary node, pre-boundary node, internal node and the cluster head. Boundary nodes do not require any recovery but pre-boundary node, internal node and the cluster head have to take appropriate actions to connect the cluster. Usually, if node energy becomes below a threshold value, it will send a fail_report_msg to its parent and children. This will initiate the failure recovery procedure so that failing node parent and children remain connected to the cluster.

3. Cluster formation

The sensor nodes are dispersed over a terrain and are assumed to be active nodes during clustering.

3.1.Problem Definition

The clustering strategy limits the admissible degree, *D* and the number of nodes in each cluster, *S*. The clustering aims to associate every node with one cluster. Every node does not violate the admissible degree constraint, *D* and every cluster does not violate the size constraint, *S* while forming the cluster. The number of clusters(C) in the network is restricted to a minimum of N/S, $N < C < N/S$, where N is the number of nodes in the terrain.

3.2.Sensor Network model

A set of sensors are deployed in a square terrain. The nodes possesses the following properties
i. The sensor nodes are stationary.
ii. The sensor nodes have a sensing range and a transmission range. The sensing range can be related to the transmission range, Rt> 2rs.
iii. Two nodes communicate with each other directly if they are within the transmission range
iv. The sensor nodes are assumed to be homogeneous i.e. they have the same processing power and initial energy.
v. The sensor nodes are assumed to use different power levels to communicate within and across clusters.
vi. The sensor nodes are assumed to know their location and the limits *S* and *D*.

3.3.Description of the Clustering Algorithm

Initially a set of sensor nodes are dispersed in the terrain. We assume that sensor nodes know their location and the limits *S* and *D*. Algorithms for estimating geographic or logical coordinates have been explored at length in the sensor network research [21,22].

In our algorithm, the first step is to calculate *Eth* and *Eic* for every node i, N <I< 1. Eth is the energy spent to communicate with the farthest next hop neighbor. *Eic* is the total energy spent on each link of its next hop neighbours. Every node *i* has an initial energy, *Einit*. A flag bit called *"covered flag"* is used to denote whether the node is a member of any cluster or not. It is set to 0 for each node initially.

3.4.Calculation of Eth and Eic

i. Nodes send a message *hello_msg* along with their coordinates which are received by nodes within the transmission range. For example in figure (1) nodes *a,b,c,d,w,x,y* are within transmission range of *v*.
ii. After receiving the *hello_msg*, the node *v* calculates the distance between itself and nodes *a,b,c,d,w,x,y* using the coordinates from *hello_msg*. It stores the distance *di* and the locations in the *dist_table*.
iii. Nodes within the sensing range are the neighbours of a node. In figure (1) nodes *w,x,y,b* are neighbours of *v*.
iv. Among the nodes within the sensing range, it chooses the first *D* closest neighbours as its potential candidates for next hop. Assuming *D*=3, in figure (1), the closest neighbours of *v* are *w,x,y*.

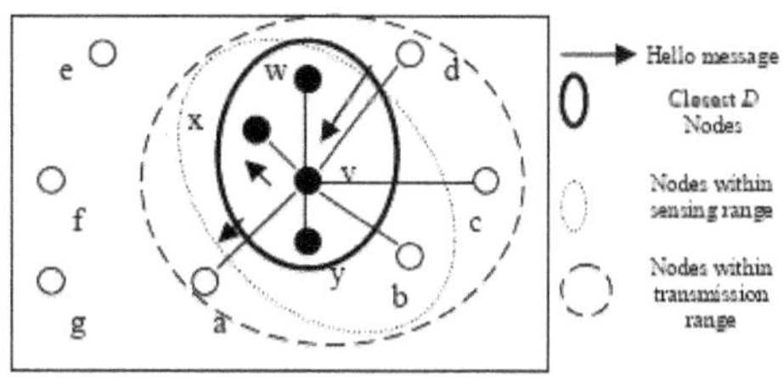

Fig1. Topology

v. Among the potential candidates, the farthest node's distance, *dmax* is taken for the calculation of *E th* .
vi. Suppose a node needs power E to transmit a message to another node who is at a distance 'd' away, we use the formula E = $E=kd^{c}$ [7,8], where k and c are constants for a specific wireless system. Usually 2<c<4. In our algorithm we assume k=1, c=2. For a node v, d $*1*d^2$max = E thv, since there are D members to which a node sends message.
vii. *Eic* is the total energy spent on each of link of the *D* closest neighbours. For a node *v*,

$$E_{ic\,v} = \sum_{i=1}^{D} d_{iv}^{2}$$

where k= 1. *div* is the distance between node *i* and node *v*. After the calculation of threshold energy *Eth*, nodes become eligible for cluster head position based on their energies. A node *v* becomes eligible for the cluster head position if its *Einit> Ethv. and.* node with the second *Einit> Ethv* becomes secondary Cluster heed.. When no nodes satisfy this condition or when there is insufficient number of cluster heads, the admissible degree *D* is reduced by one and then *Eth* is recalculated. The lowest value that *D* can reach is one. In a case where the condition *Einit> Ethv* is never satisfied at all, clustering is not possible because no node can support nodes other than itself. There may also be situations where all the nodes or more number of nodes are eligible for being cluster heads. A method has been devised by which the excess cluster heads are made to relinquish their position.

3.5. Relinquishing of the cluster head position

i. Every cluster head sends a message *cluster_head_status msg* and *Eic* to its neighbours (within sensing range).

ii. Every cluster head keeps a list of its neighbor cluster heads along with its *Eic*
iii. The nodes which receive *Eic* lesser than itself relinquishes its position as a cluster head.
iv. The cluster heads which are active send their messages to the cluster_head_manager outside the network. Thecluster_head_manager has the information of the desired cluster head count.
v. If the number of cluster heads are still much more than expected, then another round of cluster head relinquishing starts. This time the area covered would be greater than sensing range.
vi. The area covered for cluster head relinquishing keeps increasing till the desired count is reached.

3.6. Choosing cluster members

i. The cluster head select the closest *D* neighbours as next hop and sends them the message *cluster_join_msg*. The *cluster_join_msg* consists of cluster ID, *Sa*, *D*, *S*, *covered flag*. *Sa* is (*S*-1)/number of next hop members
ii. Energy is expended when messages are sent. This energy, *Eic* is calculated and reduced from the cluster head's energy.
iii. The cluster head's residual energy *Er* = *Einit* – *Eic*. *Einit* is
the initial energy when the cluster is formed by the cluster head.
iv. After receiving the *cluster_join_msg*, the nodes send a message, *cluster_join_confirm_msg* to the cluster head if they are uncovered, else they send a message, *cluster_join_reject_msg*.
v. After a *cluster_join_confirm_msg*, they set their *covered_flag* to 1.
vi. The next hop nodes now select *D*-1 members as their next hop members. *D*-1 members are selected because they are already associated with the node which selected them. For example, in figure (1) where *D* =3, a node *v* selects node *w*, node *x* and node *y*. In the next stage, node y selects
only node *a* and node *b* because it is already connected to node *v* making the *D* = 3.
vii. After selecting the next hop members, the residual energy is calculated , *Er* (new) = *Er* (old) - *Eic*
viii. This proceeds until S is reached or until all nodes have their *covered_flag* set to 1.

3.7.Tracking of the size

The size *S* of the cluster is tracked by each and every node. The cluster head accounts for itself and equally distributes *S*-1 among its next hop neighbors by sending a message to each one of them. The neighbours that receive the message account for themselves and distribute the remaining among all their neighbours except the parent. The messages propagate until they reach a stage where the size is exhausted. If the size is not satisfied, then the algorithms terminates if all the nodes have been covered. After the cluster formation, the cluster is ready for operation. The nodes communicate with each other for the period of network operation time.

4.Cluster Heed Failure recovery Algorithm

we employ a back up secondary cluster heed which will replace the cluster heed in case of failure. no further messages are required to send to other cluster members to inform them about the new cluster heed. Cluster heed and secondary cluster heed are known to their cluster members. If cluster heed energy drops below the threshold value, it then sends a message to its cluster member including secondary cluster heed. Which is an indication for secondary cluster heed to standup as a new cluster heed and the existing cell manager becomes common node and goes to a low computational mode. Common nodes will automatically start treating the secondary cluster heed as their new cluster heed and the new cluster heed upon receiving updates from its cluster members; choose a new secondary cluster heed. Recovery from cluster heed failure involved in invoking a backup node to standup as a new cluster heed.

5. PERFORMANCE EVALUATION

The energy model used is a simple model shown in [19] for the radio hardware energy dissipation where the transmitter dissipates energy to run the radio electronics and the power amplifier, and the receiver dissipates energy to run the radio electronics. In the simple radio model [19], the radio dissipates Eelec = 50 nJ/bit to run the transmitter or receiver circuitry and Eamp = 100 (pJ/bit)/m^2 for the transmit amplifier to achieve an acceptable signal-to-noise ratio. We use MATLAB Software as the simulation platform, a high performance discrete-event Java-based simulation engine that runs over a standard Java virtual machine.
The simulation parameters are explained in Table1.

Simulation parameters	Value
terrain dimensions	1 km^2
total number of nodes in terrain, *N*	100–1200
transmission range	100–450 m
cluster size limit, *S*	10–50
supportable degree, *D*	3–10

Table 1.Simulation parameters

5.1. Characteristics of the Clusters

Fig.1 depicts the percentage of cluster heads observed with varying cluster ranges. The cluster range was varied from 200 to 400. The size limit, S in our algorithm was set to 50 with admissible degree, D set to 3. The percentage of cluster heads was observed and noted for about 10 runs of the clustering algorithm. The percentage of cluster heads does not vary over various runs of the algorithm. This is because for a total number of N nodes in terrain, the limit S is set to 50 leading to N/50 cluster heads or clusters. Due to this limitation the results do not show a decrease or an increase in the cluster heads. Though the percentage of cluster heads is not changing, the role of cluster heads is exchanged among the nodes in the network.

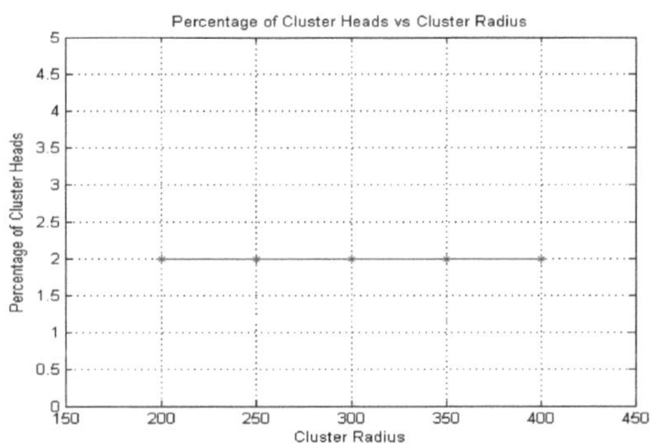

Fig2. Percentage of cluster heads with varying cluster radius

5.2. Energy characteristics in Clusters

Fig.3 depict the energy drain during the cluster formation. Energy drain refers to the loss of energy in each node after every round of cluster formation and operation. The energy loss is according to the relation in the first order radio model. Total energy loss would be the energy loss due to transmission added to the loss due to receiving. energy consumption depends on parameters used in first order radio model, distance and the number of bits, k. This energy consumption is also dependent on how many nodes would the concerned node be transmitting to and receiving from. In this clustering algorithm, the distance is sensing range, which is about 50 % of the transmission range. Also the number of nodes each node would handle is D. These two factors make the loss of energy at every stage uniform.

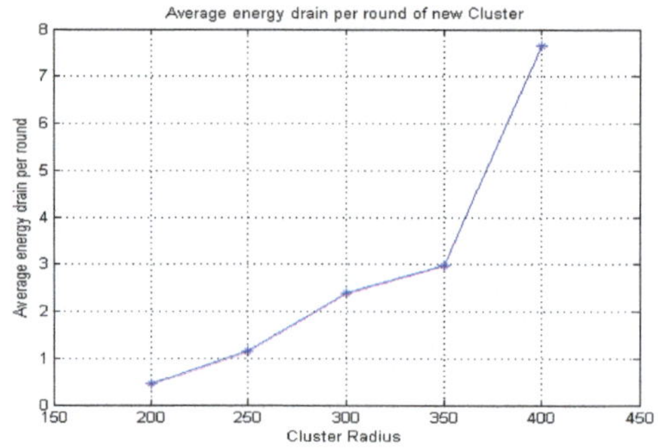

Fig 3. Ratio of average balance energy drain per round with varying cluster radius

5.3. Evalution New Algorithm

We compared our work with that of Venkataraman algorithm [10], which is based on recovery due to energy exhaustion.

In Venkataraman algorithm, nodes in the cluster are classified into four types: boundary node, pre-boundary node, internal node and the cluster head. Boundary nodes does not require any recovery but pre-boundary node, internal node and the cluster head have to take appropriate actions to connect the cluster. Usually, if node energy becomes below a threshold value, it will send a fail_report_msg to its parent and children. This will initiate the failure recovery procedure so that failing node parent and children remain connected to the cluster. A join_request_mesg is sent by the healthy child of the failing node to its neighbors. All the neighbors with in the transmission range respond with a join_reply_mesg/join_reject_mesg messages. The healthy child of the failing node then selects a suitable parent by checking whether the neighbor is not one among the children of the failing node and wEther the neighbor is also not a failing node. In our proposed mechanism, common nodes does not require any recovery but goes to low computational mode after informing their cell managers. In Venkataraman algorithm, cluster head failure cause its children to exchange energy messages. The children who are failing are not considered for the new cluster-head election. The healthy child with the maximum residual energy is selected as the new cluster head and sends a final_CH_mesg to its members. After the new cluster head is selected, the other children of the failing cluster head are attached to the new cluster head and the new cluster head becomes the parent for these children. This cluster head failure recovery procedure consumes more energy as it exchange energy messages to select the new cluster head. Also, if the child of the failing cluster head node is failing as well, then it also require appropriate steps to get connected to the cluster. This can abrupt network operation and is time consuming.

In our proposed algorithm, we employ a back up secondary cluster heed which will replace the cluster heed in case of failure.

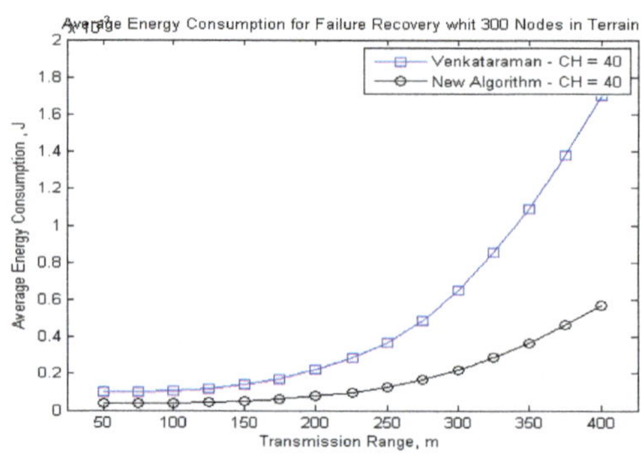

Fig4. Average time for cluster head recovery

no further messages are required to send to other cluster members to inform them about the new cluster heed Figs. 4 and 5 compare the average energy loss for the failure recovery of the three algorithms. It can be observed from Fig. 4 that when the transmission range increases, the greedy algorithm expends the maximum energy when compared with the Gupta algorithm and the proposed algorithm. However, in Fig. 5, it can be observed that the Gupta algorithm spends the maximum energy among the other algorithms when the number of nodes in the terrain increases.

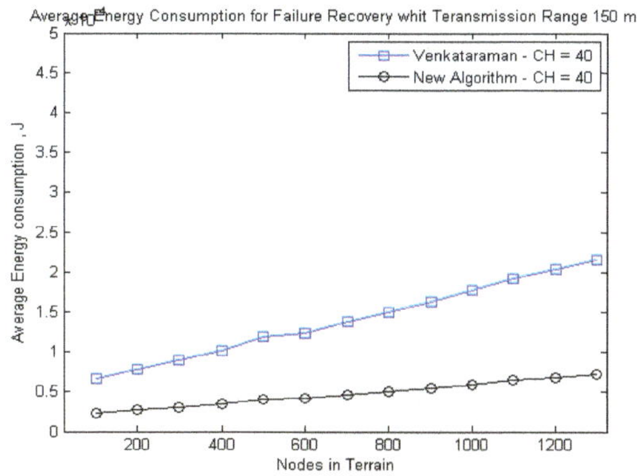

Fig5. Average time for cluster head recovery

VI. CONCLUSION

In this paper, we have proposed a cluster-based recovery algorithm, which is energy-efficient and responsive to network topology changes due to sensor node failures. The proposed cluster-head failure-recovery mechanism recovers the connectivity of the cluster in almost less than of the time taken by the fault-tolerant clustering proposed by Venkataraman . The Venkataraman algorithm is the latest approach towards fault detection and recovery in wireless sensor networks and proven to be more efficient than some existing related work. Venkataraman algorithm is more energy efficient in comparison with Gupta and Algorithm Greedy Therefore, we conclude that our proposed algorithm is also more efficient than Gupta and Greedy[20] algorithm in term of fault recovery. The faster response time of our algorithm ensures uninterrupted operation of the sensor networks and the energy efficiency contributes to a healthy lifetime for the prolonged operation of the sensor network.

VII. REFERENCES

[1] A. Bharathidasas, and V. Anand, “Sensor networks: An overview”, Technical report, Dept. of Computer Science, University of California at Davis, 2002

[2] D. Estrin, R. Govindan, J. Heidemann, and S. Kumar, "Next century challenges: Scalable coordination in sensor networks", in Proceedings of ACM Mobicom, Seattle, Washington, USA, August 1999, pp. 263-- 270, ACM.

[3] I.F. Akyildiz, W. Su, Y. Sankarasubramaniam and E. Cayirci, "A Survey on Sensor Networks", IEEE Communications Magazine, pp. 102--114, August 2002.

[4] D. Estrin, L. Girod, G. Pottie, M. Srivastava, “Instrumenting the world with wireless sensor networks”, In Proceedings of the International Conference on Acoustics, Speech and Signal Processing (ICASSP 2001.

[5] E. S. Biagioni and G. Sasaki, “Wireless sensor placement for reliable and efficient data collection”, in the 36th International Conference on Systems Sciences, Hawaii, January 2003.

[6] G. Gupta and M. Younis, “Load-Balanced Clustering in Wireless Sensor Networks”, in the Proceedings of International Conference on Communication (ICC 2003), Anchorage, AK, May 2003.

[7] J. Chen, S. Kher and A. Somani, “Distributed Fault Detection of Wireless Sensor Networks”, in DIWANS'06. 2006. Los Angeles, USA: ACM Pres.

[8] F. Koushanfar, M. Potkonjak, A. Sangiovanni- Vincentelli, "Fault Tolerance in Wireless Ad-hoc Sensor Networks", Proceedings of IEEE Sensors 2002, June, 2002.
[9] W. L. Lee, A. Datta, and R. Cardell-Oliver, "Network Management in Wireless Sensor Networks", to appear in Handbook on Mobile Ad Hoc and Pervasive Communications, edited by M. K. Denko and L. T. Yang, American Scientific Publishers.
[10] G. Venkataraman, S. Emmanuel and S.Thambipillai, "Energy-efficient cluster-based scheme for failure management in sensor networks" IET Commun, Volume 2, Issue 4, April 2008 Page(s):528 – 537
[11] L. Paradis and Q. Han, "A Survey of Fault Management in Wireless Sensor Networks", Journal of Network and Systems Management, vol. 15, no. 2, pp. 171-190, 2007.
[12] L. M. S. D. Souza, H. Vogt and M. Beigl, "A survey on fault tolerance in wireless sensor networks", 2007.
[13] W. L Lee, A.D., R. Cordell-Oliver, WinMS: Wireless Sensor Network-Management System, An Adaptive Policy-Based Management for Wireless Sensor Networks. 2006.
[14] L. B. Ruiz, I. G.Siqueira, L. B. Oliveira, H. C. Wong, J.M. S. Nigeria, and A. A. F. Loureiro. "Fault management in event-driven wireless sensor networks", MSWiM'04, October 4-6, 2004, Venezia, Italy
[15] G. Gupta and M. Younis; Fault tolerant clustering of wireless sensor networks; WCNC'03, pp. 1579.1584.
[16] M. Ding, D. Chen, K. Xing, and X. Cheng, "Localized fault-tolerant event boundary detection in sensor networks", in Proceedings of the 24th Annual Joint Conference of the IEEE Computer and Communications Societies (INFOCOM '05), vol. 2, pp. 902–913, Miami, Fla, USA, March 2005
[17] C. Hsin and M.Liu, "Self-monitoring of Wireless Sensor Networks", Computer Communications, 2005. 29: p. 462-478
[18] S. Chessa and P. Santi, "Crash faults identification in wireless sensor networks", Comput. Commun., 2002, 25, (14), pp. 1273-1282.
[19] W. R. Heinzelman, A. Chandrakasan, and H. Balakrishnan, "Energy-Efficient Communication Protocol for Wireless Microsensor Networks," Proc. Hawaii Int'l Conf. System Sciences 2000.
[20] GUPTA G., YOUNIS M.: 'Fault-tolerant clustering of wireless sensor networks'. Proc. IEEE WCNC, New Orleans, USA,March 2003, vol. 3, p. 1579 – 1584
[21] N. Bulusu, J. Heidemann and D. Estrin, "GPS-less Low Cost Outdoor Localization For Very Small Devices", IEEE Personal Communications, Special Issue on "Smart Spaces and Environments", Vol. 7, No. 5, pp. 28-34, October 2000.
[22] Radhika Nagpal, "Organizing a Global Coordinate System from Local Information on an Amorphous Computer", MIT AI Memo 1666, August 1999

DATA MINING OF NS-2 TRACE FILE

Ahmed Jawad Kadhim

IRAQ- Ministry of Education-
General Directorate of Education in Qadisiyah-
Ibn al Nafis School for Boys

ABSTRACT

Data mining is important process to extract the useful information and pattern from huge amount of data. NS-2 is an efficient tool to build the environment of network. The results from simulate these environment in NS-2 is trace file that contains several columns and lines represent the network events. This trace file can be used to analyse the network according to performance metrics but it has redundant columns and rows. So, this paper is to perform the data mining in order to find only the necessary information in analysis operation to reduce the execution time and the storage size of the trace file.

KEYWORDS

Data mining, NS-2, old trace file, new trace file.

1. INTRODUCTION

Data mining is more suitable process to find the novel, useful, and previously unknown information and knowledge from huge data [1]. This information and knowledge can be used for many applications such as fraud detection, science exploration, and customer to make certain decision to develop their businesses. Many people use term Knowledge Discovery from Data (KDD). But others see that KDD contains sequence of steps that are data cleaning, data integration, data selection, data, transformation, data mining, pattern evaluation, and knowledge presentation. They consider that data mining is an essential and required step in KDD [2]. There are many algorithms used for KDD such as classification, regression, clustering, neural networks, decision tree, association rule, genetic algorithm, and artificial intelligence [3].

2. THE NETWORK SIMULATOR (NS-2)

The network simulator (NS-2) is a discrete event simulator that has several capabilities to build the network environment with all network layers. It can simulate the wire, wireless network, sensor, satellite, and etc [4]. This simulator is building depending on two programing language that are C++ and OTCL [5]. It has many types of mobility models and traffic generators. The results of the simulation that performed based on NS-2 are NAM file (display file) and trace file (analysis file) that differ in storage size according to the network size [6]. There are two types of the trace file:

2.1 OLD TRACE FILE

This type of the trace file has about 21 columns that records each event occurs through simulation time such as event type, event time, and packet number, etc.

2.2 NEW TRACE FILE

This type takes more memory size than old trace file because it has about 51 columns [7].

3. SIMULATION AND RESULT

The simulation of the mobile ad-hoc network is performed using NS-2 with variant numbers of nodes in order to obtained many trace files that differ in size, number of columns, and number of rows. The simulation applied with old and new trace file. The resulted trace files have large storage size and take more time in the analysis operation. So that, the data mining is applied on these trace files to extract the useful and essentially information for the analysis process. The most useful columns in the analysis of the network events in old trace file only 6 columns that are: event type, event time, object type, packet id, packet type, and packet size and all other columns are often redundant. In new trace file, only 7 columns are important that are: the event type, event time, object type, packet id, packet type, packet size, and the energy. These columns are choosing because only these columns are important and useful in the analysis operation that based on NS-2 to compute the performance metrics that determine the behaviour of the network, routing protocols, traffic generation, and etc. Table1 illustrates the results of data mining for number of nodes with old and new trace file. Figure1, figure2, figure3, figure4 shows the effect of data mining on the size of the old trace file for network with 30, 45, 60, 75 nodes.

Table1: The results of data mining for old and new trace file

Number of Nodes	Trace File Type	Trace File Size		Number of Column		Number of Rows	
		Before Mining	After Mining	Before Mining	After Mining	Before Mining	After Mining
30	Old	1.8 MB	231 KB	21	6	24969	7205
45	Old	3 MB	460 KB	21	6	38843	13913
60	Old	4 MB	698 KB	21	6	49460	20770
75	Old	6.6 MB	1.1 MB	21	6	78611	32660
30	New	4.1 MB	270 KB	51	7	24969	7205

45	New	6.6 MB	537 KB	51	7	38843	13913
60	New	8.8 MB	814 KB	51	7	49460	20770
75	New	14.2 MB	1.28 MB	51	7	78611	32660

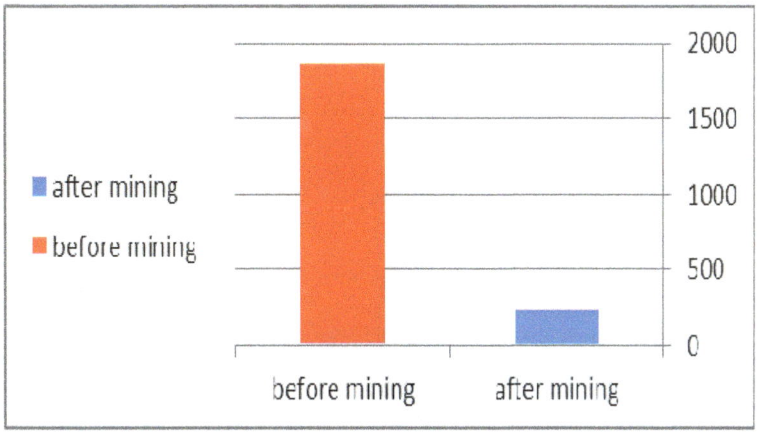

Fig1: The effect of data mining on the size of old trace file for network with 30 nodes

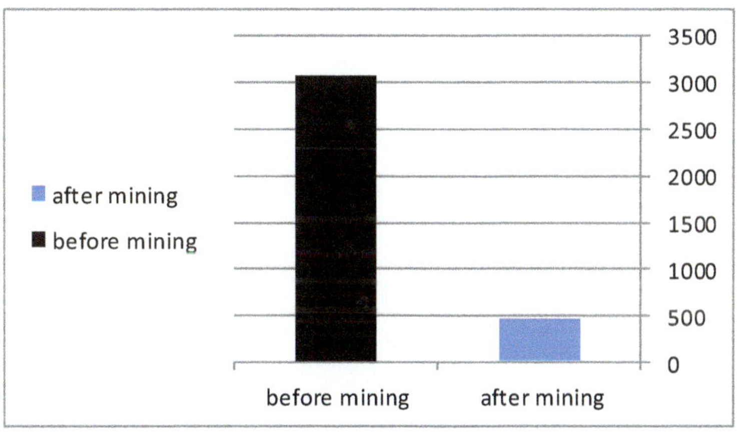

Fig2: The effect of data mining on the size of old trace file for network with 45 nodes

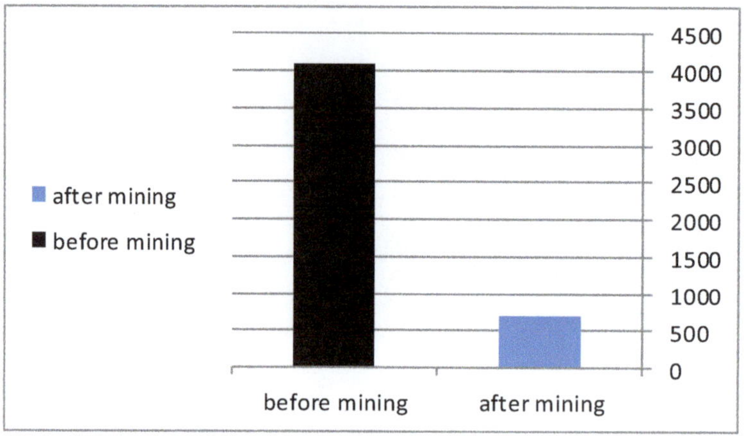

Fig3: The effect of data mining on the size of old trace file for network with 60 nodes

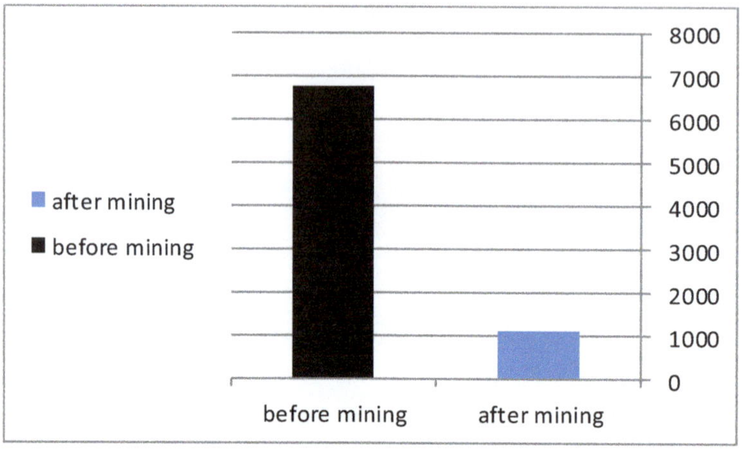

Fig4: The effect of data mining on the size of old trace file for network with 75 nodes

Figure5, figure6, figure7, figure8 illustrates the effect of data mining on the number of lines for the old trace file for network with 30, 45, 60, 75 node respectively.

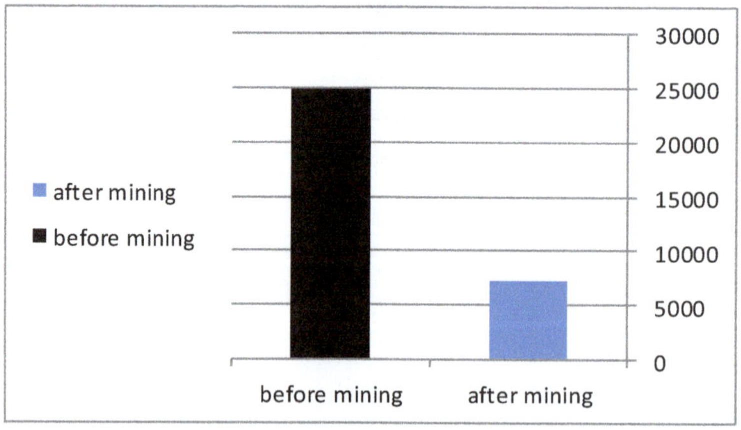

Fig5: The effect of data mining on the lines number of old trace file for network with 30 nodes

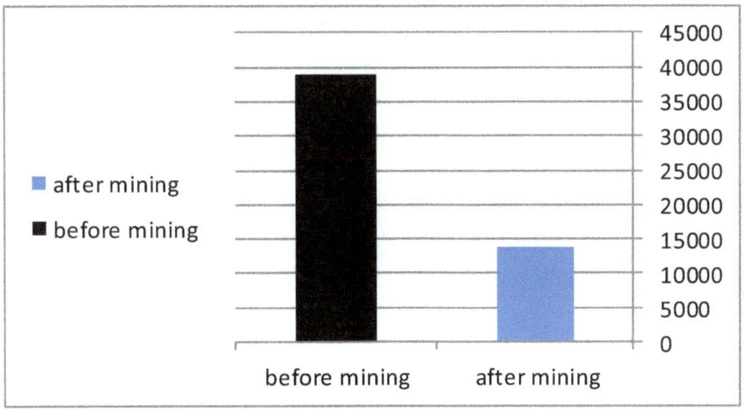

Fig6: The effect of data mining on the lines number of old trace file for network with 45 nodes

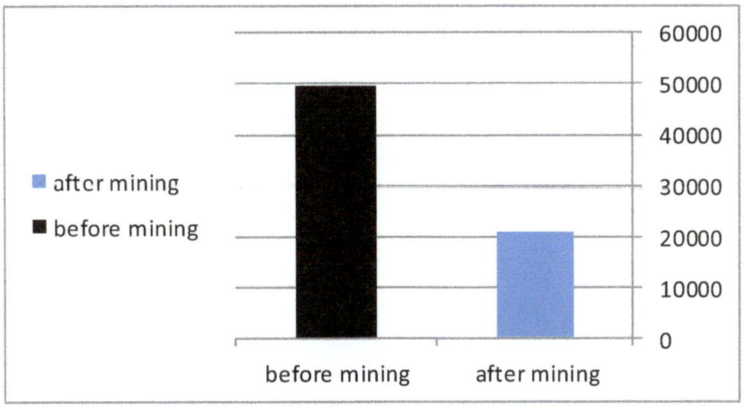

Fig7: The effect of data mining on the lines number of old trace file for network with 60 nodes

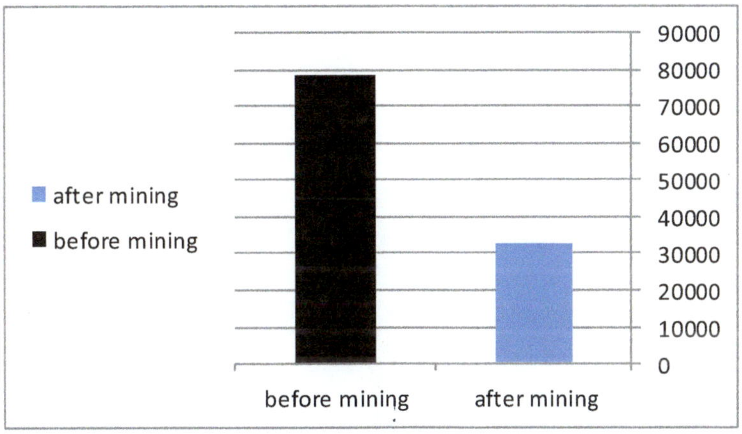

Fig8: The effect of data mining on the lines number of old trace file for network with 75 nodes

Figure9, figure10, figure11, figure12 represents the effect of data mining on the size of the new trace file for network with 30, 45, 60, 75 nodes.

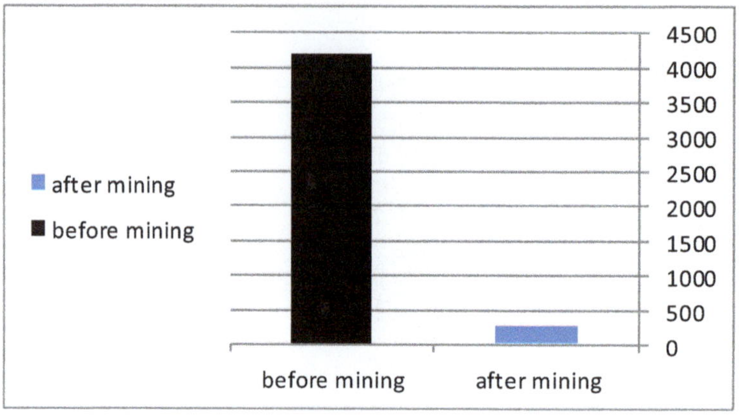

Fig9: The effect of data mining on the size of new trace file for network with 30 nodes

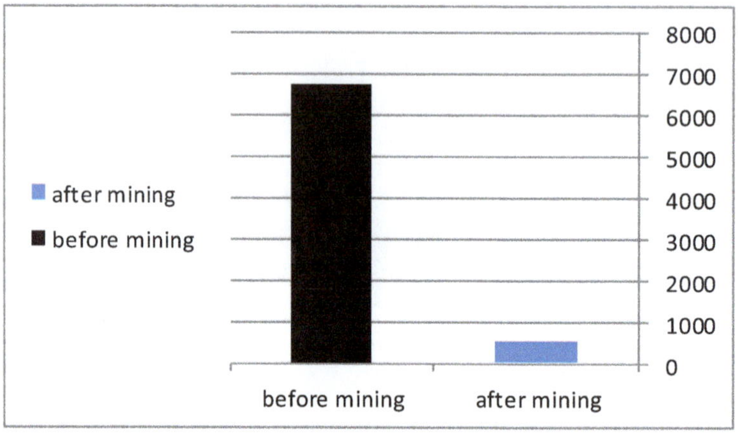

Fig10: The effect of data mining on the size of new trace file for network with 45 nodes

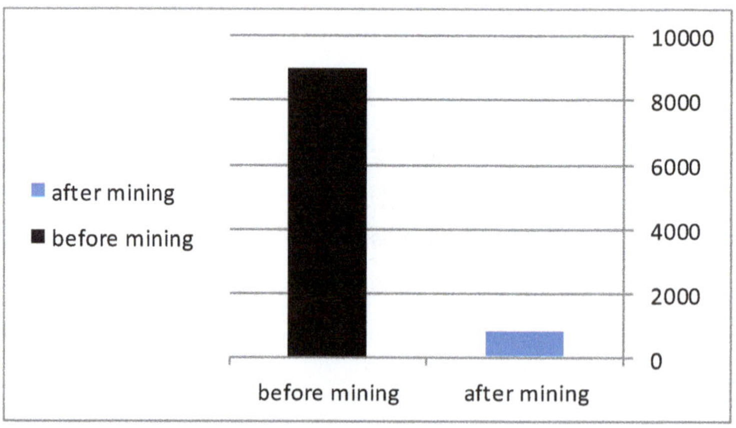

Fig11: The effect of data mining on the size of new trace file for network with 60 nodes

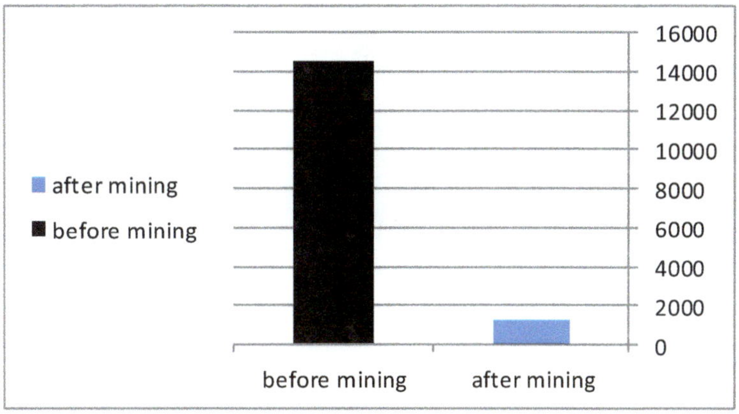

Fig12: The effect of data mining on the size of new trace file for network with 75 nodes.

Figure13, figure14, figure15, figure16 illustrates the effect of data mining on the number of lines for the new trace file to network with 30, 45, 60, 75 node respectively.

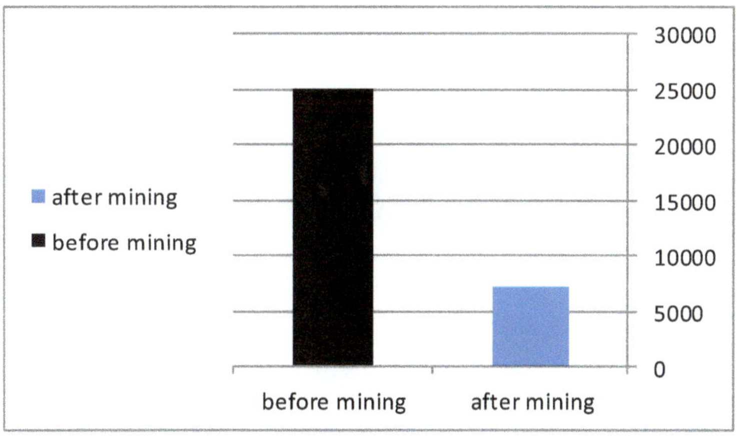

Fig13: The effect of data mining on the lines number of new trace file for network with 30 nodes

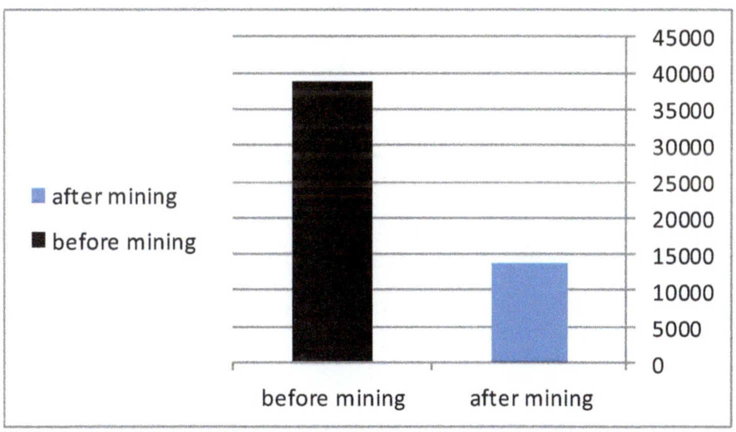

Fig14: The effect of data mining on the lines number of new trace file for network with 45 nodes

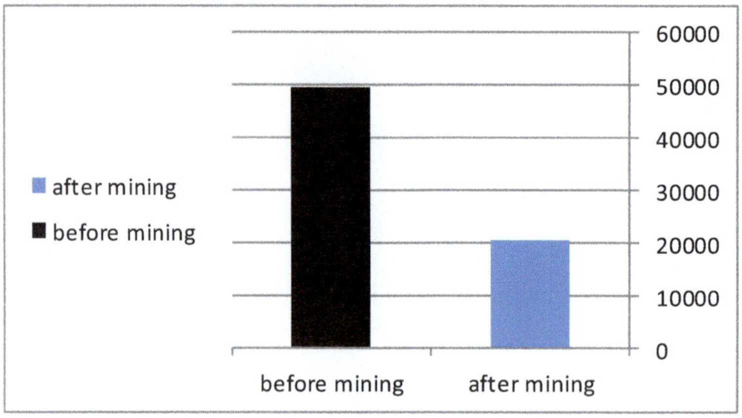

Fig15: The effect of data mining on the lines number of new trace file for network with 60 nodes

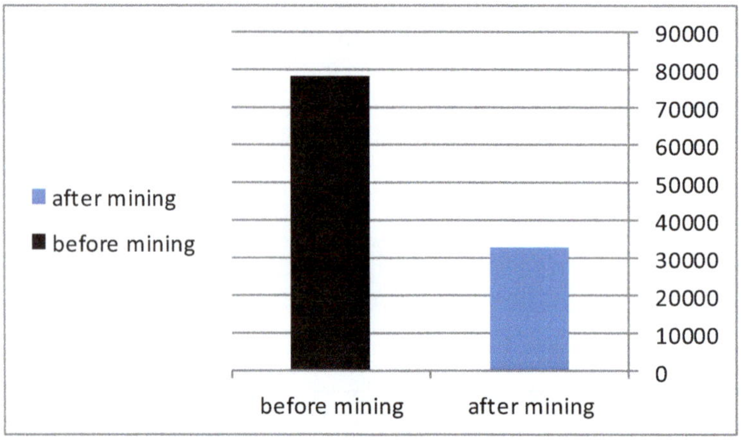

Fig16: The effect of data mining on the lines number of new trace file for network with 75 nodes

4. Conclusion

After applied the data mining on the trace files, the execution time decrease about 90% that the time for file before mining. Therefore, the computation speed increases about 90%. The size of the trace file decrease approximately 80%. These results for both old and new trace file. The data mining is a good approach to reduce time and size of NS-2 trace file.

5. References

[1] *Bharati M. Ramageri*, " DATA MINING TECHNIQUES AND APPLICATIONS", Indian Journal of Computer Science and Engineering Vol. 1 No. 4 301-305,

[2] Jiawei Han and Micheline Kamber, (2006) "Data Mining Concepts and Techniques", second edition, ISBN 13: 978-1-55860-901-3, ISBN 10: 1-55860-901-6.

[3] David Hand, Heikki Mannila and Padhraic Smyth, (2001) " Principles of Data Mining", ISBN: 026208290x.

[4] Saad Talib Al-jebori, ahmed jawad kadhim, (2012) " Design and Simulation of Optimal MANET Routing Protocol Selection System ", MSc. Thesis, science college, department of computer science, Babylon university.

[5] Jyotsna Rathee, (2009) "Simulation, Analysis and Comparison of DSDV Protocol in MANETS", MSc. thesis, Computer Science and Engineering Department, Thapar University.

[6] Teerawat Issariyakul and Ekram Hossain, (2009) "Introduction to Network Simulator NS2", springer, ISBN: 978-0-387-71759-3.

[7] Bjrn Wiberg, (2002) "Porting AODV-UU Implementation to ns-2 and Enabling Trace-based Simulation", MSc. thesis, Information Technology, Department of Computer Systems, Uppsala University, Sweden.

LOCAL DISTANCE'S NEIGHBOURING QUANTIFICATION

Meryem Saadoune[1] , Abdelmajid Hajami[2] and Hakim Allali[3]

Department of Mathematics and Computer Sciences, Hassan 1st University, Settat, Morocco

ABSTRACT

Mobility is one of the basic features that define an ad hoc network, an asset that leaves the field free for the nodes to move. The most important aspect of this kind of network turns into a great disadvantage when it comes to commercial applications, take as an example: the automotive networks that allow communication between a groups of vehicles. The ad hoc on-demand distance vector (AODV) routing protocol, designed for mobile ad hoc networks, has two main functions. First, it enables route establishment between a source and a destination node by initiating a route discovery process. Second, it maintains the active routes, which means finding alternative routes in a case of a link failure and deleting routes when they are no longer desired. In a highly mobile network those are demanding tasks to be performed efficiently and accurately. In this paper, we focused in the first point to enhance the local decision of each node in the network by the quantification of the mobility of their neighbours. Quantification is made around RSSI algorithm a well known distance estimation method.

KEYWORDS

Ad hoc, Mobility, RSSI, AODV, Localization, Distance, GPS-free.

1.INTRODUCTION

Mobile ad hoc network (MANET) is an appealing technology that has attracted lots of research efforts. Ad hoc networks are temporary networks with a dynamic topology which doesn't have any established infrastructure or centralized administration or standard support devices regularly available as conventional networks [1]. Mobile Ad Hoc Networks (MANETs) are a set of wireless mobile nodes that cooperatively form a network without infrastructure, those nodes can be computers or devices such as laptops, PDAs, mobile phones, pocket PC with wireless connectivity. The idea of forming a network without any existing infrastructure originates already from DARPA (Defense Advanced Research Projects Agency) packet radio network's days [2][3]. In general, an Ad hoc network is a network in which every node is potentially a router and every node is potentially mobile. The presence of wireless communication and mobility make an Ad hoc network unlike a traditional wired network and requires that the routing protocols used in an Ad hoc network be based on new and different principles. Routing protocols for traditional wired networks are designed to support tremendous numbers of nodes, but they assume that the relative position of the nodes will generally remain unchanged. In ad hoc, since the nodes are mobile, the network topology may change rapidly and unpredictably and the connectivity among the terminals may vary with time. However, since there is no fixed infrastructure in this network, each mobile node operates not only as a node but also as a router forwarding packets from one node to other mobile nodes in the network that are outside the range of the sender. Routing, as an act of transporting information from a source to a destination through intermediate nodes, is a fundamental issue for networks. [4]

The problem that arises in the context of ad hoc networks is an adaptation of the method of transport used with the large number of existing units in an environment characterized by modest computing capabilities and backup and fast topology changes.

According to the way of the creation and maintenance of roads in the routing of data, routing protocols can be separated into three categories, proactive, reactive and hybrid protocols. The pro-active protocols establish routes in advance based on the periodic exchange of the routing tables, while the reactive protocols seek routes to the request. A third approach, which combines the strengths of proactive and reactive schemes, is also presented. This is called a hybrid protocol.

Ad-hoc On-Demand Distance Vector routing protocol (AODV) [5] is a reactive routing protocol, who was standardized by the working group MANET [6] with IETF (Internet Engineering Task force), by the (RFC 3561).

The protocol's algorithm creates routes between nodes only when the routes are requested by the source nodes, giving the network the flexibility to allow nodes to enter and leave the network at will. Routes remain active only as long as data packets are traveling along the paths from the source to the destination .When the source stops sending packets, the path will time out and close.

In this paper we propose a solution that enables each node in the network to determine the location of its neighbors in order to create a more stable and less mobile road. For that purpose, we locally quantify the neighbor's distances of a node as the metric of mobility using AODV protocol.

The remainder of this paper is organized as follows. Section 2, describes briefly the AODV protocol. In Section 3, a summary of related work is presented. we present in Section 4 how to quantify, evaluate, estimate mobility in ad hoc network. Section 5 shows the algorithm used the quantification of the distance in AODV protocol. Section 6 presents some simulations and results. Finally Section 7 concludes this paper.

2. AD HOC ON-DEMAND DISTANCE VECTOR

AODV is an on-demand protocol which is capable of providing unicast, multicast [7], broadcast communication and Quality of Service aspects (QoS) [8], [9]. It combines mechanisms of discovery and maintenance roads of DSR (RFC 4728) [10] involving the sequence number (for maintains the consistency of routing information) and the periodic updates of DSDV [11].
At the discovery of routes, AODV maintains on each node transit information on the route discovery, the AODV routing tables contain:

- The destination address
- The next node
- The distance in number of nodes to traverse
- The sequence number of destination
- The expiry date of the entry of the table time.

When a node receives a packet route discovery (RREQ), it also notes in its routing table information from the source node and the node that just sent him the package, so it will be able to retransmit the response packet (RREP). This means that the links are necessarily symmetrical. The destination sequence number field of a route discovery request is null if the source has never been linked to the destination, else it uses the last known sequence number. It also indicates in this query its own sequence number. When an application sends a route discovery, the source

waits for a moment before rebroadcast its search query (RREQ) road, after a number of trials, it defines that the source is unreachable.
Maintained roads is done by periodically sends short message application called "HELLO" , if three consecutive messages are not received from a neighbor, the link in question is deemed to have failed . When a link between two nodes of a routing path becomes faulty, the nodes broadcast packets to indicate that the link is no longer valid. Once the source is prevented, it can restart a process of route discovery.

AODV maintains its routing tables according to their use, a neighbor is considered active as long as the node delivers packets for a given destination, beyond a certain time without transmission destination, the neighbor is considered inactive. An entered routing table is considered active if at least one of the active neighbors using the path between source and destination through active routing table entries is called the active path. If a link failure is detected, all entries of the routing tables participating in the active path are removed.

3. Related Work

In [12], a geometric mobility metric has been proposed to quantify the relative motion of nodes. The mobility measure between any pair of nodes is defined as their absolute relative speed taken as an average over time. This metric has certain deficiencies: First, it assumes a GPS like scheme for calculation of relative speeds while in a MANET, we cannot assume the existence of GPS, so we have to resort to other techniques for measuring relative mobility. Secondly, it is an "aggregate" mobility metric and does not characterize the local movement of the neighboring nodes to another particular node.

The Reference Point Group Mobility Model (RPGM) proposed in [13] was useful for predictive group mobility management. In RPGM, each group has a logical "center" and the center's motion defines the entire group's motion behavior including location, velocity, acceleration etc.

In [14], They proposed a measure of the network mobility which is relative and depending on neighboring and link state changes. Each node estimates its relative mobility, based on changes of the links in its neighboring. This measure of mobility has no unit, it is independent of any existing mobility models and it is calculated at regular time intervals.

The degree mobility used in [15] was calculated from the change of its neighboring to each node in time. The node mobility degree, represents at a given time t for each node in the ad hoc network, the change variations undergone in its neighboring compared to the previous time $t-1-1$. Thus, nodes that join or/and leave the neighboring of a given node will have surely an influence on the evaluation of its mobility.

However, the last two measures are not representative's values of a change node's motion with respect to another node.

We can see that none of the metrics described above are suitable for characterizing the relative mobility of nodes in a particular node's neighborhood in a MANET. Hence, we feel that there is a need to develop such a metric which can be used by any routing protocol.

4. Local Quantification of Neighbouring Mobility

In this section, we define how we estimate nodes mobility in ad hoc network. Mobility is quantified locally and independently of localization of a given node. We represent this local quantification node mobility, by calculating of the distance between a node and its neighbors.

The quantification of the distance can be done using 3 methods:

Calculate the exact distance: this is done by two ways:

1st way: ***The distance calculation using the GPS***: This operation is done by using a terminal capable of being localized through a positioning system by satellites: GPS. The principle of localization by the GPS system is based on the use of satellite coordinates and the estimation of distances between the receiver satellites. Distances are obtained from the estimation of the TOA (Time Of Arrival) of the signals transmitted by the satellites [16]

2nd way: ***Distance calculation function in a simulation environment:*** Like NS2, OPNET, tor other simulator.

Calculate the distance using the RSSI (Received Signal Strength Indication): in case that the absolute positioning is not accessible, dedicated equipment not available or not possible, in theory, to determine the distance between a transmitter and a receiver we can use the RSSI. RSSI is a generic radio receiver technology metric, which is usually invisible to the user of the device containing the receiver, but is directly known to users of wireless networking of IEEE 802.11 protocol family.

The distance using RSSI can be calculated using the FRIIS transmission formula:

$$P_r = \frac{P_t\ G_t\ G_r\ \lambda^2}{(4\pi)^2\ d^2\ L}$$

P_r: Receiving power.
P_t: Transmitting Power.
G_t: Gain of a transmitting antenna = ability to radiate in a particular direction in space.
G_r: Gain of a receiving antenna = ability to couple the energy radiated from a direction in space.

λ: is the wavelength.

L: is system loss factor which has nothing to do with the transmission

d: is the distance between the antennas.

Then, to calculate the distance between two nodes that are equipped by transmitting antennas, the formula is:

$$d = \sqrt{\frac{P_t\ G_t\ G_r\ \lambda^2}{(4\pi)^2\ L\ P_r}}$$

Calculate distance using GPS-free [17]**:** In case that the GPS is not accessible, we can use a GPS-free to localize the neighbor of each node. This method uses a mobile reference to calculate the coordinates of all the nodes in the network. However we can conclude the distance between any nodes. In this part, we use the distance reception power to determinate the distance between the reference and the others nodes.

To implement this method in AODV protocol, we have to choose the reference:

Choice of the reference

a and **b** are two nodes that want to communicate in a MANET network.

We put **i** the center of the coordination system.

Let **N**: The set of nodes in the network.

Pi: The set of one hop neighbor of the node **i**.

dij: the distance between nodes **i** and **j**.

Di: All distances **dij**

We choose **i / i ϵ Pa , i ≠ b and I ≠ a**

We choose **p** and **q / p,q ϵ Pa , dpq ≠ 0,**

pîq ≠ 180° , pîq ≠ 0°

Node **i** defines its system of local coordination.

We set the x-axis as the line **(ip).**

We set the y-axis as the line **(iq).**

Thus the system is defined:

The node **i** is the center of the system.

ix = 0 , iy = 0

p is the node situated on the axis of abscissas

px = dip , py = 0

The node q is located on the ordinate axis

qx = diq cos α , py = diq sinα

with **α = pîq**

Using the theorem of Al-Kashi:

$$dpq^2 = dip^2 + diq^2 - 2 * dip * diq * \cos\alpha$$

α is calculated using this formula:

$$\alpha = \arccos \frac{dip^2 + diq^2 - dpq^2}{2 * dip * diq}$$

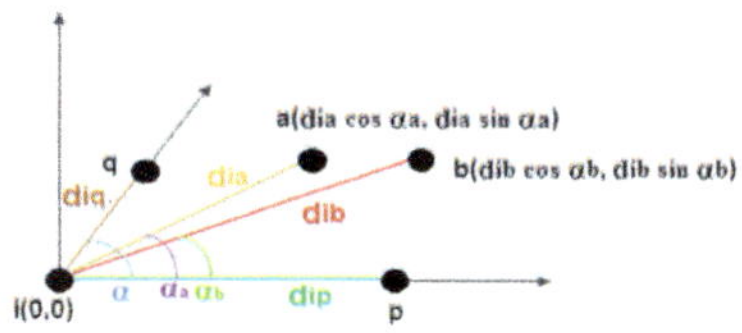

Figure 1 : The reference system

Once the reference is selected, the calculation of the coordinates of the nodes that belong to **Pi** is easy.

$$a \in Pi : \begin{cases} a_x = d_{ia} \cos \alpha_a \\ a_y = d_{ia} \sin \alpha_a \end{cases}$$

$$\alpha_a = \hat{aip} = \arccos \frac{dip^2 + dia^2 - dpa^2}{2 \cdot dip \cdot dia}$$

For **b**, the coordinates are:

$$b \in Pa \begin{cases} b_x = d_{ib} \cos \alpha_b \\ d_{ib} = d_{ia} + d_{ab} \\ b_y = d_{ib} \sin \alpha_b \end{cases}$$

$$\alpha_b = \hat{bip} = \arccos \frac{dip^2 + dib^2 - dpb^2}{2 \cdot dip \cdot dib}$$

Then the final system is as follow:

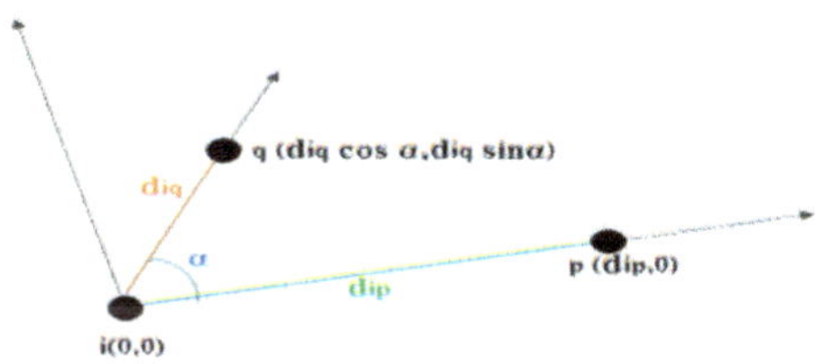

Figure 2: The system of localization local

<u>Changing the Benchmark:</u>

We propose in this part to replace the reference by another composed by the **a**'s neighbors having the smallest distances.

However, we choose **i, p, q / I,p,q ϵ Pa , i ≠ b , i ≠ a ,**

$\hat{piq} \neq 180°$, $\hat{piq} \neq 0°$

And $\forall$ k $\in$ Pi and $\forall$**k $\neq$ a,b,i,p,q**

dak > dai , dak>dap and dak>daq.

After the quantification of the distance between all nodes, we can describe the behavior of the node in the network by calculate the average of all the distances **Avg(dij).**

If the average is very high we say that the network nodes are very agitated else the network is supposedly more stable.

5. ALGORITHM OF QUANTIFICATION DISTANCE IN THE AODV ROUTING PROTOCOL

In this part, we propose to use one of those methods in the first function of a AODV protocol (rout establishment between a source and a destination).

A node **x** wants to communicate with a node **y**.

x diffuse **RREQ**.

Each node receiving **RREQ**, calculates the distance between itself and the neighbor who sent him **RREQ** (in this part we use the exact distance or the distance using the Pr) and broadcasts its table **[neighbors-distance]** to its neighbors.

To use the third method for the quantification of the distance, the algorithm has to change.

A node **x** wants to communicate with a node **y**.

x diffuse **RREQ**.

Each node receiving **RREQ**, calculates the distance between itself and the neighbor who sent him RREQ (in this part we use the exact distance or the distance using the Pr), broadcasts its table **[neighbors-distance]** to its neighbors and choose the reference who has the smallest distance and recalculate the newest distances using the third method.

N.B.: the node who receive the **RREQ** is the node a in the previous part.

6. SIMULATIONS AND RESULTS

In the following simulations, we applied our proposition to the AODV protocol .For this, we have been used the simulator NS-2 [18], with its implementation of AODV protocol of the version NS-2.35 and Ying-3D [19] to represent some results in 3D.

6.1. Environment

The network size considered for our simulations is (1000m×1000m) . The nodes have the same configuration, in particular TCP protocol for the transport layer and Telnet for the application layer. Time for each simulation is of 60s. For each simulation the mobility of the nodes is represented by the choice of an uniform speed between v_{min}= 0 and v_{max} = 100 m/s. The nodes are moved after a random choice of the new destination without leaving the network (1000m×1000m).

6.2. Discussions of results

The results present the local quantification of neighbor's distances during the simulation.

After the application of our proposition on a AODV protocol we obtain:

Table 1: Quantification's results

RREQ's source	RREQ's destination	Time	Distance	Distance/Pr
7	0	0,00128055	165,397	317,654
8	0	0,00128093	277,895	897,138
3	0	0,001281	299,47	1041,43
17	0	0,00128117	351,078	1430,93
15	0	0,00128119	356,321	1474,36
19	0	0,00128119	358,238	1490,13
11	8	0,00301544	152,84	271,179
9	8	0,00301572	238,948	662,809
1	8	0,00301579	258,118	773,428
7	8	0,00301585	276,525	887,669
0	8	0,00301585	277,731	896,079
15	8	0,00301586	280,446	913,022
10	8	0,00301598	315,33	1154,29
3	8	0,00301599	318,399	1176,86
5	8	0,00301617	371,712	1603,97
19	8	0,0030162	380,623	1681,8
1	15	0,00584584	308,585	1105,44
13	15	0,00584587	316,332	1161,63
5	10	0,00903948	103,368	124,039
4	10	0,00903976	187,032	406,083
9	10	0,00903978	194,023	437,009
11	10	0,00903982	206,903	496,956
1	10	0,00903988	224,54	585,286
18	10	0,00903999	256,821	765,673
16	10	0,00904017	309,608	1112,77
12	10	0,00904037	371,486	1602,02
13	10	0,0090404	378,926	1666,83
19	3	0,0112602	66,4831	66,4831
17	3	0,0112604	136,741	217,059
0	3	0,011261	299,018	1038,29

In the following figures, we observe the change of the distance between the node 1 and theirs neighbours in the first 3s of the simulation.

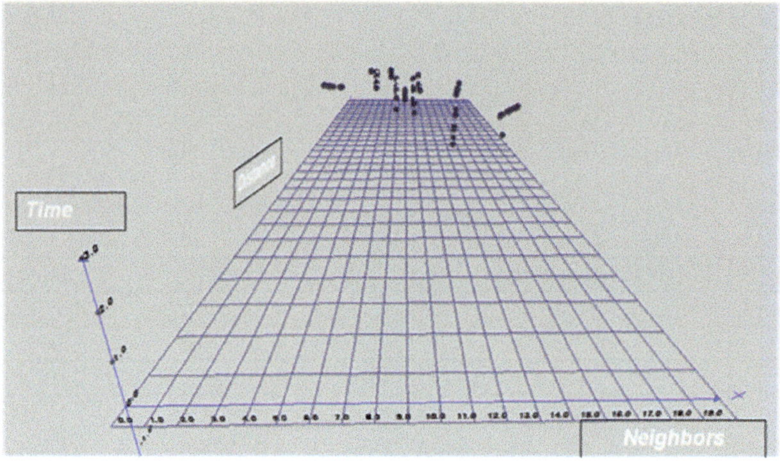

Figure 3: Quantification of the Distance between node 1 and its neighbors during the first 3s of the simulation

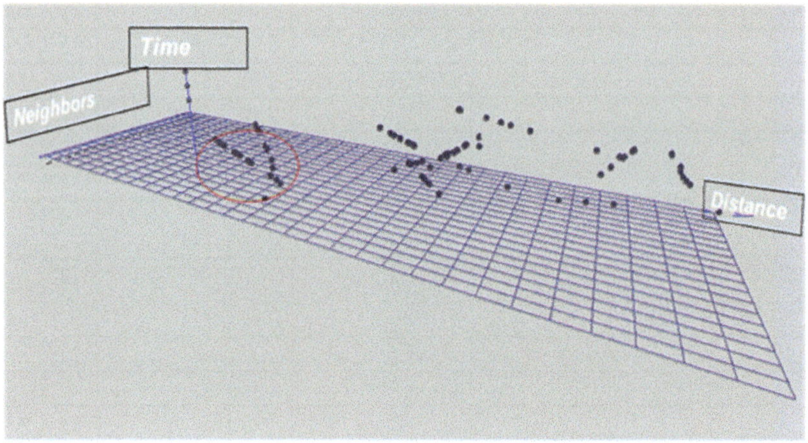

Figure 4: Quantification of the Distance between node 1 and its neighbors during the first 3s of the simulation "Another observation angle"

Figure 5: The distance between node 1 and its neighbor node 8 during the first 3s of the simulation

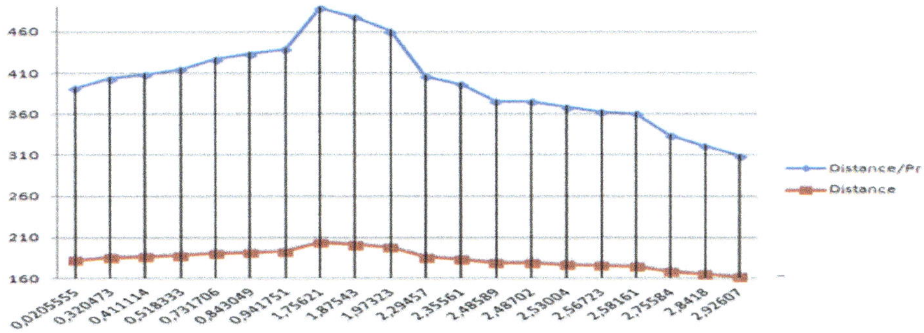

Figure 6: The distance and the distance using RSSI between node 1 and its neighbor node 8 during the first 3s of the simulation

7. CONCLUSION:

In this paper, we tried to calculate a local distance between a node and its neighbors in a AODV routing protocol for the ad hoc networks. This metric of mobility that can be used to choose a stable rout to transmit data thus ameliorate the Quality of Service in this kind of networks.

To allow this proposition more really feasible, we present the three methods to calculate the distance between two nodes. First, we use the exact distance with a GPS or using RSSI. In case that the absolute positioning is not accessible, we propose our improved GPS-free implementing in AODV protocol.

REFERENCES

[1] "Mobile Ad hoc Networking(MANET): Routing Protocol Performance Issues and Evaluation Considerations "Request for Comments 2501, IETF, January, 1999

[2] J. Jubin and J. D. Tornow, "The DARPA Packet Radio Network Protocols," Proceedings of the IEEE, Vol. 75, No. 1, pp. 21-32, Jan. 1987.

[3] B. M. Leiner, D. L. Neilson, F. A. Tobagui "Issues in Packet Radio Network esign," Proceedings of the IEEE, Vol. 75, No. 1, pp. 6-20, Jan. 1987.

[4] S. Baraković, S. Kasapović, and J. Baraković, "Comparison of MANET Routing Protocols in Different Traffic and Mobility Models", Telfor Journal, Vol. 2, No. 1, 2010.

[5] C. Perkins, B.-R. E. and D. S."Ad hoc On-demand Distance Vector routing" Request For Comments (Proposed Standard) 3561, Internet Engineering Task Force, July, 2003.

[6] http://datatracker.ietf.org/wg/manet/charter/

[7] C. Cordeiro, H. Gossain and D. Agrawal "Multicast over Wireless Mobile Ad Hoc Networks: Present and Future Directions" vol. 17, no. 1, pp. 52-59, January/February, 2003

[8] Sung-Ju Lee, Elizabeth M. Royer and Charles E. Perkins "Scalability Study of the Ad Hoc On-Demand Distance Vector Routing Protocol" In ACM/Wiley International Journal of Network Management, vol. 13, no. 2, pp. 97-114, March, 2003

[9] Ian Chakeres Elizabeth and M. Belding-Royer "AODV Implementation Design and Performance Evaluation" to appear in a special issue on Wireless Ad Hoc Networkirng' of the InternationalJournal of Wireless and Mobile Computing (IJWMC), 2005

[10] David B. Johnson, David A. Maltz and Josh Broch "DSR: The Dynamic Source Routing Protocol for Multi-Hop Wireless Ad Hoc Networks" Proceedings of INMC, 2004 - cseweb.ucsd.edu

[11] Guoyou He. "Destination-sequenced distance vector (DSDV) protocol." Technical report, Helsinki University of Technology, Finland. 2 Dec 2003

[12] P. Johansson, T. Larsson, N. Hedman, B. Mielczarek, and M. Degermark, "Scenario-based performance Analysis of Routing Protocols for Mobile Ad Hoc Networks," F'roc. ACM Mobicom 1999, Seattle WA, August 1999

[13] X. Hong, M. Gerla, G. Pei, and C.-C. Chiang, "A Group Mobility Model for Ad Hoc Wireless Networks," Proc. ACM/IEEE MSWiM '99, Seattle WA, August 1999

[14] N. Enneya, K. Oudidi and M. Elkoutbi "Network Mobility in Ad hoc Networks", Computer and Communication Engineering, 2008. ICCCE 2008. International Conference on, 13-15 May 2008, Kuala Lumpur, Malaysia.

[15] N. Enneya, M. El Koutbi and A. Berqia"Enhancing AODV Performance based on Statistical Mobility Quantification", Information and Communication Technologies, 2006. ICTTA '06, 2nd (Volume:2), Pages 2455 – 2460.

[16] A. Norhisyam Idris, A. Mohd Suldi & J. Rizal Abdul Hamid, "Effect of Radio Frequency Interference (RFI) on the Global Positioning System (GPS) Signals", 2013 IEEE 9th International Colloquium on Signal Processing and its Applications, 8 - 10 Mac. 2013, Kuala Lumpur, Malaysia.

[17] S. Capkun, M. Hamdi and J.P. Hubaux, "GPS-free positioning in mobile Ad-Hoc networks," Hawaii International Conference On System Sciences, HICSS-34 January 3-6, 2001 Outrigger Wailea Resort

[18] "The network simulator - ns-2"January, 2006

[19] http://revue.sesamath.net/spip.php?article362

Permissions

All chapters in this book were first published in IJWMN, by AIRCC Publishing Corporation; hereby published with permission under the Creative Commons Attribution License or equivalent. Every chapter published in this book has been scrutinized by our experts. Their significance has been extensively debated. The topics covered herein carry significant findings which will fuel the growth of the discipline. They may even be implemented as practical applications or may be referred to as a beginning point for another development.

The contributors of this book come from diverse backgrounds, making this book a truly international effort. This book will bring forth new frontiers with its revolutionizing research information and detailed analysis of the nascent developments around the world.

We would like to thank all the contributing authors for lending their expertise to make the book truly unique. They have played a crucial role in the development of this book. Without their invaluable contributions this book wouldn't have been possible. They have made vital efforts to compile up to date information on the varied aspects of this subject to make this book a valuable addition to the collection of many professionals and students.

This book was conceptualized with the vision of imparting up-to-date information and advanced data in this field. To ensure the same, a matchless editorial board was set up. Every individual on the board went through rigorous rounds of assessment to prove their worth. After which they invested a large part of their time researching and compiling the most relevant data for our readers.

The editorial board has been involved in producing this book since its inception. They have spent rigorous hours researching and exploring the diverse topics which have resulted in the successful publishing of this book. They have passed on their knowledge of decades through this book. To expedite this challenging task, the publisher supported the team at every step. A small team of assistant editors was also appointed to further simplify the editing procedure and attain best results for the readers.

Apart from the editorial board, the designing team has also invested a significant amount of their time in understanding the subject and creating the most relevant covers. They scrutinized every image to scout for the most suitable representation of the subject and create an appropriate cover for the book.

The publishing team has been an ardent support to the editorial, designing and production team. Their endless efforts to recruit the best for this project, has resulted in the accomplishment of this book. They are a veteran in the field of academics and their pool of knowledge is as vast as their experience in printing. Their expertise and guidance has proved useful at every step. Their uncompromising quality standards have made this book an exceptional effort. Their encouragement from time to time has been an inspiration for everyone.

The publisher and the editorial board hope that this book will prove to be a valuable piece of knowledge for researchers, students, practitioners and scholars across the globe.

List of Contributors

Berker Tasoluk
Department of Informatics, Istanbul University, Istanbul, Turkey

Zuhal Tanrikulu
Department of Management Information Systems, Bogazici University, Istanbul, Turkey

A.V. Prabu
Gandhi Institute Of Engg & Technology, Gunupur, Rayagada, Orissa-765022, India

Baburao Kodavati
Gandhi Institute Of Engg & Technology, Gunupur, Rayagada, Orissa-765022, India

Tholada Appa Rao
Gandhi Institute Of Engg & Technology, Gunupur, Rayagada, Orissa-765022, India

E. Rambabu
Gandhi Institute Of Engg & Technology , Gunupur,Rayagada,Orissa-765022,India

Shyam sundar tripathy
NTTF at J.N.TATA Technical Education Center , Gopalpur,Orissa,India

T. Linciya
P.G Scholar, Department of Computer Science and Engineering, Easwari Engineering College, Chennai, India

K. M. Anandkumar
Assistant Professor (Sl.Gr.), Department of Computer Science and Engineering, Easwari Engineering College, Chennai, India

Mohammed Azharuddin
Department of Computer Science and Engineering, M.S.Ramiah Institute of Technology, Bangalore-560054, India

Annapurna P Patil
Department of Computer Science and Engineering, M.S.Ramiah Institute of Technology, Bangalore-560054, India

Mahdieh Ghazvini
Department of Computer Engineering, Faculty of Engineering, University of Isfahan, Isfahan, Iran

Naser Movahedinia
Department of Computer Engineering, Faculty of Engineering, University of Isfahan, Isfahan, Iran

Kamal Jamshidi
Department of Computer Engineering, Faculty of Engineering, University of Isfahan, Isfahan, Iran

Suparna Biswas
Department of Computer Science & Engineering, West Bengal University of Technology, Salt Lake, Kolkata

Sarmistha Neogy
Department of Computer Science & Engineering

Priyanka Dey
Department of Computer Science & Engineering, West Bengal University of Technology, Salt Lake, Kolkata

Saad Almutairi
De Montfort University, Software Technology Research Laboratory (STRL) Leicester, United Kingdom

Hamza Aldabbas
De Montfort University, Software Technology Research Laboratory (STRL) Leicester, United Kingdom

Ala Abu-Samaha
Middle East University, Amman, Jordon

Zhang Nan
Wuhan Maritime Communication Research Institute, Wuhan, 430079, China

Gao Xiao
Wuhan Maritime Communication Research Institute, Wuhan, 430079, China

Himadri Nath Saha
Department of Computer Science and Engineering, Institute of Engineering & Management, West Bengal, India

Dr. Debika Bhattacharyya
Department of Computer Science and Engineering, Institute of Engineering & Management, West Bengal, India

Dr. P. K.Banerjee
Department of Electronics and Tele Communication Engineering, Jadavpur university, West Bengal, India

Subhankar joardar
Department of Computer Science and Engineering, Haldia Institute of Technology, Haldia, West Bengal, India

Vandana Bhattacherjee
Department of Computer Science and Engineering, Birla Institute of Technolog, Mesra, Jharkhand, India

Debasis Giri
Department of Computer Science and Engineering, Haldia Institute of Technology, Haldia, West Bengal, India

Cristian Androne
Communications Department, Technical University of Cluj-Napoca Cluj-Napoca, Romania

Tudor Palade
Communications Department, Technical University of Cluj-Napoca Cluj-Napoca, Romania

Monika Raghatate
Department of Computer Engineering, St. Vincent Pallotti College of Engineering and Technology, Nagpur, India

Prof. Dipak W. Wajgi
Department of Computer Engineering, St. Vincent Pallotti College of Engineering and Technology, Nagpur, India

Chunming Liu
T-Mobile, Bellevue, USA

Bryan Fleming
T-Mobile, Bellevue, USA

Eleazar Chukwuka
School of Engineering, University of Greenwich Chatham Maritime ME4 4TB United Kingdom

Kamran Arshad
School of Engineering, University of Greenwich Chatham Maritime ME4 4TB United Kingdom

Yassine MALEH
Emerging Technologies Laboratory (VETE), Faculty of Sciences and Technology Hassan 1stUniversity, Settat, MOROCCO

Abdellah Ezzati
Emerging Technologies Laboratory (VETE), Faculty of Sciences and Technology Hassan 1stUniversity, Settat, MOROCCO

Abolfazl Akbari
Dept. of Computer Engineering, Islamic Azad University Ayatollah Amoli Branch, Amol, Iran

Arash Dana
Islamic Azad University, Central Tehran Branch, Tehran, Iran

Ahmad Khademzadeh
Iran Telecom Research Center, Tehran, Iran

Neda Beikmahdavi
Islamic Azad University Ayatollah Amoli Branch, Amol, Iran

Ahmed Jawad Kadhim
IRAQ- Ministry of Education-General Directorate of Education in Qadisiyah-Ibn al Nafis School for Boys

Meryem Saadoune
Department of Mathematics and Computer Sciences, Hassan 1st University, Settat, Morocco

Abdelmajid Hajami
Department of Mathematics and Computer Sciences, Hassan 1st University, Settat, Morocco

Hakim Allali
Department of Mathematics and Computer Sciences, Hassan 1st University, Settat, Morocco